3D Imaging in Medicine

Algorithms, Systems, Applications

NATO ASI Series

Advanced Science Institutes Series

A series presenting the results of activities sponsored by the NATO Science Committee, which aims at the dissemination of advanced scientific and technological knowledge, with a view to strengthening links between scientific communities.

The Series is published by an international board of publishers in conjunction with the NATO Scientific Affairs Division

A Life Sciences B Physics	Plenum Publishing Corporation London and New York
C Mathematical and Physical Sciences D Behavioural and Social Sciences E Applied Sciences	Kluwer Academic Publishers Dordrecht, Boston and London
F Computer and Systems Sciences G Ecological Sciences H Cell Biology	Springer-Verlag Berlin Heidelberg New York London Paris Tokyo Hong Kong

Series F: Computer and Systems Sciences Vol. 60

3D Imaging in Medicine

Algorithms, Systems, Applications

Edited by

Karl Heinz Höhne

Institute of Mathematics and Computer Science in Medicine
University Hospital Hamburg-Eppendorf
Martinistraße 52, D-2000 Hamburg 20, FRG

Henry Fuchs

Stephen M. Pizer

Department of Computer Science
University of North Carolina
Chapel Hill, NC 27599, USA

Springer-Verlag Berlin Heidelberg New York
London Paris Tokyo Hong Kong
Published in cooperation with NATO Scientific Affairs Division

Proceedings of the NATO Advanced Research Workshop on 3D Imaging in Medicine, held in Travemünde, Germany, June 25–29, 1990.

Directors:
Karl Heinz Höhne, University of Hamburg
Henry Fuchs, UNC, Chapel Hill
Stephen M. Pizer, UNC, Chapel Hill

Scientific Committee:
Jean-Louis Coatrieux, University of Rennes-I
Alan C. F. Colchester, Guys Hospital, London
François Hottier, Philips, Paris
Olaf Kübler, ETH Zürich
David N. Levin, University of Chicago
Wilfried K. Löffler, Siemens, Erlangen
Richard A. Robb, Mayo Clinic, Rochester
Wolfgang Schlegel, DKFZ, Heidelberg

Workshop Coordinators:
Michael Bomans, University of Hamburg
Linda Houseman, UNC, Chapel Hill
Andreas Pommert, University of Hamburg

ISBN-13:978-3-642-84213-9 e-ISBN-13:978-3-642-84211-5
DOI: 10.1007/978-3-642-84211-5

Softcover reprint of the hardcover 1st edititon 1990

2145/3140-543210 – Printed on acid-free-paper

Preface

The visualization of human anatomy for diagnostic, therapeutic, and educational purposes has long been a challenge for scientists and artists. In vivo medical imaging could not be introduced until the discovery of X-rays by Wilhelm Conrad Röntgen in 1895. With the early medical imaging techniques which are still in use today, the three-dimensional reality of the human body can only be visualized in two-dimensional projections or cross-sections. Recently, biomedical engineering and computer science have begun to offer the potential of producing natural three-dimensional views of the human anatomy of living subjects.

For a broad application of such technology, many scientific and engineering problems still have to be solved. In order to stimulate progress, the NATO Advanced Research Workshop in Travemünde, West Germany, from June 25 to 29 was organized. It brought together approximately 50 experts in 3D-medical imaging from all over the world.

Among the list of topics *image acquisition* was addressed first, since its quality decisively influences the quality of the 3D-images. For 3D-image generation – in distinction to 2D-imaging – a decision has to be made as to which objects contained in the data set are to be visualized. Therefore special emphasis was laid on methods of *object definition.* For the final visualization of the segmented objects a large variety of *visualization algorithms* have been proposed in the past. The meeting assessed these techniques. Their practical application in medicine depends heavily on the availability of suitable tools for *object manipulation and interaction* and the corresponding *hardware systems.* Therefore these topics were included as important aspects. Although the meeting was dominated by the algorithmic and systems aspects, the discussion of their relevance to the *applications* was considered indispensible.

We chose to publish the proceedings before the meeting. We are aware of the fact that new ideas gained following stimulating discussions at the meeting could not be incorporated into articles of this book. On the other hand the field is developing so rapidly that a timely publication could only be provided by asking the authors to meet a deadline prior to the meeting. We are highly indebted to the authors for having accepted the heavy workload we have put on them.

The workshop would not have been possible without the help of many people in the Department of Computer Science in Medicine at the University of Hamburg and the Department of Computer Science at the University of North Carolina at Chapel Hill. Our special thanks go to Linda Houseman, Michael Bomans, and Andreas Pommert for their enthusiastic support. And last but not least we thank NATO for providing the financial basis for this meeting.

Hamburg and Chapel Hill
June 1990

Karl Heinz Höhne
Henry Fuchs
Stephen M. Pizer

Table of Contents

Image Acquisition

Object Definition

Visualization

Object Manipulation and Interaction

Systems

Applications

Image Acquisition

Magnetic Resonance Imaging

Wilfried K. Loeffler

Siemens AG, Med
Henkestrasse 127
D-8520 Erlangen, FRG

Abstract

Magnetic Resonance Imaging within the past 10 years has become the primary medical imaging modality for many neurological and orthopedic applications. Starting from the basic phenomena of Nuclear Magnetic Resonance, the principles of 2D and 3D Magnetic Resonance Imaging are described. As a new and possibly important future application the basics of Magnetic Resonance Angiography allowing the display of vascular structures without the use of contrast agents is being elaborated.

Keywords: Nuclear Magnetic Resonance (NMR) / Magnetic Resonance Imaging (MRI) / magnetic field / magnetic field gradients / relaxation times / 3D-Magnetic Resonance Imaging / Bloch Equations / Radio-Frequency (RF) pulses / Magnetic Resonance Angiography

Introduction

In principle, all fields or particles can be employed for medical imaging, if they fulfill the three following conditions: First, the fields or particles have to interact with the body tissue, second, they still have to penetrate the body to a sufficient degree, and third, a spatial association of the interaction must be possible. Looking at the electromagnetic spectrum for candidates for medical imaging only two wavelength ranges allow a sufficient penetration of the human body. This is, on the one hand, the well known range of ionizing radiation with x-rays, still being utilized in most medical imaging systems. On the other hand, also electromagnetic fields with wave lengths above about 20 cm will show a penetration depth in tissue sufficient for medical imaging use. Unfortunately, the

NATO ASI Series, Vol. F 60
3D Imaging in Medicine
Edited by K. H. Höhne et al.

spatial resolution required for a meaningful diagnosis must be much higher than the wavelength limit in this VHF or UHF range would allow. Therefore, other than quasi-optical means of imaging are required, if this range of electromagnetic fields is to be used for medical imaging. Magnetic Resonance Imaging (MRI) represents such an imaging method.

Nuclear Magnetic Resonance (NMR)

All atomic nuclei with an odd number of neutrons or protons (that is about two thirds of all stable isotopes) possess an angular momentum or spin. Due to the electrical charge of the nuclei their spin always is associated with a magnetic moment. This is usually written as:

$$\underline{\mu} = \gamma \hbar \, \underline{I}$$

$\underline{\mu}$ = magnetic moment
γ = gyromagnetic ratio (characteristic for each type of nucleus)
$\hbar$ = Planck's constant
$\underline{I}$ = nuclear spin

The interaction of these nuclei with an external magnetic field can be described by:

$$\begin{aligned} W &= \langle \underline{\mu} \cdot \underline{B} \rangle \\ &= \gamma \hbar B \langle I_Z \rangle \end{aligned}$$

W = energy of a nucleus in the magnetic field
$\underline{B}$ = magnetic field
I_Z = component of $\underline{I}$ parallel to $\underline{B}$

Quantum mechanics allow I(I + 1) values of $\langle I_Z \rangle$:

$$\langle I_Z \rangle = -I, -I+1, \ldots, I-1, I$$

For an ensemble of nuclei the relative number of spins in each energetic level corresponding to a certain value of $\langle I_Z \rangle$ is governed by the Boltzmann equation:

$$\frac{N_{n+1}}{N_n} = e^{-\frac{\gamma \hbar B}{k T}}$$

N_n = number of nuclei in level n
k = Boltzmann's constant
T = absolute temperature

As a consequence of the competing thermal and magnetic interaction as described by the Boltzmann equation, the spins in an external magnetic field will not be aligned uniformly. A small preponderance of nuclei will be oriented in the energetically more favorable direction. As a result, a small macroscopic magnetization can be associated with an ensemble of nuclei in a magnetic field:

$$M_o = \frac{N \gamma^2 \hbar^2 I (I+1)}{3 k T} B$$

M_o = equilibrium magnetization (magnetic moment / volume)
N = number of nuclei / volume

The quantitative description of the motion of the macroscopic magnetization M does no longer require the use of quantum mechanics; instead classical mechanics can be employed. Therefore, the macroscopic angular momentum associated with the magnetization M can be written as:

$$\underline{P} = \frac{1}{\gamma} V \underline{M}$$

V = volume

In an external magnetic field $\underline{P}$ will change as a consequence of the magnetic torque $V\underline{M} \times \underline{B}$:

$$\frac{d\underline{P}}{dt} = V \cdot \underline{M} \times \underline{B}$$

Inserting for $\underline{P}$ leads to the well known Larmor equation which describes the motion of a magnetization in a magnetic field:

$$\frac{d\,\underline{M}}{d\,t} = \gamma\,\underline{M} \times \underline{B}$$

As can be easily verified the solution of the Larmor equation is a type of motion where the component of $\underline{M}$ parallel to $\underline{B}$ remains steady while the component perpendicular to $\underline{B}$ rotates at an angular frequency of

$$\omega = \gamma B$$

about the axis of $\underline{B}$. This type of motion is known as precession.

This simple equation in fact correctly describes the results of NMR in thin gases. In condensed matter, however, interactions between the nuclei have to be taken into account additionally. This can be done by introducing empirical relaxation factors into the equation of motion

$$\frac{d\,M_x}{d\,t} = \gamma\,[\underline{M} \times \underline{B}]_x - \frac{M_x}{T_2}$$

$$\frac{d\,M_y}{d\,t} = \gamma\,[\underline{M} \times \underline{B}]_y - \frac{M_y}{T_2}$$

$$\frac{d\,M_z}{d\,t} = \gamma\,[\underline{M} \times \underline{B}]_z + \frac{M_z - M_o}{T_1}$$

This system of equations first has been introduced by Felix Bloch to describe the results of his early NMR-experiments (1). It has since been termed Bloch's equations. The time constant T_1 is called longitudinal or Spin-Lattice Relaxation Time, while T_2 is called transverse or Spin-Spin Relaxation Time. The following is a simple solution for a static magnetic field along the z-direction

$$\underline{B} = (0, 0, B_o)$$

and a nuclear magnetization starting perpendicular to $\underline{B}$

$\underline{M}_{(o)} = (M_\perp, 0, 0)$

$$M_x = M_\perp \, e^{-t/T_2} \cos \omega_o t$$

$$M_y = M_\perp \, e^{-t/T_2} \sin \omega_o t$$

$$M_z = M_o \, (1 - e^{-t/T_1})$$

$$\omega_o = \gamma B_o$$

In order to describe the results of experiments in the presence of high frequency electromagnetic fields it has proven useful to transform the equations of motion into a coordinate system that rotates with the Larmor Frequency ω_o. This transformation can be described by:

$$\left(\frac{d\,\underline{M}}{d\,t}\right)_{\text{fixed}} = \left(\frac{\partial\,\underline{M}}{\partial\,t}\right)_{\text{rotating}} + \underline{\omega} \times \underline{M}$$

Neglecting the relaxation terms for a moment, this leads to the following equation of motion:

$$\left(\frac{\partial\,\underline{M}}{\partial\,t}\right)_{\text{rotating}} = \gamma\,\underline{M} \times \underline{B}_{eff}$$

with $\underline{B}_{eff} = \underline{B} + \underline{\omega} / \gamma$

This means that the magnetization in a frame of reference rotating at an angular frequency of $\underline{\omega}$ will precess about an effective field $\underline{B}_{eff}$. The correct choice of $\underline{\omega}$ will dramatically simplify the equations of motion if both a static and high frequency magnetic field are applied. This can be shown with three simple examples:

1) Static magnetic field $\underline{B} = (0, 0, B_o)$

A choice of $\omega = -\gamma B_o$ leads to

$$\frac{\partial \underline{M}}{\partial t} = 0$$

This simply means that the magnetization in the rotating frame of reference remains static.

2) Inhomogeneous static field $\underline{B} = (0, 0, B_o + \Delta B)$

With the same choice of ω this leads to

$$B_{eff} = \Delta B$$

Therefore, the magnetization in the rotating frame of reference will precess slowly at an angular frequency corresponding to the local field offset ΔB.

3) Static field plus alternating field $\underline{B} = (B_1 \cos \omega_o t, B_1 \sin \omega_o t, B_o)$:

With the usual choice of ω this leads to

$$B_{eff} = (B_1, 0, 0)$$

Therefore, if an alternating magnetic field with a frequency corresponding to the Larmor Frequency is applied, the nuclear magnetization in the rotating frame of reference will precess about the direction of the alternating field with an angular frequency of

$$\omega_1 = \gamma B_1$$

If B_1 - as in our example - is perpendicular to B_o, the magnetization in the rotating frame of reference will be turned away from the direction of the static field B_o. In the fixed frame of reference this corresponds to a precession about the axis of $\underline{B}_o$ with the angle between the static field and the magnetization changing at the rate of ω_1.

In most NMR experiments the alternating magnetic field B1 is not applied continuously but rathed in a pulsed mode. Starting an NMR experiment, for example, with an RF-pulse (alternating field in the radio-frequency range) of the duration

$$\Delta t = \frac{\pi}{2 \gamma B_1}$$

will tilt the magnetization starting from its equilibrium direction parallel to $\underline{B}_0$ through 90°. An RF-pulse like this, therefore, is termed 90°-pulse. As seen from the fixed laboratory frame of reference a 90° pulse will tilt the magnetization into a plane perpendicular to $\underline{B}_0$ and thus start its precession about the axis of $\underline{B}_0$. Similarly RF-pulses of different duration Δt or strengths B_1 are termed according to the angle through which they turn the magnetization in the rotating frame of reference (e.g. 180° pulse). Figure 1 shows a simple pulsed NMR-experiment. The same coil is employed to generate the excitation field B_1 and, after the RF-transmitter is turned off, to detect the precession of the nuclear magnetization.

Imaging with Nuclear Magnetic Resonance

In an experiment as shown in Figure 1 the NMR signal received will be a superposition of signals from all the material inside the test tube within the pick-up coil. In order to employ this type of measurement for imaging a mechanism must be found which allows to distinguish contributions to the NMR-signal coming from spatially different parts of the sample. For this purpose, a simple trick can be used: If the static magnetic field B_0 is made spatially nonuniform, the Larmor Frequency or NMR frequency ω_0 will vary correspondingly. Therefore, contributions to the NMR signal from different regions can be distinguished by means of a simple frequency analysis. Provided the spatial dependence of the magnetic field strength is known and unambiguous a precise correlation between the NMR-spectrum and the position in space is possible (2).

The most straight forward correlation between field strength and space (in one dimension) is represented by a magnetic field the strength of which varies linearly with one spatial coordinate:

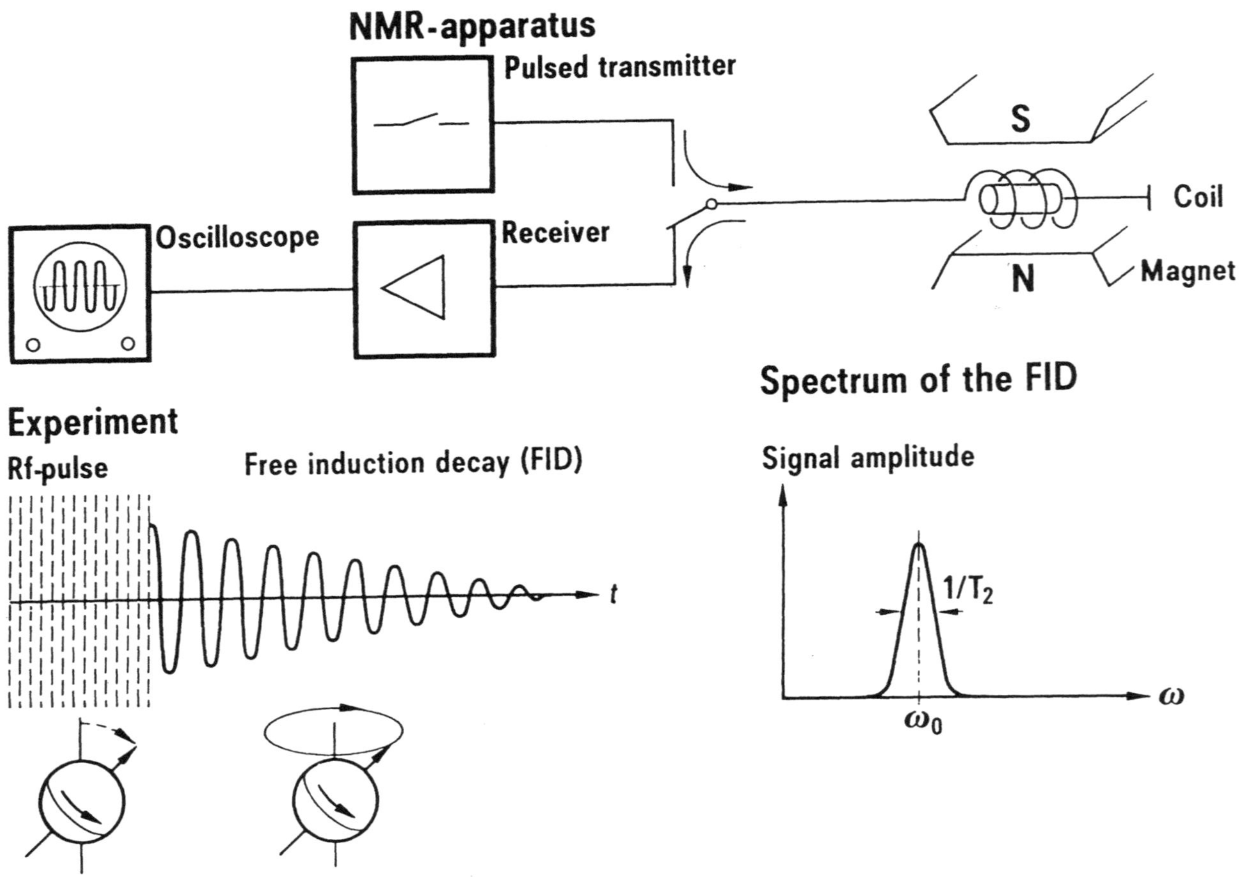

Figure 1: Set-up and time course of a simple pulsed NMR-experiment

$$\underline{B} = (0, 0, B_z)$$

$$B_z = B_o + \underline{G} \cdot \underline{r}$$

$$\text{with} \quad \underline{r} = (x, y, z)$$

$$\underline{G} = \left(\frac{\partial B_z}{\partial x}, \frac{\partial B_z}{\partial y}, \frac{\partial B_z}{\partial B_z} \right)$$

In this case, the magnetic field B can be considered as the superposition of a homogeneous field with a linear gradient field. In imaging systems usually three separate coil systems are employed which allow the generation of gradient fields in the x, y, and z-directions. The Larmor Frequency in an NMR-experiment then will be spatially dependent with

$$\omega = \gamma (B_o + \underline{G} \cdot \underline{r})$$

$$\omega = \omega_o + \gamma \underline{G} \cdot \underline{r}$$

As an example let us consider an NMR-experiment like the one described in Figure 1 with the x-gradient field of an imaging system activated. After an initial 90° RF-pulse which starts the precession of the magnetization, the resonance frequency will be:

$$\omega (x, y, z) = \omega_o + \gamma G_x \cdot x$$

In a frame of reference rotating with the angular frequency ω_o this simplifies to (observation in a rotating frame of reference is equivalent with using a phase sensitive demodulator tuned to ω_o):

$$\omega_r (x, y, z) = \gamma G_x \, x$$

The signal contribution from the location (x, y, z) then can be written as

$$A (x, y, z) \, e^{i \omega_r t} \, dx \, dy \, dz$$

with A representing the local contribution to the NMR signal as a result of the density of nuclei and - depending on the pulsing scheme that is used - also as a function of the local relaxation parameters T_1 and T_2. The total signal in the receiver coil then can be seen as the result of the integration:

$$S(t) = \iiint A(x, y, z)\, e^{i\gamma G_x x t}\, dx\, dy\, dz$$

A Fourier Transform of the received time domain signal yields:

$$s(\omega_r) = \int_{-\infty}^{\infty} S(t)\, e^{-i\omega_r t}\, dt$$

$$= \iint A'\left(x = \frac{\omega r}{\gamma G_x}, y, z\right) dy\, dz$$

This represents a projection of the "NMR-density" A of the sample on the axis of the gradient field x. One way of obtaining three dimensional information of the sample, therefore, could be to repeat the above experiment for a sufficient number of different projection directions and then use the well known mathematics for reconstruction of 3D objects from projections (3). Although this idea has marked the beginning of NMR-imaging in 1973, in the meantime a slightly different approach has established itself as the standard for 3D-imaging (4).

Figure 2 shows the time course of a 3D-Fourier-Transform NMR-imaging experiment. The leading RF-pulse starts the precession of the magnetization within the sample. As will become obvious, the two following gradient field pulses G_y and G_z imprint a typical spatially dependent phase characteristic on the magnetization. These pulses, therefore, are called phase encoding gradient pulses. Finally, the NMR-signal is sampled while a third gradient field in x-direction is activated. The negative x-gradient lobe in front of the sampling interval is added only to effectively shift the time domain zero into the middle of the sampling interval to simplify the time domain Fourier Transform. As a result of the phase encoding gradient pulses the magnetization will experience a phase shift

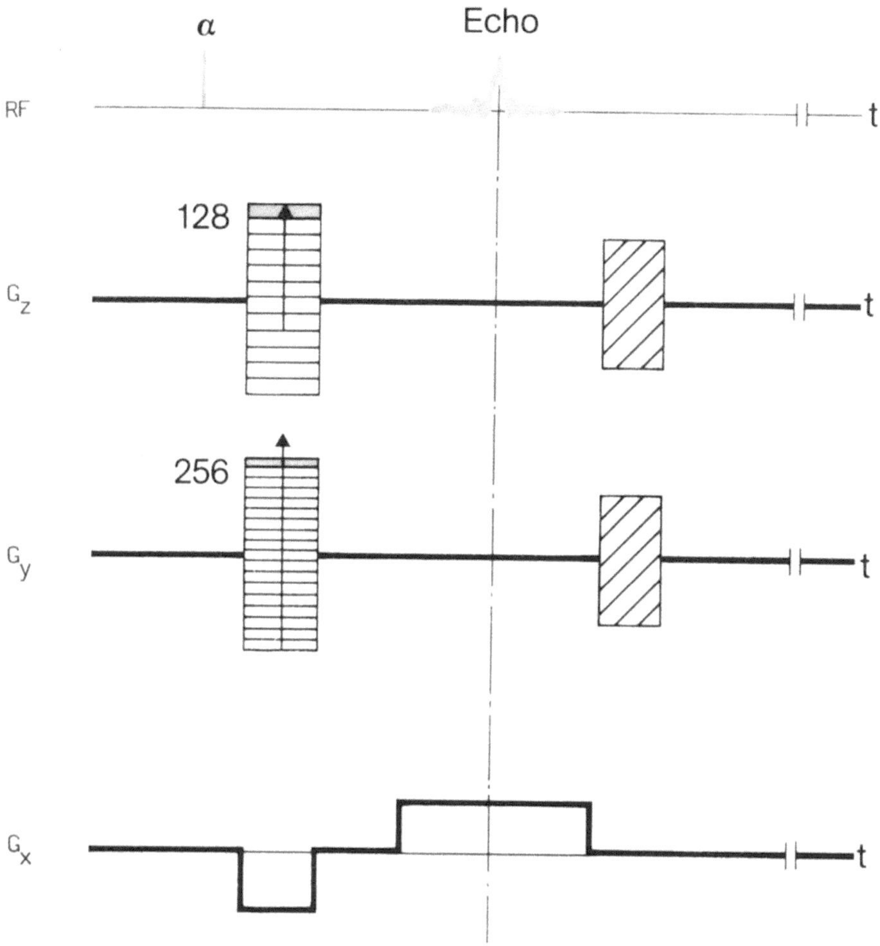

Figure 2: Nonselective 3D-Fourier-Transform MR-Imaging sequence

$$\varphi = \gamma\,(G_y \cdot y + G_z \cdot z)\,\Delta t$$

or $$\varphi = k_y \cdot y + k_z \cdot z$$

with $k_y = \gamma\, G_y\, \Delta t, \qquad k_z = \gamma\, G_z \cdot \Delta t$

The signal contribution from a location (x, y, z) during the sampling interval, therefore, can be written:

$$A\,(x, y, z)\, e^{i(k_x\, x + k_z\, z)}\, e^{i\gamma\, G_x\, x\, t}\, dx\, dy\, dz$$

Again the total signal in the receiver coil can be considered the result of the integration

$$S(k_x, k_y, k_z) = \iiint A(x, y, z)\, e^{i(k_x\, x + k_y\, y + k_z\, z)}\, dx\, dy\, dz$$

with $k_x = \gamma\, G_x \cdot t$

A three-dimensional Fourier Transform of the NMR signal in the experiment described, obviously will result in a 3D representation of the "NMR-density" A(x, y, z). Since only the time domain parameter $k_x = \gamma \cdot G_x \cdot t$ can be adequately sampled in a single experiment, the pulse sequence shown in Figure 2 has to be repeated with different settings of the phase encoding gradients G_y and G_z in order to acquire a sufficient number of samples in k-space. The total data acquisition time T_A required to build up a 3D image matrix of the dimensions l x m x n, therefore, is given by:

$$T_A = l \times m \times T_R$$

T_R represents the pulse sequence repetition time. Further it is assumed that in each of the data sampling periods n samples of k_x are taken.

If for reasons of image contrast, a long repetition time T_R is required, a variant of the imaging method described is frequently employed. This method circumvents the excessive data acquisition time demanded by a long repetition time T_R in 3D mode if only a limited number of crossectional 2D images are required. Here the 90° RF-pulse is modified such that only the magnetization within a sufficiently thin slice is flipped out of its equilibrium state. Therefore, only areas within such a slice will contribute to the NMR signal received in the pick-up coil. To achieve this kind of slice selection, a narrow band 90° RF-pulse is applied while one of the three orthogonal gradient fields is switched on. As a result, the NMR-frequency will only match the frequency of the RF-pulse in a narrow spatial band defining the desired slice. Subsequent to the leading slice selective RF-pulse a gradient pulse sequence similar to the one shown in Figure 2 is used except that phase encoding gradient pulses are employed only in one direction. In order to build up an m x n 2D image, therefore, only a time

$$T_A = m \times T_R$$

is required. For reasons of image contrast, frequently a repetition time T_R much longer than the duration of the pulse sequence shown in Figure 2 is used. In these cases the excessive waiting time can be employed to acquire information from other 2D slices parallel to the first one. This 2D multislice mode is used frequently to acquire up to 20 or 30 parallel slices with moderately long repetition times T_R. A disadvantage compared to true 3D methods is relative smooth slice profiles and cross-talk between adjacent slices. Still the 2D multislice mode is the method most frequently used in routine diagnostic imaging.

Magnetic Resonance Angiography

Techniques for magnetic resonance angiography are based on two physical principles: Inflow and phase effects (5, 6). Inflow effects are due to motion of spins with a different history into the slice or volume to be investigated. Phase effects are a consequence of motion of spins along the direction of the field gradients employed for imaging.

In any imaging sequence the nuclear magnetization is exposed to a large number of RF-pulses repeated at fixed intervals T_R. To shorten the total data acquisition time, T_R is usually chosen to be of the order of or even shorter than T_1. Therefore, the magnetization has not enough time between two successive RF-pulses to completely reestablish its full equilibrium value. Instead a dynamic equilibrium - the so-called steady-state magnetization - will be established after the first couple of RF pulses. For a sequence of α-pulses the value of the steady-state magnetization can be derived from the Bloch equations

$$M = M_o \frac{(1 - e^{-T_R/T_1})}{1 + (1 - e^{-T_R/T_1}) \cos \alpha}$$

As a consequence, the NMR signal available in an imaging sequence, will be reduced, particularly if short T_R values and flip angles α approaching 90° are being used. This phenomenon has been termed saturation. By using slice or slab selective RF-pulses or RF-transmitter coils which encompass only parts of the human body, like the head, for example, it is possible to limit the volume which shows a reduced steady-state magnetization to the corresponding areas. If blood flow now moves spins from outside the volume exposed to the RF-pulses into this area, this unsaturated blood volume possesses the full equilibrium longitudinal magnetization and, therefore, upon entering the area of interest produces a much stronger signal than stationary spins. This effect has also been termed "entry slice phenomenon".

The second physical phenomenon employed for MR-angiography, the phase effect, is a consequence of the phase memory of the magnetization. When travelling along the direction of magnetic field gradients, the spins experience a motion dependent phase shift. In general, this phase shift will give rise to flow artifacts frequently seen as repeated parallel "ghosts" of vessels with blood flow. The physical origin of the flow-related phase shift lies in the fact that moving spins, as long as a magnetic field gradient is turned on, will experience different local magnetic fields on their way along the direction of the gradient. A quantitative description of flow-related phase effects can be based on the Larmor Equation. The phase of the spin system can be written as:

$$\varnothing (t) = \gamma \int_o^t B (x, t') \, dt'$$

If x(t) represents the position of the spins along the x-axis, the magnetic field can be rewritten:

$$B (x, t) = B_o + G_x (t) \cdot x (t)$$

In case of a motion of spins at a constant velocity v this yields

$$x(t) = x_o + v \cdot t$$

and for the phase

$$\emptyset(t) = \gamma \int_o^t (B_o + G_x(t')\, x_o + G_x(t')\, vt')\, dt'$$

The motion dependent phase term, therefore, is represented by:

$$\emptyset_v(t) = \gamma\, v \int_o^t G_x(t')\, t'\, dt'$$

This type of integral is also known as the first moment of the gradient wave form.

Some methods for magnetic resonance angiography make use of this flow-related phase effect to allow the display of blood vessels. Moreover, the above relationship allows to quantitate the flow velocity in blood vessels using the known time dependence of the magnetic field gradients. A problem common to all of these methods is their sensitivity to the pulsatile flow in arteries. In order to obtain consistent data it is generally necessary to use cardiac synchronization for data acquisition. This severely increases the data acquisition times particularly if three-dimensional data are to be obtained.

The problem of pulsatile flow can be overcome by compensating the flow dependent phase term derived above, for example by the use of additional negative gradient pulses which allow to reduce or even null the first moment of the gradient wave form at the time of data acquisition T_E:

$$\emptyset_v(T_E) = \gamma\, v \int_o^{T_E} G_x(t')\, dt' = 0$$

This equation can be fulfilled with a number of gradient wave forms. A pulse sequence like this will be called flow compensated along the x-direction. In general, flow compensation can be employed along the three orthogonal directions of the gradients used in imaging. These types of pulse sequences allow the elimination of flow artifacts without cardiac synchronization. The inflow phenomena described initially can only be used for an enhanced display of blood vessels if the flow related phase effects for the pulsatile flow in arteries are completely compensated.

In principle, with this technique for an enhanced display of blood flow, any vessel segment can be imaged by cutting through the vessel with 2D slice selective sequences perpendicular with regard to the flow direction. With repetitive increments of the slice position a complete vessel tree can be reconstructed with appropriate postprocessing techniques. Still, good resolution will require relatively narrow slices with the corresponding loss in signal-to-noise ratio.

Improvement of the spatial resolution without a reduction of the signal-to-noise can be achieved with 3D techniques. Also here complete phase compensation is required to avoid flow artifacts in the form of signal loss or ghosting. In order to increase the signal from flowing blood above that of stationary tissue slab selective RF-pulses perpendicular to the direction of arteries feeding the area of interest are employed. This allows to make optimally use of the inflow enhancement.

For an improved visualization of the three dimensional flow enhanced data one particular postprocessing technique has proven to be especially useful. Since the data acquisition techniques described make sure that flowing blood will show a higher signal than stationary tissue, a simple maximum intensity projection allows to extract vessel information in projective images like those commonly used in conventional angiography. A good spatial impression can be achieved by rapidly showing a sequence of projective images with different projection angles obtained from a single 3D data set.

Contrary to conventional angiography magnetic resonance angiography is completely noninvasive and does not require the use of contrast agents. The 3D technique described above allows a complete examination in about 15 minutes and, therefore, appears well suited for routine clinical use. A major advantage over two dimensional projective data acquisition techniques are the postprocessing capabilities which allow a representation of vascular structures at any projection angle. Clinical studies have shown that pathological conditions such as stenoses, aneurysms, and arteriovenous malformations can be visualized.

References

1) Abragam, A.: The Principles of Nuclear Magnetism. Oxford Universiy Press (Clarendon), London and New York (1961)

2) Mansfield, P., and Morris, P. G.: NMR Imaging in Biomedicine. Academic Press, London and New York (1982)

3) Lauterbur, P. C., Nature 242 (1973) 190 - 191

4) Kumar, A., Welti, D., and Ernst, R. R., J. Magn. Reson. 18 (1975) 69 - 83

5) Dumoulin, C. L., Souza, S. P., Walker, M. F., and Wagle, W., Magn. Res. in Med. 9 (1989) 139 - 149

6) Masaryk, T. J., Modic, M. T., Ruggieri, P. M., Ross, J. S., Laub, G. L., Lenz, G. W., Tkach, J. A., Haacke, E. M., Selman, W. R., and Harik, S. I., Radiology 171 (1989) 801 - 806

3D Echography : Status and Perspective

F. Hottier, A. Collet Billon

Laboratoires d'Electronique Philips. 3 Av. Descartes, 94450 Limeil-Brévannes, FRANCE

Abstract

3D echography is still in its infancy but recent preliminary clinical results reveal exciting potentialities. Echography presents specific characteristics which have to be exploited when designing a 3D ultrasound system. After reviewing the various schemes which have been followed to reliably acquire 3D data, the possible exploitation of these volumic data will be described, insisting on the advantages / limitations of the various possibilities. Finally, a survey of the still meager clinical investigations will be made.

Keywords : 3D medical imaging, echography, ultrasound

1. Introduction

All medical imaging modalities are able to generate 3D data [Freiherr 1987, Hemmy 1988]. 3D display of CT data is routinely used in some advanced clinical institutions to evaluate hip displasy or for preoperative surgical planning. Large research and development efforts are also devoted to propose 3D acquisition and reconstruction systems for other types of imaging such as MRI and now, ultrasound [Greenleaf 1982].

Echography presents specific characteristics with respect to other imaging modalities. Its geometrical resolution is clearly poorer than those obtained with CT or MRI imaging although high-end echographic systems now offer good performances (1 mm is a typical resolution). An other difficulty is that ultrasound images are corrrupted by a rather strong texture noise, due to a speckle phenomenon as in the case of coherent optics, and this impedes the differentiation of weakly contrasted regions [Smith 1983]. Moreover, velocity fluctuations in biological tissues may cause ultrasound ray bending which distorts the image.

On the advantage side, the image frame rate typically ranges between 10 and 50 images per second, depending on the type of probe and the scanning mode ; this allows to image moving tissue, in particular the beating heart. As it will be shown later, this also ensures an acquisition time for a 3D echography of a few seconds, to be compared to the several minutes needed in the case of CT or MRI. When the acquisition is performed with annular arrays [Pini 1987], echography also provides thinner tomographic slices, which reduces the danger of merging adjacent structures in the

NATO ASI Series, Vol. F 60
3D Imaging in Medicine
Edited by K. H. Höhne et al.

direction perpendicular to the scan. Let us recall also that echography is a an innocuous, and relatively inexpensive, technique.

Extending medical ultrasound imaging from 2D to 3D raises problems at several levels. We will first review different ways to design and optimize the acquisition unit for getting the most accurate data. Then several methods to exploit these 3D data will be presented, mentioning also the difficult areas of data segmentation due to the peculiar characteristics of ultrasonic images. Finally, clinical interest of 3D echography will be discussed, based on still preliminary clinical studies.

2. The Data Acquisition Problem

2.1 Generalities

Designing a reliable 3D data acquisition system is probably the most crucial step when extending 2D imaging to 3D since the accuracy of the recorded data will greatly condition the quality of the final results. That step is also completely imaging modality dependent and tools to acquire 3D ultrasound data are not at all related to those used for CT or MRI.

Two parameters are especially important when specifying the acquisition system :

- the 3D scanning of the ultrasound beam is to be such that the 3D volume is correctly sampled ;
- the acquisition time has to be reduced as much as possible to minimize possible artefacts introduced by natural tissue motion ; even a minor organ motion can lead to severe misregistrations of the 2D acquired images, blurring then the reconstructed 3D image.

In addition to these two requests, shaping the ultrasound beam in order that its point spread function is homogeneous to get a similar geometrical resolution in the three directions is also important. Some considerations have also to be given to the footprint of the 3D probe, especially when the body acoustic window has a limited extent (for heart exams specially).

We will start with the two following characteristics :

- size of the volume to be scanned,
- sampling pitch,

to define the main specifications of a 3D scanner.

Let us consider a very simple type of apparatus, as shown in fig. 1, consisting in fixing a conventional linear array (the ultrasound beam, always perpendicular to the array front face, is electronically scanned in the X direction) to a mechanical set-up translating it in the Z direction. For a depth of exploration equal to 20 cm, the maximum pulse repetition frequency is 3.8 kHz, and for a sampling pitch fixed at 1 mm (a rather conservative specification which is however justified by the resolution performances of high-end systems), the frame rate is 38 images per second for a 10 cm wide image. Scanning a volume of 10 cm × 10 cm × 20 cm thus leads to a total acquisition time of about 3 seconds.

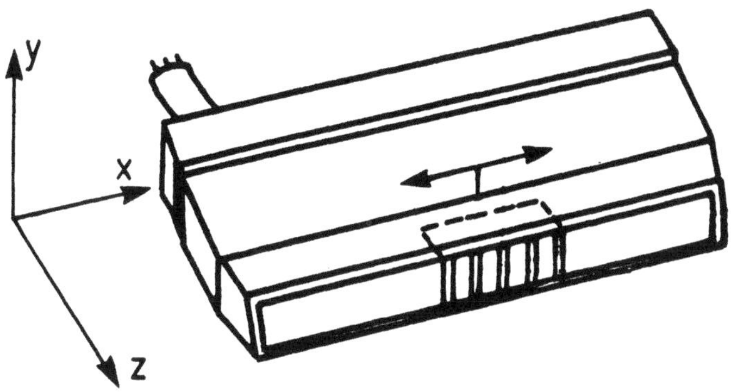

Fig. 1. Principles of a 3D ultrasound scanning system.

These figures have to be adapted, depending on the specific type of scanning and probes used. The three main classes of 3D probes reported in literature, i.e. hand controlled scan head, mechanical 3D probe and 2D arrays, will be now described.

2.2 Hand Controlled 3D Scanner

A simple arrangement [Geiser 1982, Nikravesh 1984, Levaillant 1989] consists in fixing the probe of a conventional real time echographic system at the end of a pantographic arm, for example a B-scan arm of a manual echographic scanner which was used some ten years ago for radiological exams (fig. 2). The practician scans the patient and the spatial position of the probe is indicated by the arm. Note that the scanning plane orientation is constrained to be perpendicular to the B-scan arm movement. This set-up can be readily adapted to any commercial echographic system but the volume scanning is never perfectly regular and strongly depends on the ability of the practician ; this implies a non even sampling of the space. The scanning requires generally ten seconds, which can bring artefacts associated to natural or scanning induced tissue motions.

Due to the large volume of data to acquire (a typical 10 seconds acquisition corresponds to 250 images with a resolution of 512 ×512 pixels, which amounts to a 64Mbyte data volume if the pixel is coded on 8 bits), a video recorder is used as an intermediate storage unit. Synchroneously with the video recording, the position of the probe is picked up by measuring the voltage of the potentiometer based coders of the

B-scan arm, digitizing their values and storing the result on an IBM PC. After the acquisition phase, image and probe position data have then to be transferred to a computer for off-line processing, the video data on the video tape recorder being first digitized using a real time video disk .

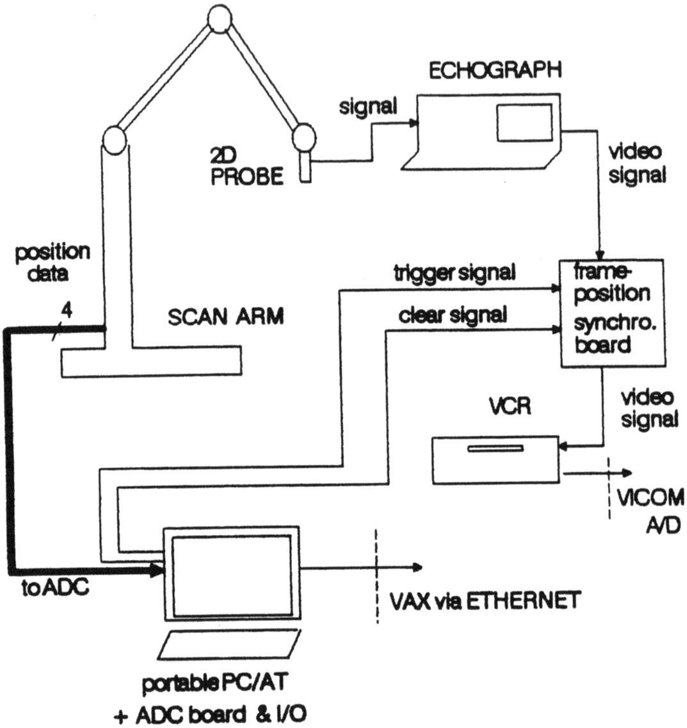

Fig. 2. A simple set-up for 3D echographic data acquisition.

Several investigators [Geiser 1982, Brinkley 1982, Moritz 1983] have also reported the use of a method based on a direct localization of the transducer through the acoustic range principle illustrated in fig. 3. An array of three spark gap sources is attached to the mechanical sector scan head ; spark sources have been chosen as sound sources because their size is small, ensuring an adequate localization of the sound source, and the amplitude of the generated sound is reasonably high. On the other hand, an array of three hemispherical microphones is mounted over the patient. The operator holds the transducer and, when the orientation of the plane is to be recorded, a switch is depressed causing each spark gap source to be activated in turn. Knowing the sound velocity in air, the position of each spark gap source can be successively deduced, with elementary geometry, from the time of arrival of the sound at the three microphones. Since the orientation and the position of a plane are completly defined once the location of three of its points is known, the ultrasound scan plane is straightly deduced from the spark gap array plane. Designing such a system requires some care, in particular when positioning the microphones : the spark gaps must be within the range of the microphones and the operator should ensure than the sources have an unobstructed line of sight to each receiver ; it is also needed to take into account the air temperature since a 10°C variation results in about a 2 % change in ultrasound velocity. Reported precision is of the order of 1 mm.

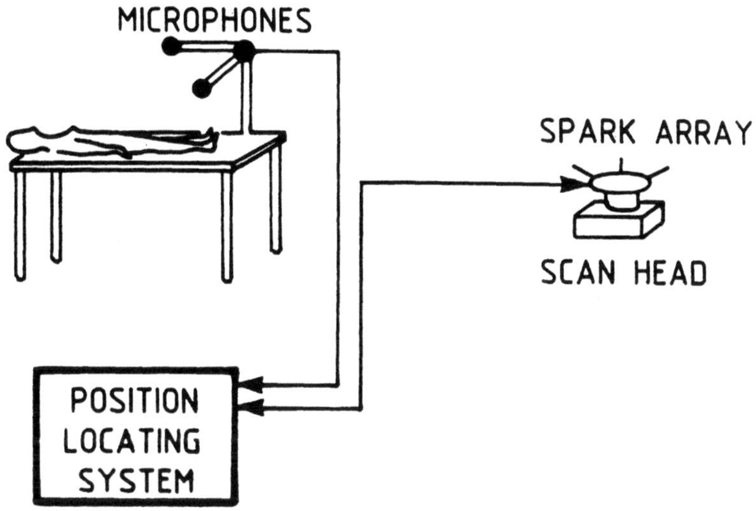

Fig. 3. Direct localization of the transducer probe through an acoustic range principle.

This localization principle is attractive because it offers a maximum of flexibility. The transducer probe is hand held, as for normal 2D echography, and ultrasound slices with any orientation of the scan plane can be recorded, their absolute spatial localization being given by the acoustic range device. However, this flexibility has its drawback which is that the recorded sections may have completely different orientations ; there is no mechanism which partly constrains the scanning of the volume as when employing the previously described B-scan arm based mechanism, or completely controls the scanning order as explained in the next paragraph for mechanical probes or 2D arrays. This completely uncontrolled sampling strategy of the volume may possibly lead to grossly undersampled regions. As in the previous arrangement, images are stored on videorecorders, while the position of the probes, as deduced from the sound time arrival on the microphones, is computed and stored on a microcomputer.

Other approaches can be suggested to position the scan head without direct contact, using a set of laser diodes with suitably located receptors or filming with one or several cameras the hand held moving scan head. For example [Watkins 1989], 3D acquisition of the tongue is performed by painting landmarks on the transducer and the face of the patient and recording with a TV camera the motion of the hand held tranducer, synchroneously with the B-scan images .

2.3 Mechanical 3D Probes

Starting from the probe and the type of scanning used in present real time echographic systems, the idea is to add some computer controlled additional degree of movement to progressively shift the ultrasound scanning plane for volume exploration. There are several possibilities at this level, consisting in combining different types of 2D images (sectorial scanning as in the case of mechanically driven 2D probes, or phased array; parallel scanning for linear array) with a third direction of scanning. Let us take two examples.

A 3D acquisition system for the ultrasound diagnosis of breast tumors has been built [Itoh 1979]. It simply consists in fixing a probe (either a linear array or a mechanical 2D probe) to a computerized translation unit actuated by a stepping motor : the probe assembly can be shifted with regular steps over the breast ; a plastic water bag is placed over the patient's breast and the probe is immersed in the water bath ; the experimental arrangement is quite similar to the one sketched out in fig. 1.

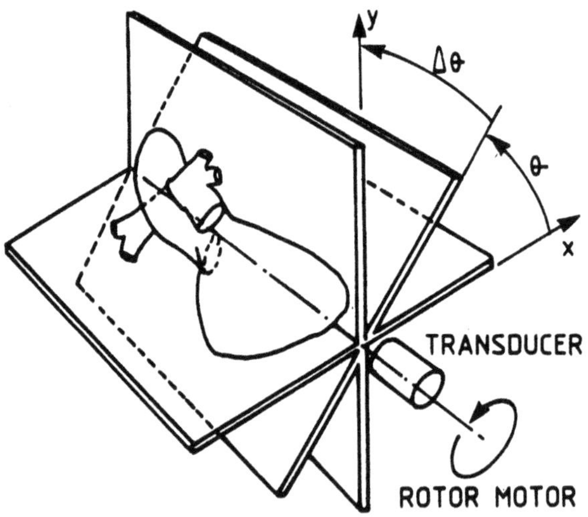

Fig. 4. A small footprint probe for imaging heart.

Imaging heart, in vivo, with relatively narrow acoustic windows (because of the ribs and the gas filled lungs) implies that the 3D probe footprint size has to be small. The approach followed at the Mayo Clinic [McCann 1988] consists in setting up a commercial phased array (type of probe mostly used in clinical echocardiography) on a rotating stage, in such a way the scanned plane can be turned around the phased array axis as shown in fig. 4 ; a 180° field of view can then be explored to obtain 50 views at 3.6° increments. This type of scanning does not provide an even sampling, since the areas close to the rotation axis are sampled with a much smaller pitch than those near the external boundaries of the sectorial images. Recording heart data raises other issues, since the image acquisition has to be triggered with respect to the natural heart motion (caused by the heart natural beating and the patient breathing), in order to get coherent 3D data corresponding to the same position of the heart within the chest. The trick is to gate the image acquisition at the end of the expiration period, when the acoustic window size is the largest. The images are then recorded on a video tape recorder (with the phased array orientation and the ECG), and cardiac gating is then performed by a manual selection of the frames corresponding to a specific phase of the heart cycle. Similar arrangements have also been reported by others [Sohn 1989].

2.4 2D Arrays

This is certainly a very attractive way and will likely be the ultimate solution. The challenge however is formidable. Let us take a phased array-like configuration, adapted to heart exams. A 1D phased array is made of typically 64 elemental transducers with dimensions such as 10 mm length and 0.15 mm width. Such a fine pitch for the array is needed to avoid the presence of grating lobes inducing artefacts on the image ['tHoen 1982, Kino 1987] ; it corresponds to a half-wavelength spacing between elements.

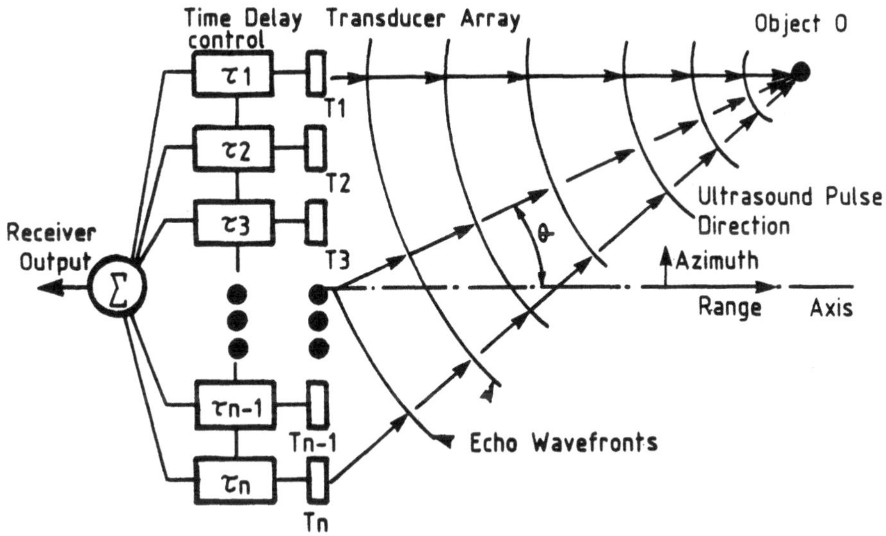

Fig. 5. Beam formation for a 1D phased array.

During the transmission phase, transmission times from the transducers are delayed to focus and steer the beam in a given direction ; these delays are used to compensate the different path lengthes between each of the transducers and the point focused on. During the reception phase (fig. 5), the echoes are dynamically delayed to optimize the focalisation when the wave is coming back to the array. A complex front-end electronics controls the transmission/reception of signals to properly steer and focus the beam. A typical delay accuracy of $\lambda/8$ (λ being the ultrasound wavelength) is generally implemented.

The straightforward extension of this 1D to a 2D phased array would involve the design of a 64 × 64 arrays (to avoid grating lobes), see fig. 6. Making this type of transducer raises very serious issues : the geometry of the elemental transducer is to be such that it generates a unimodal vibration, there must be no cross-coupling between elements, electrical connection to each of these elements has to be provided [Pappalardo 1981, Kino 1987]. However, the main issue here remains the design of the associated front-end electronics which must control in parallel these 4096 elemental transducers !

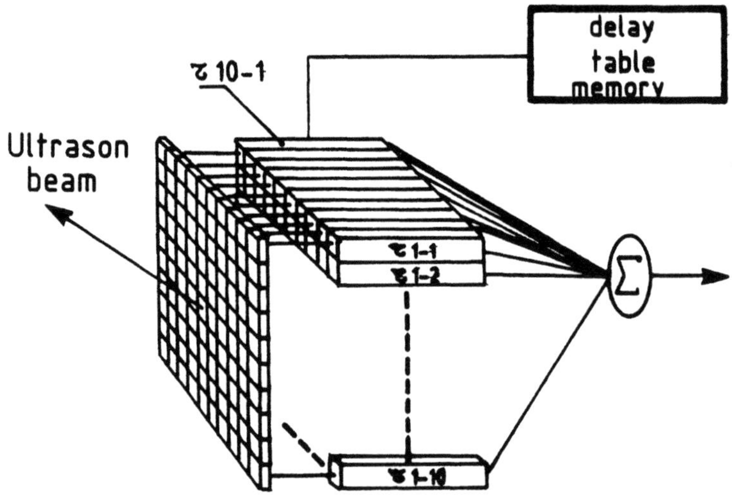

Fig. 6. A 2D phased array.

These characteristics are clearly out of grip with the presently available technology but the realization of a simpler array is now undertaken at Duke University [Poutton 1988]. It consists in a 16 × 16 (256 elements) 2D phased array for echocardiography. Although the image quality will not be optimum, this system will allow to explore new and very exciting potentialities of echocardiography.

There are two possible exploitations of such arrays:

- perform a 2D beam formation in transmission and reception to steer the beam in one particular direction of the 3D space. If the beam former architecture authorizes a full control of the applied delays, any type of scanning can then be programmed. Some examples are indicated in the fig. 7 : the 3D space can be explored by a succession of sectorial images to get a partial or full acquisition of the volume ; this exploration can be effected by rotating the sectorial plane around the symmetry axis of the array, or by modifying the azimuth angle of the sectorial plane. Depending on the required frame, compromise on the data sampling can be useful, going for example towards the acquisition of two orthogonal views which could provide a first estimation of the heart ventricle volume.

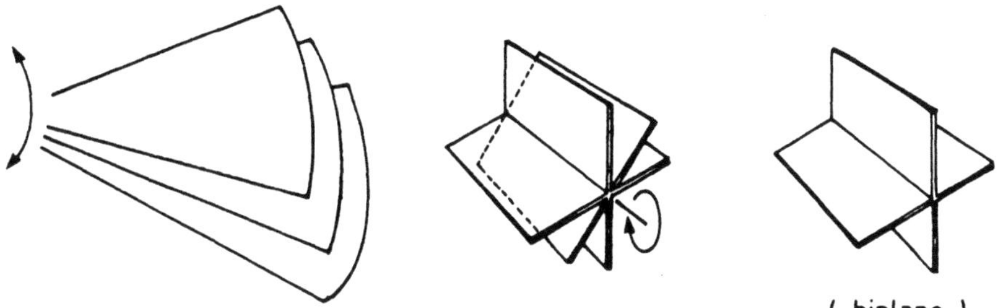

Fig. 7. Different scanning strategies.

- realize a true real time 3D imaging system which is the option chosen by the Duke University (fig. 8). The basic principle here is to transmit an unfocused wide beam which insonifies the whole volume of interest and form in parallel during the reception phase many focused ultrasound beams for adequately sampling the volume of interest ; this requires a number of beam formers equal to the one of the ultrasound beams generated in parallel. They call this concept *Explososcan* Galloway 1986] and its main force is that it is imaginable to have a real time 3D imaging which will be extremely valuable for heart exams (no need to trigger data acquisition, no blurr,...) ; however there are two problems : i) the system obviously is very complex since it requires the realization of a series of beam formers ii) there is no focalisation during the transmission phase, meaning that the geometrical resolution will be poorer than when using more conventional techniques where some focusing is also done during transmission.

The projected front-end electronics is to be made of 64 independant beam formers allowing to generate 9600 separate directions at a 20 cm range in 1/30th second corresponding typically to 100 sectorial planes made of 96 ultrasound lines ; this is quite a reasonable spatial sampling pitch of the 3D space and the 1/30th second time sampling is adequate to *freeze* the heart when capturing 3D data. To achieve this goal, an implementation based on dedicated digital integrated circuits is under way, each chip containing essentially the circuitry for a complete analog to digital conversion, programmable digital delay lines and adders (partial beam formation [Poutton 1988]).

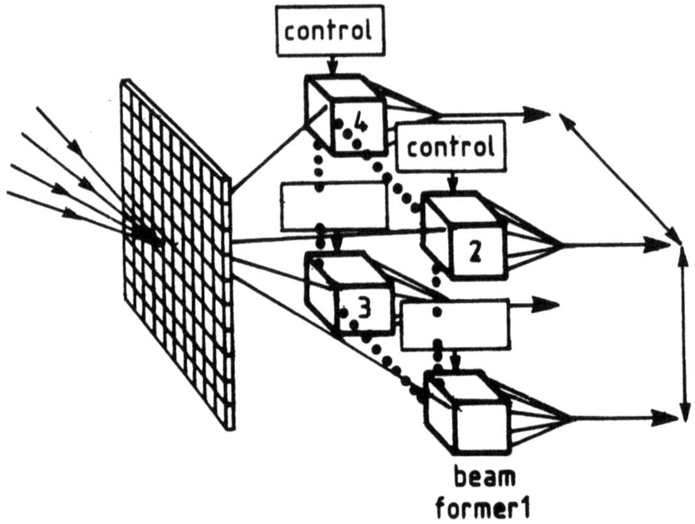

Fig. 8. Parallel beam formers for a 2D phased array.

These specifications in spatial / time sampling lead to a data flow of more than 60 million points per second to transfer, store or process ; this is also a major challenge.

To make easier a practical implementation of 2D arrays, the use of sparse 2D arrays could be envisaged. Such arrays contain only a limited number of elements with a very carefully chosen distribution avoiding in particular any periodic arrangement. It is a well known technique in radar antenna design. Preliminary simulations [Turnbull 1989] for ultrasounds indicate that the number of elements could be reduced by a factor of 6 while keeping side lobes at a sufficiently low level.

3. Data Set Manipulation

At the end of the acquisition phase, a certain number of tissue slices have been recorded with information on their spatial localization. A 3D digital scan conversion has first to be done to obtain a regular 3D sampling of the volume. Then two modes of visualization are possible, multiplanar reformatting and solid 3D rendering.

3.1 3D Digital Scan Converter and Preprocessing

In the case of a hand controlled scanning, with a B-scan arm set-up for example, the scan conversion can be performed as follows (fig. 9) : each pixel of the recorded B-scan images is mapped on the nearest voxel in the 3D cartesian data set, thanks to the probe position derived from the voltage outputs of the B-scan arm. The vertical axis can be chosen, for instance, as the average of the angles of the B-scans.

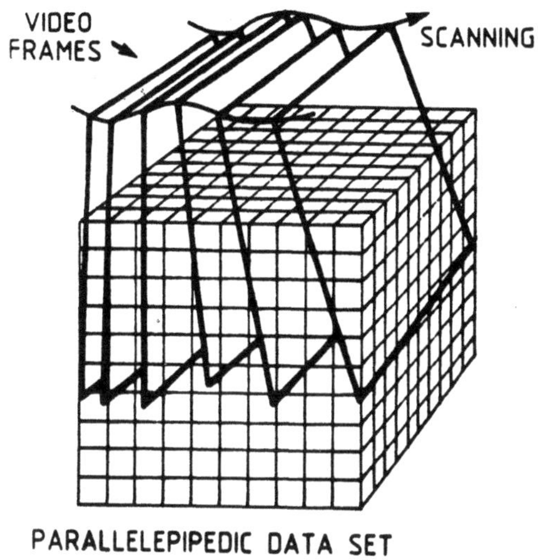

Fig. 9. Transformation of the acquisition data into a regular 3D grid.

As the scanning is irregular, some voxels are oversampled and others are *missed*. When several pixels fall in the same voxel, the average of their value is retained resulting in an efficient speckle smoothing. Missed voxels are filled by linearly interpolating the two closest non-empty voxels in the transverse direction.

In the case when the scanning is automatic, conventional 3D interpolation, linear or cubic, which resamples the data at the location of the voxels can be used.

The last step before visualization is to perform appropriate editing of the 3D data to eliminate overlying structure which might otherwise hide or obscure the object of interest. Ideally automatic segmentation would be a great help at that level, but the present performances of these techniques on echographic data (difficult to segment because of the strong texture pattern) are so poor that the work is most generally done interactively by the operator.

3.2 Multiplanar Data Reformatting

The present 2D echographic systems provide tissue slices which are roughly perpendicular to the body wall ; this is a severe limitation. Selecting one or several views which have a clear anatomical reference would certainly be useful to improve diagnosis accuracy.

Starting from the reconstructed cartesian voxel volume, a procedure to extract a selected plane with a given orientation is easily implemented on a workstation with human interactive time response. The position and orientation of the plane can be controlled by using an interactive device such as the mouse of the workstation. To get the voxel value of the selected section, a tri-linear interpolation from the nearest neighbour voxels is generally sufficient. In fig. 10 are shown three orthogonal views (sagittal, lateral, coronal) of the head of a 16 weeks hydrocephalous fetus. Such scans are impossible to visualize with present echographic systems. Note that, by averaging parallel planes that generates *thick* slices, it is also possible to improve the image contrast at the expense of the out-of-plane resolution.

Other modes of representation such as cutting planes or wedge tools displaying simultaneously several slices representing the visible faces of the object with their respective orientation can also be proposed.

3.3 Surface Modelling and Volume Rendering

A direct understanding of a complex 3D structure may be obtained by rotating and viewing the volume from different angles. A number of techniques, developed first for computer graphics, have been used to display body images.

Surface modelling. In 3D computerized tomography for imaging bone structures, surface volume rendering techniques have been extensively employed [Vannier 1983]. The first step consists in segmenting the data in two classes, those corresponding to soft tissues and those considered as bones, with an adapted thresholding technique, and then extract the surface of the bones. This very simple automatic segmentation procedure works rather well because the contrast between bones and soft tissues is important. Note that the segmentation step is critical since it must not remove diagnostically important information or add artefacts. After that, well known computer graphic techniques, based either on a 1D (wireframe) surface reconstruction or on 2D primitives (patches), model the boundaries of the objects.

This approach has been used also in 3D echography by several experimentators mainly to display the surface of heart ventricles or large vessels. Most often the cavities are segmented manually, isolating for example, in the case of heart, the endocardial surface. This is a lengthy procedure and developing methods to ease that step is important. Segmentation by a simple thresholding technique is clearly unsufficiently reliable. More robust semi-automatic (human expert guided) techniques are a critical research field and partial answers have been reported.

Volume rendering. Surface modelling is not always applicable to 3D ultrasound images because the boundaries between soft tissues are often quite fuzzy and somewhat smoothed out by the ultrasound speckle. Volume rendering algorithms which are largely employed for representing MRI 3D data are also powerful tools in 3D echography. These techniques process the volume directly ; there is no need for binary operation to segment data, making sure that no image information is lost.

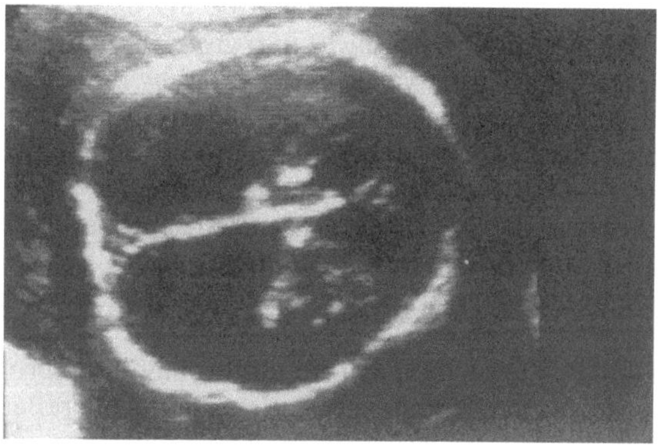

a

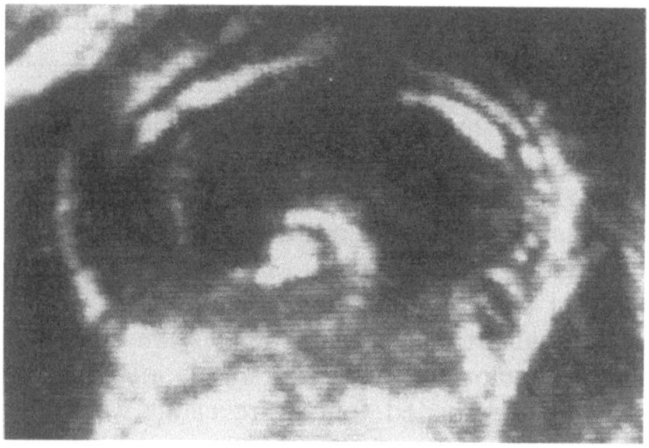

b

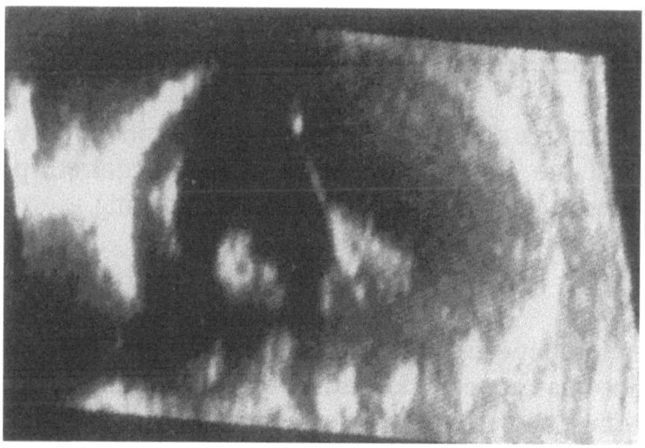

c

Fig. 10. Coronal (a), lateral (b) and sagittal (c) views of the brain of a hydrocephalous fetus.

The simplest technique consists in reprojecting the 3D data : the observer looks at the 3D volume and the intensity he perceives in a given direction (corresponding to a specific ray) is a weighted sum of the voxels intercepted by the ray ; to get some depth cues, voxel values are weighted according to the distance to the observer's eye, simulating a medium with a certain attenuation. The so obtained reprojected views, akin to a radiography of the volume, are much smoother than the original 2D echographic images because the speckle has been efficiently averaged by the process. For a good appraisal of the 3D structure, it is important to dynamically change the viewpoint, for example by continuously rotating the 3D object in front of the observer.

Editing the data volume is to be done before performing the reprojection ; to display a fetus, it is important to segment the data to cut out the highly reflecting placenta part which would impede a proper fetus visualization. Some additional processings, prior to reprojection, to enhance the contrast (3D filter, grey scale optimization) are also very useful to improve the visibility of specific anatomical features. Breast lesions, shown in fig. 11, which generally have a lower reflectivity than the surrounding healthy tissue, can be highlighted through an adaptive contrast reversal algorithm. Yet, details at the surfaces are hard to discern due to the absence of shading.

More sophisticated volume rendering techniques which involve at the same time volumic and surface representations are presently under evaluation for 3D echographic display [Drebin 1988, Levoy 1988]. For ultrasounds, a classification assigns to each voxel a partial opacity and color according to the local image reflectivity and to a possible region segmentation. Computation of a local 3D gradient estimates the amount of surface present and its orientation, relative to an external source of light. Computation of the 3D gradient magnitude estimates the amount of surface present. These three informations (color for tissue differentiation, opacity and surface) are finally combined to obtain the shaded volume, revealing weak boundary details (fig. 12).

4 . Preliminary Clinical Evaluation

Although clinical evaluation of 3D echography is still in its infancy, it is worthwhile to have a brief survey of the clinical experience already acquired, pointing out the the strength of this new technique with respect to the 2D imaging.

4.1 Breast lesions

A better differentiation between breast adenofibromas and adenocarcinomas has been reported. Typical fibromas appear like rather regular volumes presenting well defined boundaries with normal tissues ; they extend more or less parallel to the breast wall. Typical adenocarcinomas have a much more intricate structure interfering with the normal tissue ; their orientation is often rather orthogonal to the breast interface [Levaillant 1989].

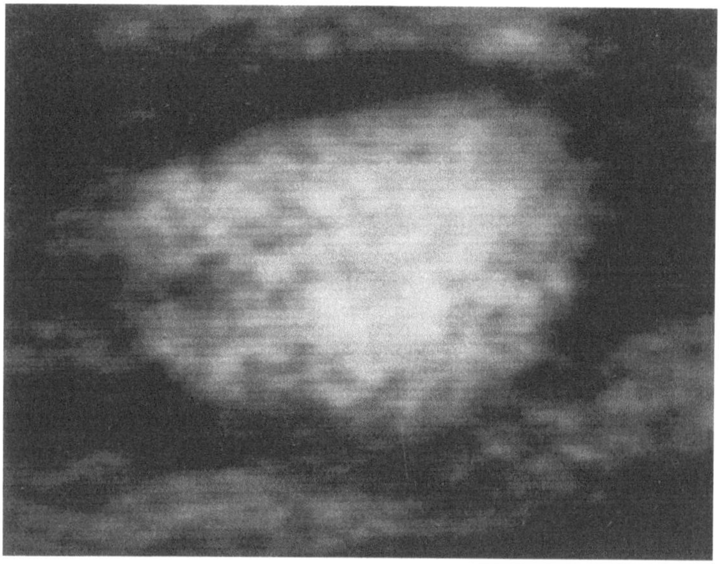

a

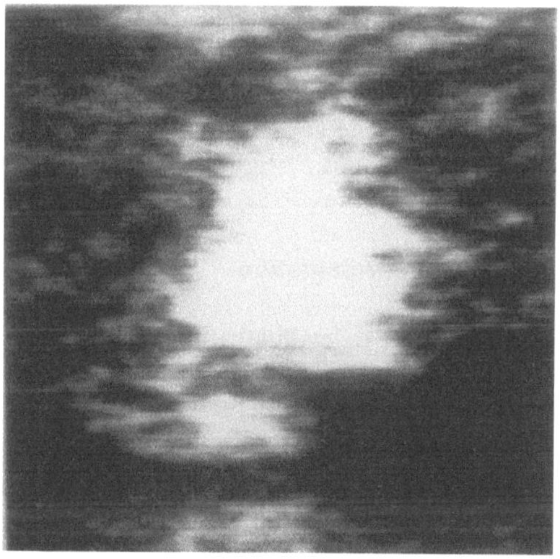

b

Fig. 11. Breast lesions : adenofibroma (a) and adenocarcinoma (b).

4.2 Fetus

- Fetal weight : fetus growth rate is generally estimated by combining a series of fetal size measurements such as circumferences (skull,...) and length (tibia,...). The reliability of such estimates are far from satisfactory [McCallum 1979, Brinkley 1984]. Measuring volume by counting the number of voxels in the area of interest should bring substantial improvements. More generally, by selecting the right anatomical view, it allows an exact anatomical measurement of distances, areas and volumes.

- Fetal morphology : a 3D image of a 12 week old fetus is shown in fig. 13 : a continuous rotation allows to appreciate its general morphology and readily detect possible abnormalities. Reslicing [Levaillant 1989] gives access to a fine morphological study of the fetus internal structure.

4.3 3D Echocardiography

Although 3D echocardiography [Matsumoto 1981, Eiho 1981, Samada 1983, Linker 1986, McCann 1987, Pini 1989] raises very difficult issues, it has attracted a lot of attention simply because 2D echocardiography is presently a major technique, routinely used to assess heart functions and performances. 2D echocardiography is a non invasive technique, portable and relatively inexpensive. Two of its features are of special interest :

- it is a real time imaging modality of the beating heart with a typical frame rate of 25 images per second, allowing to follow the evolution of chambers and stroke volumes, evaluate the ejection fraction, visualize regional wall motions indicative of possible infarcted zones, control valve efficiency ;
- blood flow within the heart chambers and the main vessels is readily assessed by conventional Doppler techniques giving the local blood velocity in selectable areas. The more recently introduced Color Flow Mapping technique (CFM) is also more and more routinely used ; such a technique supplies a real time map of the blood velocity and direction (coded with different hues / brightness) superimposed to the usual grey scale B-mode image of the soft tissues.

Which benefits can bring a 3D representation ?

2D images are successfully used to evaluate the cardiac function and show a wide variety of vascular disorders or myocardial disease but it is far from ideal to understand the 3D morphology of the beating heart and the estimation of the heart cavities (left ventricle in particular) is far from accurate. Generally volume estimation is derived from the areas of two cross-sections of the ventricle, modelling it by a crude shape such as an ellipsoid. The shape of a normal heart ventricle is far more complex, not speaking of the diseased heart. Deriving the exact volume from 3D echography will certainly largely improve the accuracy of this important parameter ; validation of this technique, based on in vitro experiments, has already been reported. The two ways to acquire the data needed to perform the reconstruction have already been described in the data acquisition section :

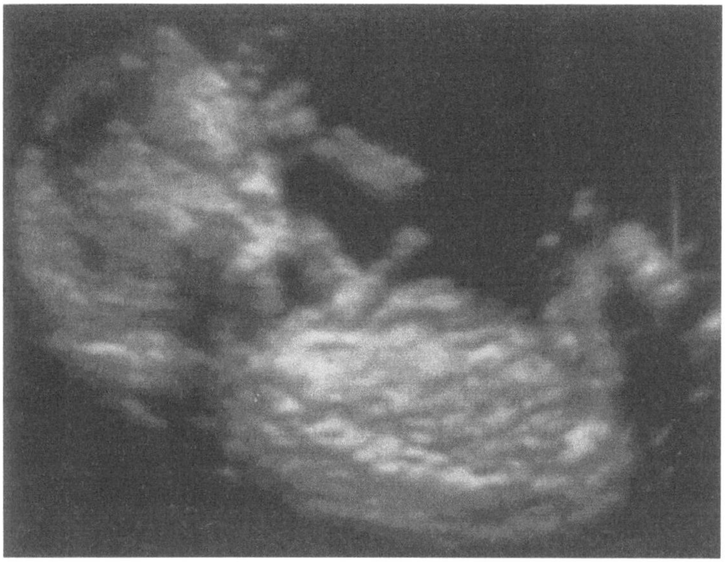

Fig. 12. 3D representation of a 12 week old fetus by a volume rendering technique.

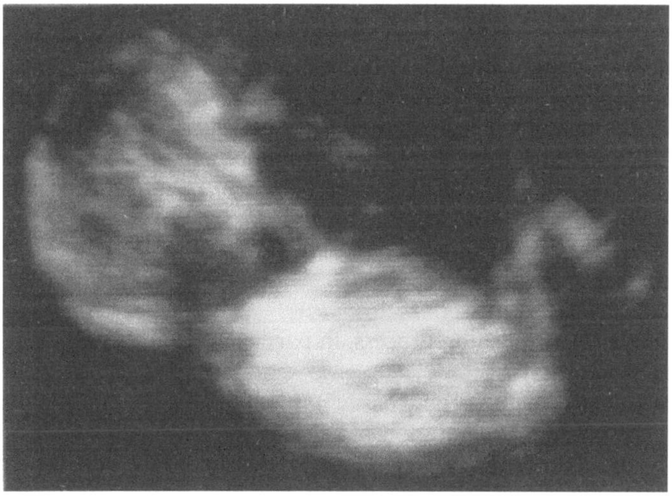

Fig. 13. 3D representation of a 12 week old fetus by reprojection.

- use short-axis images obtained by changing the position of the probe on the chest wall
- grasp a series of apical long axis views by rotating the probe around its axis.

Assessing local heart disease should also be made much easier by the 3D representation. Currently the cardiologist examines each part of the moving heart muscle in turn, to detect possible abnormalities indicative of ischemy or infarctus. Greenleaf [1982] has shown that surface modelling techniques provide an accurate representation of the anatomic structures and that the 3D complete view of the endocardial surfaces allows a much better appreciation of wall motion abnormalities. By using a color mapping representation similar to the technique employed for blood flow imaging, variations of wall thickness during a cardiac cycle can be evaluated.

Volume rendering technique are also considered to be important since they keep the real backscattering structure of the myocardium in the form of texture patterns and this type of information is considered to be useful by cardiologists.

5. Conclusion

Potentialities of 3D echography are significant :

- the acquisition time is of the order of a few seconds, which guarantees a high patient throughput ;
- reslicing the 3D volume along an arbitrary orientation relaxes the constraint of having access to tissue slices nearly perpendicular to the body wall. The most significant orientations at the anatomy level can then be selected ;
- visualizing 3D structures joined to the possibility of making accurate measurements of distances, areas and volumes provides useful tools to quantify the evolution of lesions [Dumke 1989].

It also raises issues :

- how to optimize the acquisition process, in particular when the organ is moving (heart for sure, but also fetus,...) ?
- the interacive data reslicing is now easily implementable on modern workstations. On the contrary, the volume computation needed to offer realistic volume rendering with the possibility of rotating the object is not yet feasible with non-dedicated hardware. However, system designers strive to design graphics systems for interactive volume display [Robbe 1986, Williams 1988] ;
- a segmentation/editing of the original data must be done prior to the 3D representation. Most of the time, this task is performed manually, since automatic segmentation procedures are not sufficiently reliable. This is a very tedious process.

Considering the more and more important research and development effort devoted to 3D echography, these problems should progressively be solved, making this new technology transferable into the clinical environment.

Acknowledgements

3D echography is a three years old project at LEP. We thank Y. Le Guerinel and P. Rua who, with A. Collet Billon, have developed the data acquisition system and the visualization tools reported in this paper. We also thank our colleagues MD Levaillant and Rotten who constantly indicate us what is clinically important in 3D echography.

6. References

Brinkley, J.F., McCallum, W.D., Muramatsu S.K., Liu D.Y. : Fetal weight estimation from lengths and volumes found by three-dimensional ultrasonic measurements. J. Ultrasound Med. 3, 163-168 (1984)

Brinkley, J.F., Muramatsu, S.K., McCallum, W.D., and Popp, R.L. : In vitro evaluation of an ultrasonic three dimensional imaging and volume system. Ultrasonic Imaging 4, 126-139 (1982)

Drebin, R.A., Carpenter, L. and Hanrahan, P.: Volume rendering. Computer graphics 22, n°4, 65-74 (1988)

Dumke, A.E., Lizzi, F., Coleman, D.J., Rosado, A.L. and Silvermann, R. : Three-dimensional imaging of tissue structures for treatment planning and monitoring. Ultrasonic Imaging 11, 150 (1989)

Eiho, S. : Reconstruction of 3D-images of pulsating left ventricle from two-dimensional sector scan echocardiograms of apical long axis. In : 1981 Computer Cardiology, 19-24 (1981)

Freiherr, G. : 3D imaging in medicine : synthesizing the third dimension. Diagnostic Imaging 9, 190-203 (November 1987)

Galloway, L. and Thurstone, F.L. : Recent applications of parallel processing techniques to improve ultrasound B-mode images. Ultrasonic Imaging 8, n1, p 69 (1986)

Geiser, E.A., Christie, L.G., Conetta, D.A., Conti, R. and Grossman, G.S. : A mechanical arm for spatial registration of two-dimensional echocardiographic sections. Cathet. Cardiovasc. Diagn. 8, 89-101 (1982)

Geiser, E.A., Lupkiewicz, S.M., Christie, L.G., Ariet, M., Conetta, D.A. and Conti, C.R.: A framework for three-dimensional time-varying reconstruction of the human left ventricle: sources of error and estimation of their magnitude. Computers and Biomedical Research 13, 225-241 (1980)

Greenleaf, J.F. : Three-dimensional imaging in ultrasound. J. Med. Systems 6, 579-589 (1982)

Hemmy, D.C. : Three dimensional imaging, the first decade. Special symposium on maturing technologies and emerging horizons. IEEE, 61-64 (1988)

't Hoen, P.J. : Systematic analysis of grating lobe influence on ultrasonographic array directivity. ISM III Proc., Berlin, pp 94-97 (1982)

Itoh, M. and Yokoi, H. : A computer-aided three-dimensional display system for ultrasonic diagnosis of a breast tumor. Ultrasonics, 261-268 (Nov. 1979)

Kino, G.S. : Acoustic waves : Devices,Imaging , and Analog Signal Processing. New-York, NY : Prentice Hall 1987

Levaillant, J.M., Rotten, D., Collet Billon, A., Le Guérinel, Y. and Rua, P. :

Three-dimensional ultrasound imaging of the female breast and human fetus in utero : preliminary results. Ultrasound Imaging 11, 149 (1989)
Levoy, M. : Display of surfaces from volume data. IEEE Computer Graphics and Applications (May 1988)
Linker, D.T, Moritz, W.E. and Pearlman A.S. : A new three-dimensionnal echocardiographic method of right ventricular volume measurement : in vitro validation. J. Am. Coll. Cardiol. 8, 101-106 (1986)
McCallum, W.D. and Brinkley, J.F. : Estimation of fetal weight from ultrasonic measurements. Amer. J. Obstetrics Gynecology 133, 195-200 (1979)
McCann, H.A. : A method for three-dimensional ultrasound imaging of the heart in vivo. Dynamic Cardiovascular Dynamics 1, 97-109 (1987)
McCann, H.A., Sharp, J.C., Kinter, T.M., McEwan, C.N., Barillot, C. and Greenleaf, J.F. : Multidimensional ultrasonic imaging for cardiology. Proc. IEEE 76, 1063-1072 (1988)
Matsumoto, M., Inoue, M., Tamura, S., Tanaka, K. and Abe, H. : Three-dimensional echocardiography for spatial visualization and volume calculation of cardiac structures. J. Clin. Ultrasound 9, 157-165 (1981)
Moritz, W.E., Pearlman, A.S., McCabe, D.H., Medema, D.K., Ainsworth, M.E. and Boles, M.S. : An ultrasonic technique for imaging the ventricle in three dimensions and calculating its volume. IEEE Trans. on Biomedical Engineering 30, 482-492 (1983)
Nikravesh, P.E., Skorton D.J., Chandran K.B., Attarwala Y.M., Pandian N. and Kerber R.E.: Computerized three-dimensional finite element reconstruction of the left ventricle from cross-sectional echocardiograms. Ultrasonic imaging 6, 48-59 (1984)
Pappalardo, M. : Hybrid and matrix acoustic arrays. Ultrasonics, 81-86 (March 1981)
Pini, R., Ferruci, L., DiBari, M., Greppi, B., Cerofolini, M., Masotti, L. and Devereux , R.B. : Two-dimensional echocardiographic imaging : in-vitro comparison of conventional and dynamically focused annular array transducers. Ultrasound in Med. and Biol. 13, 643-650 (1987)
Pini, R., Monnini, E., Masotti, L., Greppi, B., Carofolini, M. and Devereux, R.B. : Echocardiographic computed tomography of the heart : preliminary results. J. Am. Coll. Cardiol. 13, 224A (1989)
Poutton, J., Von Ramm, O. and Smith, S. : Integrated circuits for 3-D medical ultrasound imaging. MCNC Technical Bulletin, 6-7 (Jul./Aug. 1988)
Robb, R.A., Heffernan, P.B., Camp, J.J. and Hanson, D.P. : A workstation for interactive display and quantitative analysis of 3-D and 4-D biomedical images. In : Proc. Computer Applications in Medical Care 1986, Washington DC, pp 240-256 (1986)
Sawada, H., Fujii, J., Kato, K., Onoe, M. and Kuno, Y. : Three-dimensional reconstruction of the left ventricle from multiple cross sectional echocardiograms value for measuring left ventricular volume. Br. Heart J. 50, 438-442 (1983)
Smith, S.W., Wagner R.F., Sandrik J.M. and Lopez H : Low contrast detectability and contrast/detail analysis in medical ultrasound. IEEE Trans. on Sonics and Ultrasonics 30 (1983)
Sohn, C., Grotepass, J., Schneider, W., and Ameling, W.: Dreidimensionale organdarstellung mittels Ultraschall. Fortschr. Med. 107, no.13, 281-284 (1989)
Turnbull, D.H. and Foster, F.S. : Theoretical steered beam profiles from a two dimensional transducer array. To be published in 1989 IEEE Ultrasonics Symposium Proc.

Vannier, M.W., Marsh, J.L. and Warren, J.O. : Three dimensional computer graphics for craniofacial surgical planning and evaluation. Computer Graphics (SIGGRAPH'83 Proc.) 13, 182-189 (1983)
Watkin, K.L. and Rubin, J.M. : Pseudo three dimensional reconstruction of ultrasonic images of the tongue. J. Acoust. Soc. Am. 85, 496-499 (1989).
Williams, T. : Graphics system designers strive for photorealism. Computer Design, 50-61 (August 1988)

Object Definition

NETWORK REPRESENTATION OF 2-D and 3-D IMAGES

A.C.F.Colchester

Department of Neurology, Guy's Hospital, London Bridge SE1 9RT

Abstract

Much of the data and knowledge representation in image interpretation needs to be in the form of networks. We propose that a network representation of the grey level changes in an image should be constructed as early as possible during bottom-up processing. We refer to this low level description as the image structure representation. The image is treated as a continuous surface made up of triangular facets (in 2-D) or as a continuous volume of tetrahedrons (in 3-D). The image is completely segmented into non-overlapping slope districts. Each slope district is anchored between one peak and one pit and (usually) two or more saddle points. Slope districts can easily be grouped in several ways to form more meaningful entities such as edge support regions, ridges, convex corners etc.

Keywords: Segmentation, image structure, grey level morphology, edges, connectivity, symbolic, data representation, knowledge representation.

1. INTRODUCTION

1.1. The Need for a General Purpose Segmentation System

In the past ten years increasing attention has been paid to the value of using domain-specific knowledge in the analysis of medical images. Such knowledge is employed in two distinct ways [Ritchings et al 1985]. Firstly, factual knowledge is used in the specification of models. Secondly, strategy knowledge is used to guide bottom-up processing (including segmentation) and also top-down processing. With regard to bottom-up processing, selection of image processing operators using expert knowledge, by the human user if not by the computer, is usually taken for granted. However, in most image interpretation domains there is a requirement to be able to describe novel entities which cannot be matched to any models and which must be extracted during the segmentation process even if no prior knowledge of their probable size, shape, position, contrast etc is available. Therefore, even in a restricted domain such as the interpretation of magnetic resonance

NATO ASI Series, Vol. F 60
3D Imaging in Medicine
Edited by K. H. Höhne et al.

images of the brain, it is important that segmentation operators allow the compilation of general descriptions of salient features as well as domain-tuned descriptions of features which are candidates for matching to specific models.

The approach we have taken to segmentation at Guy's Hospital can best be understood in the context of the data and knowledge representation schemes used in the computer vision system which we are developing.

1.2. Data and Knowledge Representation Schemes for an Intelligent Vision System

There are three major sets of representations: image, model, and current patient representations. Each set consists of multiple subdivisions. The standard data structure for each representation is a semantic network. Most links in the networks imply spatial relations, particularly (1) connectivity or adjacency, or (2) a part hierarchy.

The **model representations** are of several types. The present discussion is restricted to factual knowledge rather than strategy knowledge. Factual knowledge is organised into topics such as "brain anatomy", "CT numbers", "conventional radiographic projections" etc. This is efficient for summarising general principles that enable the knowledge to be used in a variety of contexts, but the corollary is that the representations will usually have to be modified to incorporate constraints that make model matching feasible in a particular context. These modifications constitute top -down processing.

There are several **image representations** at different levels of abstraction. The lowest level consists of the input image where the primary index for each entity (pixel) is its position. At the highest level a symbolic representation refers to entities composed of linked entities from lower levels. For example, a high level entity might refer to a structure occupying a large part of the image, formed by linking together several edge sections and several regions contained in its interior. The image representations are initially constructed by bottom-up processing. They form an increasingly refined image description. Entities in the higher level descriptions may be formed by universal rules which extract salient features in any context. In addition, entities may be formed by the application of domain-specific rules which tune the bottom-up processes so that features which are likely to match model entities are sought in the image. A subset of such image entities is used to seek initial matches with certain model entities; these entities are referred to as cues.

The **current patient representations** are constructed from three types of entity which already exist in the image and/or the model representations. Firstly, there are entities which exist in both the image and the model representations and have been linked to each other. Secondly, there are model entities which are not matched to any specific image entity but whose presence in the current patient is implied by other entities for which there is direct evidence in the current image. Thirdly, there are salient features which exist in the image but have not been matched to any model entities.

1.3 Network Data Structures

Higher level, symbolic representations refer to entities that may occur at arbitrary locations and take the form of a network. The management of transformations (bottom-up and top-down) between these high level representations of irregularly spaced entities and low level arrays of regularly sampled pixel intensities is hindered by the fundamental difference in their data structures. During bottom-up processing, we argue that the transformation to a network data structure which makes key spatial relations explicit but is not dependent on regular sample spacing should be carried out as early as possible. Nevertheless, the earliest such representation must provide a detailed description of grey level changes if loss of potentially important information is to be avoided. Thus, our aim is to construct a symbolic but low level description which avoids premature interpretation. Subsequent transformations of the image representations all involve fundamentally similar processes and data structures. Furthermore, these data structures are the same type as those used in all but the lowest level of model and current patient representations.

1.4. Formation of Districts where Changes of Grey Level are Uniform

Our initial approach to the construction of a symbolic, low level description of image structure was developed in collaboration with Tim Ritchings and Nehal Kodikara at the University of Manchester Institute of Science and Technology [Colchester et al 1990a,b]. Edge support regions were constructed from maximum gradient profiles (MGPs). MGPs are polylines linking pixels along lines of maximum grey level gradient in the image.

The approach of Burns et al [1986] for constructing districts by identifying areas of uniform grey change in their interior had some similarities to ours. They grouped local gradient values into straight edge support regions without creating an intermediate representation. In contrast, we first constructed MGPs from local gradient values, and then grouped the MGPs into edge support regions. This allowed us to form regions supporting modestly curving as well as straight edges. Both methods were developed for the identification of high gradient areas of the image and involved the use of a threshold which is a potential disadvantage.

Several conclusions can be drawn from these approaches. Edges can be successfully treated as **regions** of high and/or consistent slope. Defining a region in which there is a boundary may be as important as defining the correct linear locus of the boundary. It is clear that in some domains, including quantitative angiography, the accurate localisation of boundaries requires the use of special operators which depend on domain knowledge. Yet many aspects of processing do not depend on accurate localisation of boundaries. For example, even if boundary localisation has not been refined, successful identification of vessels is possible, and accurate centre line localisation may be possible because of the symmetry of the boundary regions.

1.5. Gradient Thresholds and Plateaus

In the method described in section 1.4, we used a local gradient threshold (calculated from the current image data) below which pixels were not grouped into MGPs. Pixels where the gradient was less than this were ignored. This was appropriate for the images under examination, where the larger vessels were reliably identified. However, in some parts of the image, MGPs constructed from low gradient areas still corresponded to significant features. Furthermore, following Marr's principle of least commitment [1976], we wished to define districts of coherent grey level change which would help to establish a low level image description rather than a - possibly premature - image interpretation. Such a general description should include low gradient as well as high gradient districts. Some districts might during subsequent processing be identified as due to noise but would be merged together in a multi-scale hierarchy. Others could be used as a component of a structural texture classification. Only selected districts would be treated as edge components. Areas of zero gradient, i.e. plateaus, would merely be extreme cases of the general type of district.

1.6. Defining Districts from their Boundaries

To define a general type of district, it was necessary to establish a consistent criterion for the boundaries. Singular points such as local maxima and minima would be important anchor points on ridges and troughs which would form the upper and lower boundaries. Appropriate criteria for the sides of a district were not as easy to identify. Continuing to draw on the analogy with a terrain surface, we explored the use of watersheds and watercourses to define district boundaries.

More than a hundred years ago, Maxwell [1870] proposed decomposing a landscape into "natural districts" consisting of a dual representation of hills and dales, and more recently this has been applied to water surfaces [Longuet-Higgens, 1960]. Hills and dales correspond to parabolic sectors in mathematical topology. A hill is the set of points from which maximum uphill paths lead to one particular peak, and a dale is the set of points from which maximum downhill paths lead to one particular pit. Maxwell also showed that these regions can be defined by their boundaries in terms of maximum uphill and downhill paths emanating from saddle points and terminating at peaks and pits respectively. These uphill paths correspond to watersheds and the downhill paths correspond to watercourses. Hills are bounded by watercourses, enclosing one peak, and dales are bounded by watersheds, enclosing one pit.

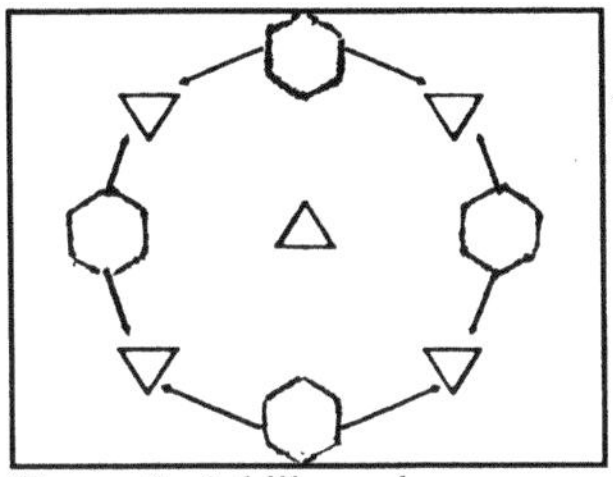

Figure 1. A hill graph

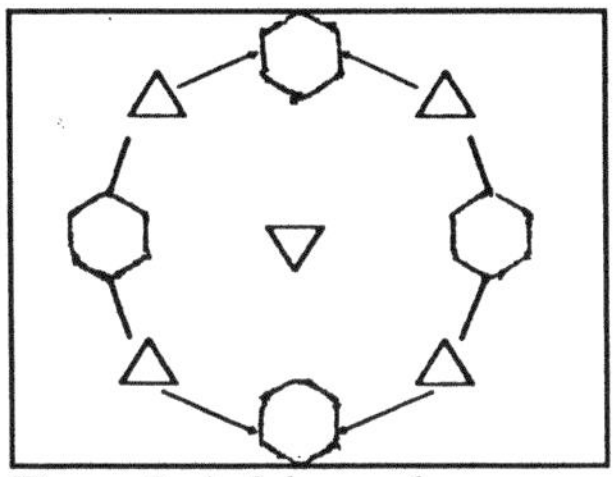

Figure 2. A dale graph

Thus, hills and dales can be represented by separate graphs whose nodes are singular points, and whose directed arcs are watersheds or watercourses. In this paper peaks are shown as ▵, pits as ▿, and saddles as hexagons. Arrows point downhill. Maximum uphill paths are shown by solid lines and maximum downhill paths by dashed lines.

Figures 1 and 2 show a hill and a dale diagrammatically. A watershed separates two districts both of whose maximum uphill paths reach the same peak, and whose maximum downhill paths terminate at two different pits. Likewise a watercourse separates two districts both of whose maximum downhill paths terminate at the same pit, and whose maximum uphill paths terminate at two different peaks.

It is possible to construct a unified representation which incorporates both watersheds and watercourses [Rosin et al 1990]. Within the basic district, in general all maximum uphill paths lead to one particular peak, and all maximum downhill paths lead to one particular pit. These regions usually have boundaries formed by connecting singular points as follows: saddle point 1 - peak - saddle point 2 - pit - saddle point 1. The interiors of each district would in principle contain sets of maximum gradient profiles running approximately parallel to each other, but by defining the sides of a district in relation to saddle points, only the MGPs linking the saddle points to peaks and pits would actually need to be constructed.

2. METHOD: 2-D SEGMENTATION

2.1. Construction of the Basic Representation of Image Structure

2.1.1. Data Structures and Representation of the Image as a Connectivity Network. Input data points in general can be specified with arbitrary sample spacing in x & y and an intensity or z value which is treated as a point height. Near neighbour data points are linked to form triangles using Delaunay triangulation. With this method of triangulation, the circumcircle of each triangle does not enclose any other data points. This is one way of producing triangles which are as regular as possible. We treat images as continuous surfaces. That is, the triangles are treated explicitly as inclined planes. At any point on one of these planes, height values are the linearly interpolated intensity values between the three neighbouring sample points.

Where data points form a symmetric pattern such as a square or hexagon, a procedure for choosing between the alternative, equally good triangulations must be established. Many implementations choose one triangulation arbitrarily. We avoid this problem by interpolating a new data point at the centre of any symmetric pattern. Our normal input image data of course form a symmetric set of points where four adjacent pixel-centres form the corners of a square. A grey value is linearly interpolated in the middle of each square which is thus always split into four triangles. The discrete input image can now be considered as a continuous surface made up of triangular planes whose vertices consist of two original pixel values and one interpolated point.

2.1.2. Classification of 2-D Data Points. The first stage of processing of the surface representation involves finding the singular points. For every pixel and interpolated point we compare its grey value with the grey value of each of its neighbours in rotation. Peaks and pits are pixels that are greater or smaller respectively than their neighbours. The interpolated points can never be local maxima or minima. Saddles are either pixels or interpolated points where the gradient changes from uphill to downhill or vice versa four or more times when the point is compared with each neighbour in rotation.

2.1.3. Construction of 2-D Maximum Gradient Profiles between Singular Points. From every saddle, MGPs are formed by tracking the maximum uphill paths to peaks, and maximum downhill paths to pits. The paths are not constrained to run along the borders of the triangles but are allowed to cross their interior (figure 3).

2.1.4. Construction of 2-D Slope Districts. Districts are formed by connecting MGPs in sequence so that they enclose an area which is not crossed by any other path. Starting from an arbitrary path, the sequence of forming the district to its right hand side is as follows. Points at the beginning and end of the path are listed in the path's data structure. The data structure for the singular point at the end of the path is inspected. The next path to which this point is connected in anticlockwise rotation is linked to the district's boundary list. The data structure of the point at the far end of this latest path is next inspected, and the next path is connected in the same way. This process is repeated until the singular point at the end of a linked path is the same as the singular point at the beginning of the first path.

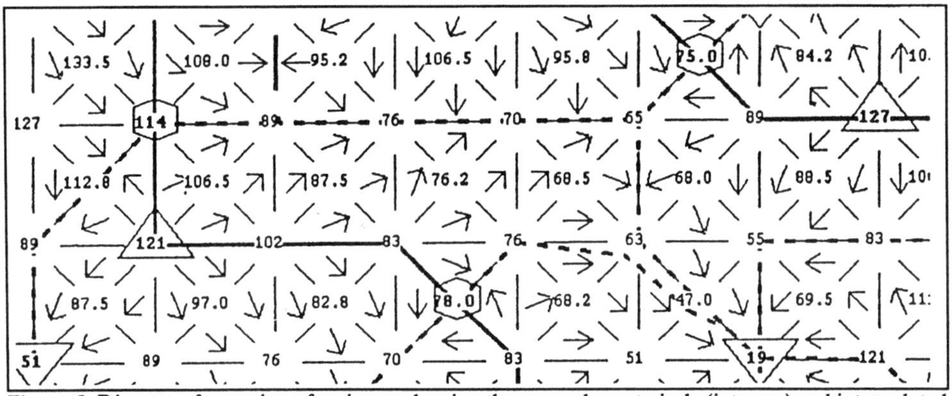

Figure 3. Diagram of a portion of an image showing the grey values at pixels (integers) and interpolated points (decimals), triangular planes with their directions of maximum gradient, singular points, and maximum gradient paths.

Figure 3 shows a small portion of an image which has been processed. The pixels and interpolated points are shown connected to form triangular planes. In each triangle an arrow shows the direction of maximum downhill gradient (quantized to sixteen directions in the display, but calculated and stored with a higher precision). The peaks, pits, and saddles are marked by ▵, ▿, and hexagons respectively, the maximum uphill paths by continuous lines, and the maximum downhill paths by dashed lines.

2.1.5. Zero Gradient Paths and Plateaus. The above rules do not cover groups of points with zero gradient. If groups of contiguous points have the same value they are grouped to form either lines or areas of zero gradient. A plateau is considered as one entity, and its neighbours are checked by tracking round the region's perimeter. If a plateau region satisfies the appropriate conditions, it will be treated as a peak, pit, or saddle in the same way as if it were a point. A saddle is defined as a point or flat region with four or more sign changes of gradient when tracking round the perimeter. The number of sign changes is always even within the image but not on its borders. Since pixels and interpolated points within the image have eight and four-way connectivity respectively, they are restricted to a maximum of four and eight sign changes. Plateau regions, however, will be connected to more points, and so the larger a region, the more sign changes of gradient round its perimeter are possible. All maximum uphill and downhill paths between sign changes are generated, and regions are formed as before.

2.1.6. District Attributes. A set of attributes is calculated for each district. The attributes allow the districts to be combined in various ways by the grouping rules described below. Currently these attributes are: the XY centroid; average grey-level value; the grey-level difference from peak to pit; gradient magnitude; and gradient direction. For ease of computation, the following approximations are made: the average grey-level value is taken as the average of the peak and pit values, the gradient direction is taken as the direction from the peak to pit, and the gradient magnitude is taken as the grey-level difference from peak to pit divided by the distance from the peak to pit.

The segmentation of the complete image into districts forms the basic representation of image structure.

2.2. District Grouping

In most application areas it is to be expected that the slope districts in the low level representation of image structure outlined above would usually require further processing to construct candidates for matching to model entities. During the initial phase of bottom-up processing, the basic operation consists of grouping near neighbour districts if they share certain characteristics. We employ two types of grouping process which must be clearly distinguished. The first is a general grouping process which may be based on any district attribute. For example, if districts have a similar mean gradient direction, they may be linked to form a longer hillside section or edge component, or if they have a similar within- district variance of gradient magnitude they might be linked to form a region of similar texture. The second is a specialised grouping process which is equivalent to a type of adaptive smoothing. It is used to create our multi-scale representation. In this case, the only district attribute used is the grey value and position of its centroid. The two types of grouping are contrasted in the table.

COMPARISON OF DIFFERENT TYPES OF DISTRICT GROUPING PROCESS USED DURING INITIAL BOTTOM-UP PROCESSING

	General Grouping	**Grouping for Multi-Scale Representation**
Criteria for Defining the Candidate Set of Districts for Possible Grouping	Near neighbours defined by districts having boundaries touching at least at a point	Near neighbour defined by Delaunay triangle links between centroids of districts
Criteria for Grouping Districts	Any type of intra-district att-ribute and any type of between-district relationship	Centroids of districts treated as new input data points
Type of Region Formed by Grouping Districts	Any type. Regions may correspond to entities of Marr's primal sketch	Coarser scale slope district

General Grouping According to Gradient Direction. One of the most important attributes of a district for general grouping is the direction of maximum gradient. We have experimented with grouping neighbouring districts using this attribute alone. Pairs of regions are classified into three major categories: straight hillsides, and convex and concave sections. Region pairs are classed as straight hillside sections if the angle between the directions of maximum gradient is <45°. If the directions of maximum gradient diverge by more than 45° the regions form a convex section. Convex sections are classified as a ridge if the directions of maximum gradient diverge by 135° to 180°. If they diverge by 45° to 135° then their classification lies between a ridge and a straight hillside and they can either be thought of as a steeply ascending ridge, i.e. a buttress, or a sharply bending hillside or convex corner. If the directions of maximum gradient converge by more than 45° the adjacent regions form a concave section or gully. Concave sections are classified as a trough if they converge by 135° to 180°. If they converge by 45° to 135° then their classification lies between a trough and a straight hillside, and they can either be thought of as a steeply descending trough, i.e. a gully, or a sharply bending hillside or concave corner.

3. WORKED EXAMPLES: 2-D SEGMENTATION

3.1. Synthetic Image to Show one Hill overlapping one Dale

Figure 4 shows an enlarged synthetic image generated to demonstrate our segmentation. When plotted as a 3D landscape it can readily be seen to contain within its interior three peaks, three pits, and four saddles (figure 5). The direction of maximum gradient within each of the original triangle planes is shown by an arrow. The peaks, pits, and saddles are marked as before. The peak right of centre is the summit of a hill whose borders are delineated by the four watercourses passing through the pits and saddles around the peak.

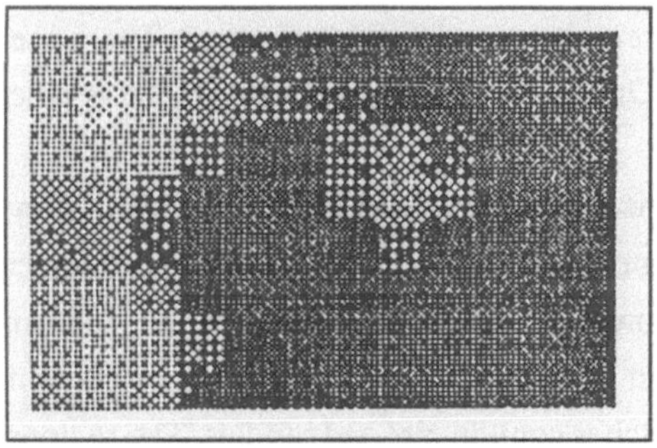

Figure 4. Synthetic 12x8 pixel image

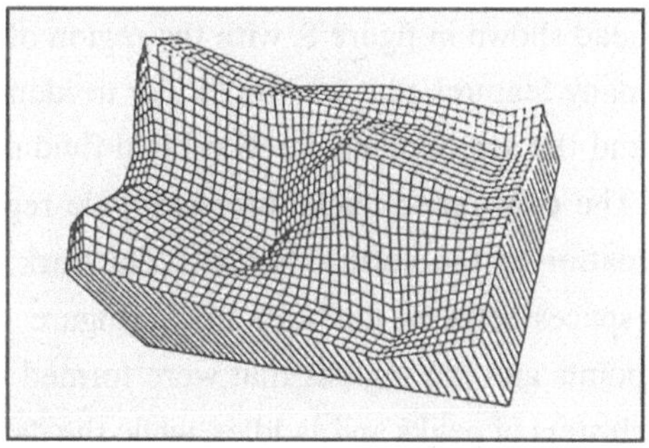

Figure 5. Synthetic image plotted as a 3D landscape

Overlapping the hill is a dale, centred on the pit left of the centre, and delineated by the four surrounding watersheds. Notice that the local gradients radiate out from the peak and converge at the pit (figure 6).

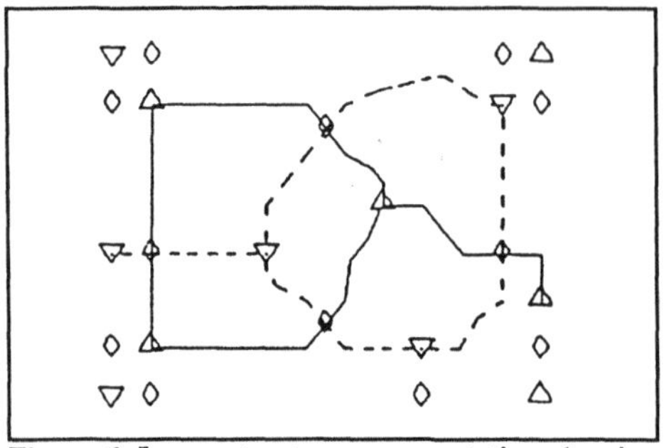

Figure 6. Image structure representation showing singular points, MGPs and districts.

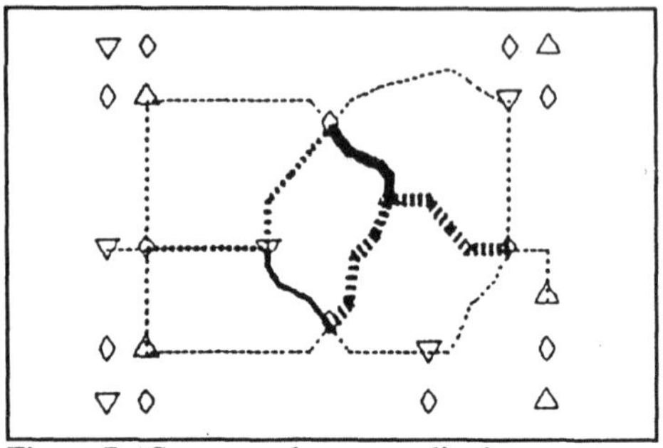

Figure 7. Convex and concave district groups.

The results after general grouping using only the district attribute of mean gradient direction are shown in figure 7. In the following figures where region boundaries have been classified as described above, districts which form straight hillside borders are separated by thin lines. Ridges and convex corners are denoted by thick continuous and dashed lines respectively, and troughs and concave corners by mid-thickness continuous and dashed lines respectively. Figure 7 shows that the three borders between regions within the interior of the hill are convex, and those within the dale are concave. The convex border point at ten o'clock (the thick continuous line) is a ridge while the other two convex borders are buttresses. Likewise, the concave borders are separated into one trough and two gullies.

3.2. MRI Image

The segmentation and grouping algorithms were also tested on a sagittal MRI slice of the head shown in figure 8, with the region of interest shown enlarged in figure 9. There are many features that we would like to identify. On this image the brain substance is pale and the surrounding cerebrospinal fluid dark.

The convoluted cauliflower-like pale regions correspond to gyri or folds of brain grey matter on the cortical surface. The dark channels between the gyri correspond to sulci (spaces between the brain folds). Figure 10 shows the calculated location of the singular points and the regions that were formed. It can be seen that the bright patches contain clusters of peaks and saddles, while the dark patches contain pits and saddles. The regions enclose areas of continuously increasing intensity lying between the bright and dark patches at either end, and are bordered by concave or convex folds at each side. Figure 11 shows the classification of adjacent districts. There is a good correspondence with the

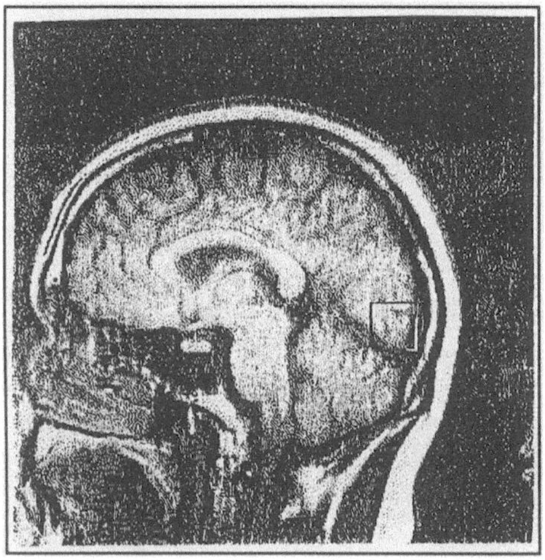

Figure 8. MRI brain image. Sagittal slice near midline

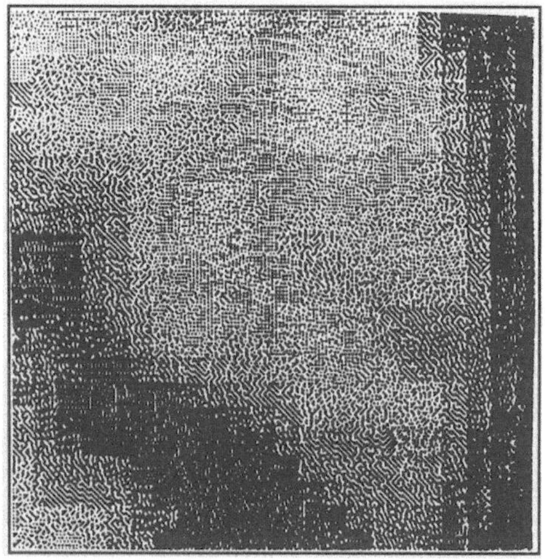

Figure 9. MRI region of interest showing the visual cortex.

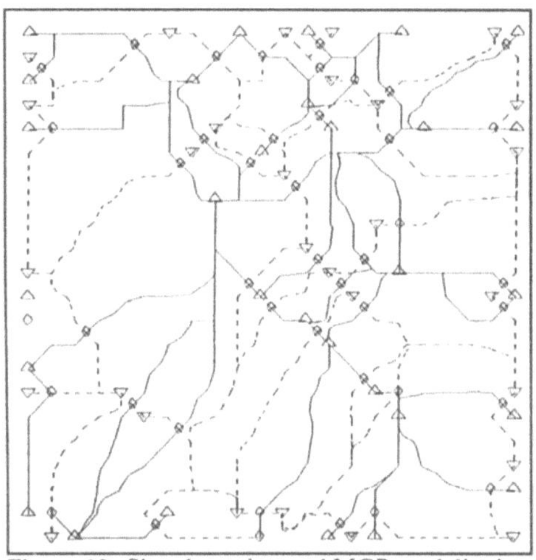

Figure 10. Singular points and MGPs and districts on MRI image

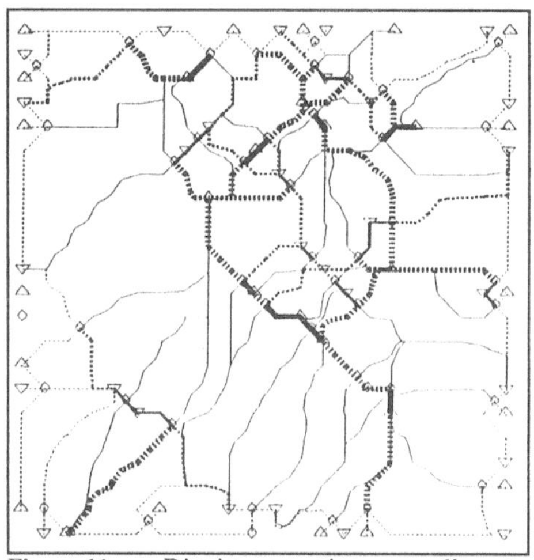

Figure 11. District grouping according to difference in gradient directions

anatomical features. The major ridges follow the centre of the gyri, the major troughs follow the sulci, and the straight hillsides or edge sections correspond to the sides of the gyri.

3.3. Image Detail to Illustrate the Effect of Warping.

The next example shows the effects of warping an image. Isotropic transformations of the XY plane have no effect, while non-isotropic scaling or more general rubber sheet transformations usually alter only a small proportion of region boundaries since the singular points and hillsides remain intact (although deformed). This is advantageous if the X and Y axes are incommensurate, as in parametric images for example. Since a non-isotropic transformation will alter the surface patch gradients unevenly, segmentation is not identical. At any point the direction of the maximum gradient path may be altered, causing different boundary paths to be followed. Figure 12 shows the original image and figure 13 shows it after it has been stretched (and resampled) along the X axis. Many corresponding paths have identical connectivity, either following an identical (stretched) path or deviating slightly along the way before being pulled back on course. For a few the connectivity is altered.

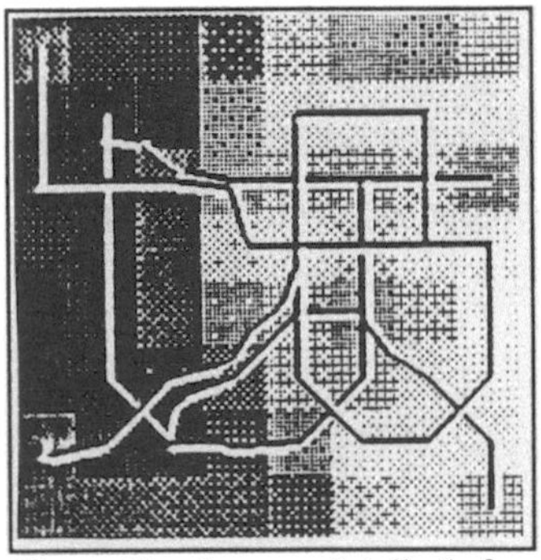

Figure 12. Image detail before stretching

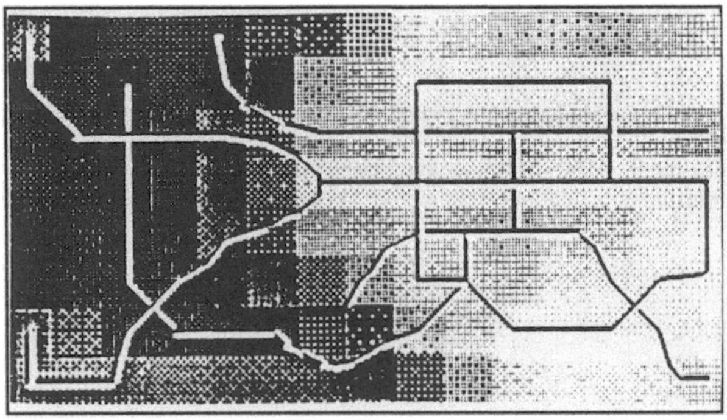

Figure 13. Image detail after stretching

When treating a 2-D grey-scale image as a 3-D surface the spatial and intensity dimensions are incommensurate. That is, there is no relationship between the spatial and intensity dimensions. The regions defined are not affected since the segmentation is unchanged under any monotonic grey-level transformations of the image.

The segmentation is fairly robust. Slight perturbations usually only cause small quantitative changes. However, depending on its topology, certain points on a surface may be unstable. An example is a point at which a maximum gradient path forks into two paths of unequal but similar gradient. A small change may be sufficient to make the less steep path the steeper, diverting the maximum gradient path, causing a different set of regions to be formed in the neighbourhood. However, these local instabilities are not likely to have much effect on the mean attributes of the district, and are even less likely to affect the grouping of districts when larger, more meaningful entities are formed.

The hillside sections correspond to edges for which basic attributes of mean gradient, direction, and amplitude are immediately available. Hillside sections could be selected on the basis of these attributes and subjected to conventional edge detectors, polynomial surface fits, or other algebraic operations if required for specific applications. The same is true for more complex topographical elements such as corners, ridges, hills, etc.

4. EXTENSION OF 2-D IMAGE STRUCTURE ANALYSIS TO 3-D

The principles of our approach to 2-D image structure analysis extend naturally to higher dimensions and are discussed in this section.

4.1. Construction of links between neighbouring 3-D data points. Where 3-D data sets are input without explicit links to define the neighbourhood set for each point, 3-D Delaunay triangulation is used to define these links. The entire 3-D volume is thus composed of tetrahedrons. Input data sets with existing links forming tetrahedrons can be accepted even if these links do not strictly join the nearest neighbours.

4.2. Classification of 3-D points. For every 3-D point a "neighbourhood polyhedron" whose apices are formed by the point's neighbourhood point set is analysed. The initial 3-D point lies in the interior of the polyhedron and the links between the initial point and the neighbourhood point set divide the polyhedron into a set of tetrahedrons. The outer triangular faces of the polyhedron form a continuous surface, a 2-D manifold in 3-space. This surface is referred to as the neighbourhood surface. Figures 14 to 17 show examples of neighbourhood polyhedrons. To assist comprehension, an icosahedron has been used as an example. Irregular polyhedra with non-equilateral triangular facets can be analysed in the same manner.

The intensity gradient between the initial point and its neighbouring points is calculated as the ratio of the difference in intensity and the length of the link in 3-space. Each gradient value can be considered as an attribute of the neighbouring points on the neighbourhood surface. Gradient values can now be interpolated within the triangular facets of the neighbourhood surface so that all points on the

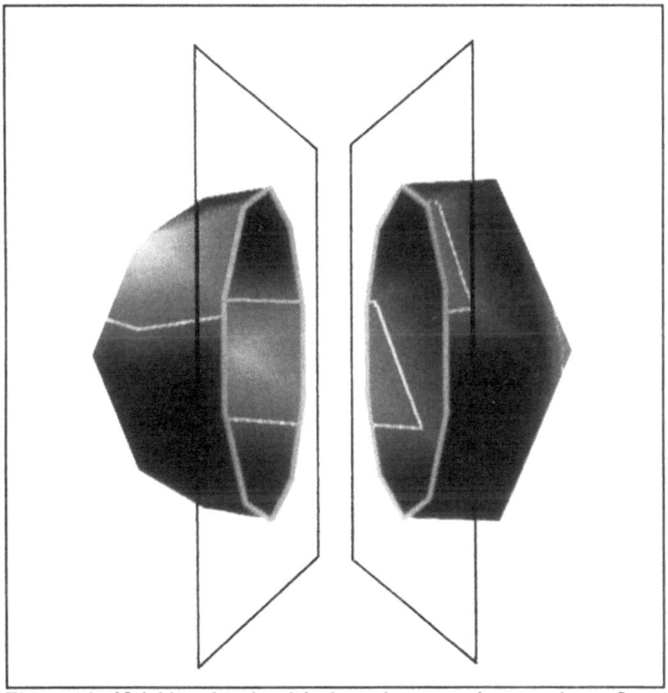

Figure 14. Neighbourhood polyhedron shown partly opened out. Same surface as Figure 17: three zero crossing paths

neighbourhood surface can be given a gradient value. Figure 14 shows a view of a neighbourhood polyhedron (an icosahedron) as if split open so that part of the inner and outer surfaces can be seen. The point in the centre of the polyhedron, which is the initial point whose neighbourhood is given by the polyhedron, is not shown. In figures 15 to 17, the split polyhedra have been opened out further so that almost all the interior can be seen.

The analogy of terrain height is again applied in describing the geometry of the types of gradient distributions that may be encountered on the neighbourhood surface. However, it should be recalled that in the present context terrain height represents first derivative values and not absolute intensity values. All areas of positive gradient (where the neighbouring point has a higher intensity value than the initial point) can be regarded as lying above sea level and are shown as pale shades on the diagrams. All areas of negative gradient can be regarded as lying below sea level and are shown as dark shades. The shore line thus represents the zero crossing of the gradients on the neighbourhood surface. On the same intensity scale this would be a mid grey shade, but to highlight the zero values they are shown as pale lines.

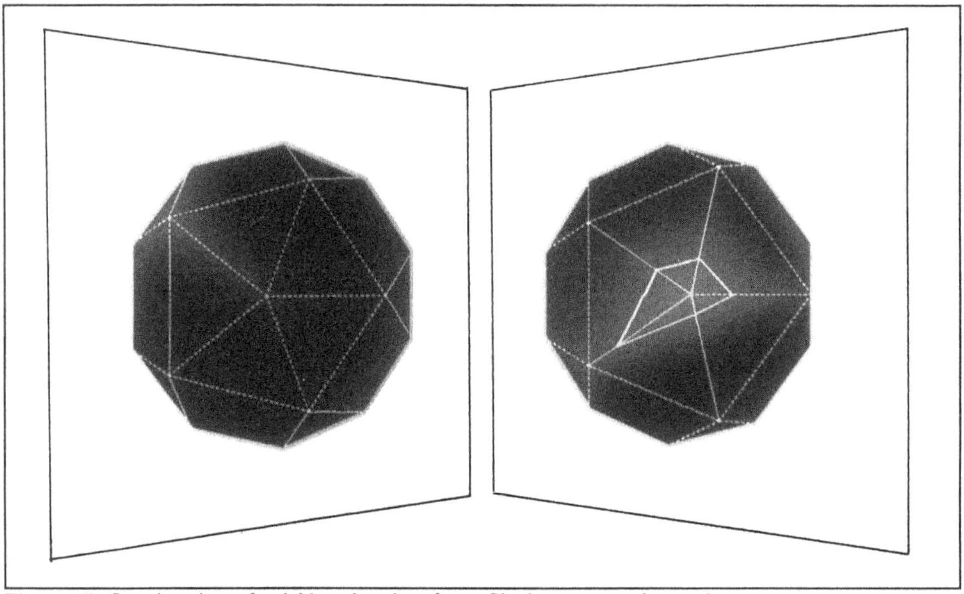

Figure 15. Interior view of neighbourhood surface. Single zero crossing path.

A neighbourhood surface which has all zero gradients means that the points all form part of a zero gradient district in 3-space. A neighbourhood surface which has all positive or all negative gradients (all land or all sea) means that the initial point is a local minimum or local maximum respectively. The commonest type of point has a neighbourhood surface with a single island surrounded by sea as in figure 15, or a single sea surrounded by land which covers the rest of the surface. This surface type thus has

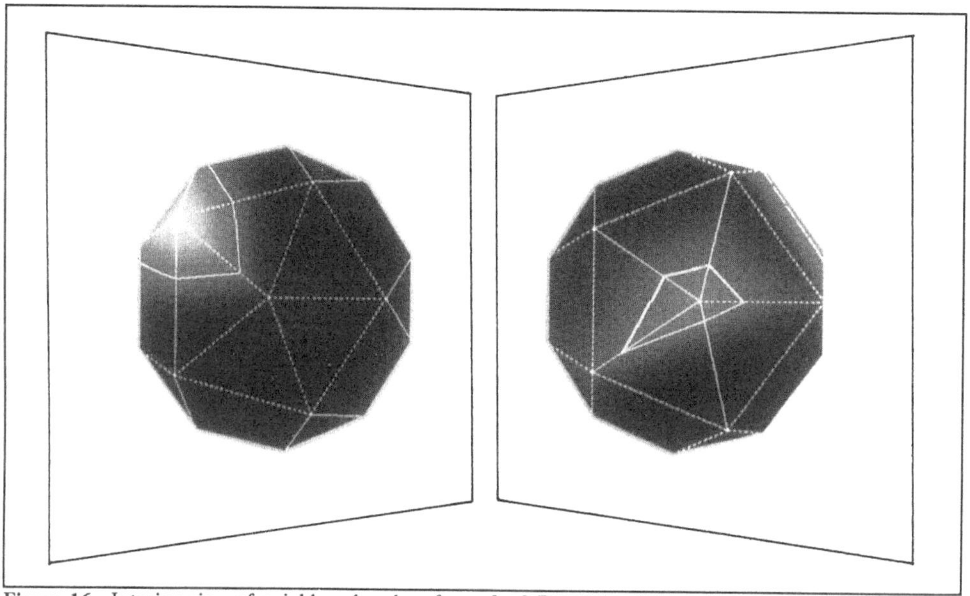

Figure 16. Interior view of neighbourhood surface of a 3-D saddle with two zero crossing paths.

a single continuous zero crossing path (shoreline). Points with such neighbourhood surfaces lie within a 3-D slope district.

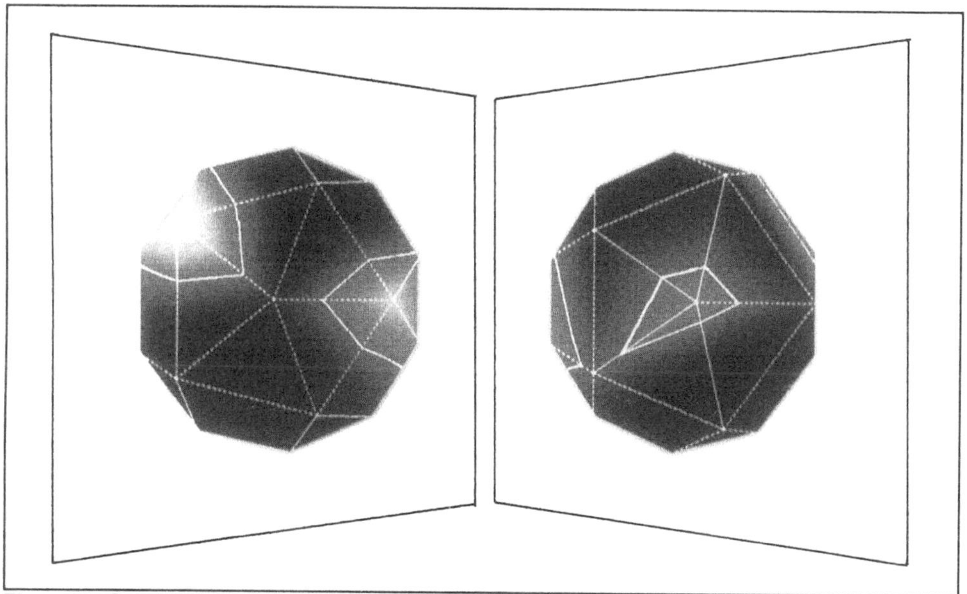

Figure 17. Interior view of a neighbourhood surface of a 3-D saddle with three zero crossing paths

More complex surfaces all indicate different types of 3-D saddle. The essential characteristic of a 3-D saddle is that there is more than one continuous zero crossing path on its neighbourhood surface. Two islands surrounded by a continuous sea (figure 16), or two seas surrounded by continuous land, form the simplest type of 3-D saddle. As in the 2-D domain, the number of points in the neighbourhood point set (polyhedron) limits the possible complexity of the saddle. The most complex saddle which can be formed by a 20-facet neighbourhood polyhedron has three separate zero crossing paths (figures 14 & 17). Rare points with a large neighbourhood set can form a saddle with several islands and/or seas.

4.3. Construction of 3-D MGPs. As in the 2-D case, each saddle point seeds maximum gradient profiles which are grown in 3-space via the highest point of each island on its neighbourhood surface (maximum uphill paths) till they reach a local maximum image intensity, and from the deepest point of each sea (maximum downhill paths) till they reach a local image intensity minimum.

4.4 Construction of 3-D slope districts. The basic concept of a 3-D slope district is that following a line of maximum uphill gradient from any point within it will lead to the specific local maximum of the district, and following a line of maximum downhill gradient will lead to its local minimum. In 3-D, multiple adjacent saddle points exist which can be linked to form a polyline running along the locus of contact of four or more adjacent 3-D slope districts. From these saddle points, multiple MGPs are generated forming a maximum gradient surface. On one side of a maximum **uphill** surface, travelling **downhill** would lead to one local minimum. From the other side of this maximum **uphill** gradient surface, travelling **downhill** would lead to a different local minimum. In the same way, on one side of a maximum **downhill** surface, travelling **uphill** would lead to one local maximum. From the other side of this maximum **downhill** gradient surface, travelling **uphill** would lead to a different local maximum.

5. DISCUSSION

Like Pizer and co-workers at the University of North Carolina (see for example Nackman 1984), we use the analogy between an image data set and a 3-D surface terrain map to describe image structure. Many other authors have applied this analogy both in computer vision and other domains. It is useful because significant topographical features such as peaks, ridges, and hillsides are likely to correspond to significant features in the data sets of the problem domain. Example data types include image, range, demographic, and economic trend data, as well as actual terrain data. In image analysis this approach has been used for a variety of purposes such as stereo matching [Pong et al 1989], 3-D shape estimation [Pong et al 1985], image approximation [Hsu et al 1978], segmentation [Lee and Fu 1980; Besl and Jain 1986], texture analysis [Ehrich and Foith 1978], edge detection

[Ehrich and Foith 1978; Colchester et al 1990a,b], and estimation of centre-lines of ribbon like-structures such as images of blood vessels [Colchester et al 1990a,b].

There are important consequences of representing surfaces as a connectivity network. Disadvantages include the need to transform standard image data structures at the very start of processing, and the inability to execute conventional image processing procedures. However, we believe that the disadvantages are outweighed by significant advantages. Much real world data is inherently irregularly sampled. This is easily handled in a connectivity representation. 3-D image data sets are usually acquired as multiple slices with different resolution in different dimensions. A 3-D connectivity network representation, constructed from 2-D representations by linking adjacent points in adjacent slices, is fundamentally the same regardless of slice separation. Scale can be represented very simply in network representations as a part hierarchy which makes explicit the connections between coarse and fine entities. Network representations provide a direct route for communication with computer aided design and vector graphics procedures. Insertions or deletions can easily be handled as local operations in the data structure. This is relevant for the handling of artefacts in sensor data, such as missing sample points, as well as the more general issue of user interaction and editing. Higher level, symbolic representations are usually already in the form of a semantic network. Thus, communication between feature-based representations (high level) and image intensity-based representations (low level) is simplified.

6. CONCLUSIONS

Our low level representation of images provides an accurate description of images at fine detail by well defined regions that can be easily and robustly generated. The segmentation is unaffected by monotonic grey-level transformations, and is insensitive to rubber sheet transformations of the spatial co-ordinates. The data structure is a connectivity network. The nodes of the network are slope districts. The links between the districts code adjacency. Each node also has pointers to the components from which it was constructed and these form a part hierarchy. We refer to this representation as image structure. It is the lowest level of symbolic representation, and provides a framework to simplify the interaction between bottom-up and top-down processes. Using simple region grouping rules we have shown the usefulness of the representation in the construction of more complex elements like those in the primal sketch.

7. REFERENCES

Besl PJ, Jain RC (1986). Invariant Surface Characteristics for 3D Object Recognition in Range Images. Computer Vision, Graphics, and Image Processing 33: 33-80.

Burns JB, Hanson AR, Riseman EM (1986). Extracting Straight Lines. IEEE PAMI 8: 425-455.

Colchester ACF, Ritchings RT, Kodikara ND (1990, in press). A Method for Multi-Scale Representation of Data Sets based on Maximum Gradient Profiles: Initial Results on Angiographic Images. In The Formation, Handling and Evaluation of Medical Images (A. Todd-Pokrapek and M. Viergever, eds). Berlin: Springer.

Colchester ACF, Ritchings RT, Kodikara ND (1990, in press). A new approach to image segmentation using maximum gradient profiles orthogonal to edges. Image & Vision Computing vol 8.

Ehrich RW, Foith JP (1978). Topology and Semantics of Intensity Arrays. In : Computer Vision Systems (EM Riseman and A Hansen A ,eds), Academic Press, New York. pp 111-127.

Hsu S, Mundy JL, Beaudet PR (1978). Web Representation of Image Data. 4th International Joint Conference Pattern Recognition: 675-680.

Lee HC, Fu KS (1981). The GLGS Image Representation and its Application to Preliminary Segmentation and Pre-attentive Visual Search. Proc. 1981 Conference Pattern Recognition Image Processing: 256-261.

Longuet-Higgens MS (1960). Reflection and Refraction at a Random Moving Surface. I: Pattern and Paths of Specular Points. Journal Optical Society America. 50: 838-844.

Marr D (1976). Early Processing of Visual Information. Philosophical Trans. Royal Society London B 275: 483-524.

Maxwell JC (1870). On Hills and Dales, The London, Edinburgh, and Dublin Philosophical Magazine and J Sci., 4th Series, 40(269): 421-425; reprinted in Niven W.D (ed) "The Scientific Papers of James Clark Maxwell", New York: Dover, 1965.

Nackman LR (1984). Two-Dimensional Critical Point Configuration Graphs, IEEE PAMI 6: 442-450.

Pong T-C, Shapiro LG, Haralick RM (1985). Shape Estimation from Topographic Primal Sketch. Pattern Recognition 18(5): 333-347.

Pong T-C, Haralick RM, Shapiro LG (1989). Matching Topographic Structures in Stereo Vision. Pattern Recognition Letters 9: 127-136.

Ritchings RT, Colchester ACF, Wang H-Q (1985). Knowledge based analysis of carotid angiograms. Image & Vision Computing 3:217-222.

Rosin PL, Colchester ACF, Hawkes DJ. (in press). Early visual representation using regions defined by maximum gradient profiles between singular points. In Information Processing in Medical Imaging (Ortendahl D, ed): Proceedings of XIth International Meeting, Berkeley, California, June 1989.

ACKNOWLEDGEMENTS

Part of this work was carried out under Alvey Project MMI 134. Paul Rosin and Glynn Robinson have made a major contribution to the material in this paper and I am very grateful for their help.

Segmentation and Analysis of Multidimensional Data-Sets in Medicine

Olaf Kübler and Guido Gerig

Institute for Communication Technology, Image Science Division
ETH-Zentrum, CH-8092 Zurich, Switzerland

Abstract

Segmentation and analysis of 2-D and 3-D CT and MR data sets is illustrated on examples covering renal function studies in MR time sequences, discrimination of brain matter in multispectral 3-D MRI, extraction of brain tumor by structure segmentation methods, and model guided analysis of hip-joint and pelvic anatomy. It is argued in favor of a tightly integrated analysis sequence which exploits the characteristics of the original measurements and all available anatomic information. Knowledge-based analysis has produced some preliminary results but must be considered still in its infancy. Novel ideas for describing and representing natural structures in conjunction with AI tools are needed for generally applicable analysis methods.

Keywords: medical image processing, nonlinear smoothing, classification, edge detection, grouping, structure segmentation, computer vision, signal-to-symbols paradigm, knowledge-based interpretation

1. Introduction

Computer-assisted analysis of multidimensional medical images - according to established practice - is specified by the sequence: data-acquisition, data-processing, segmentation, and model-guided interpretation, where the implicit assumption is made that the steps in this analysis-chain may be regarded as distinct modules executing in a strict hierarchy and, in essence, without feedback. This embodies the famous "from-signals-to-symbols" (SS) paradigm which has had its decisive and eminently productive influence on computer vision

NATO ASI Series, Vol. F 60
3D Imaging in Medicine
Edited by K. H. Höhne et al.

research and which has only recently come under critical scrutiny partly in favor of a more integrated, monolithic approach [Fischler 1987].

Applying the SS paradigm to multidimensional images in medicine has led to spectacular results such as 3-D visualizations of segmented structures like brain or blood vessels displayed within the outer anatomy. Some inherent drawbacks have also become apparent, perhaps best exemplified by the legal arguments raised against 'post-processing' of the 'original' data. Although the physical measurements have to be processed in any case to transform them into useable images and the data-format is never changed in what is normally called post-processing, a logical distinction is introduced which may, however, be regarded as largely arbitrary. The discussion about a change of paradigms in computer-vision and the technical limitations will not bring about an immediate and complete revision of current strategies but it should teach us not to overestimate the beautiful and spectacular static and animated 3-D renditions of anatomical structures which have been such a success with research oriented physicians and medical equipment manufacturers. In terms of clinical potential and achievement, computer-assisted analysis of medical imagery will have to be examined with care to indicate where the traditional approach is bringing significant progress and where a judicious combination and integration of methods appears particularly promising.

In assessing segmentation and analysis of multidimensional image data in medicine we want to demonstrate results which can be obtained by transcribing established computer-vision methods from the normal 2-D domain to higher dimensions and we will discuss some initial results in support of an integrated approach. The order of presentation is arranged according to the following objectives of medical image analysis: quantitation by interactive and automated measurements and minimizing tedium by partly automated analysis of large amounts of similar data, data-processing and improved statistical classification, determination of structure and function, integration with other data and descriptive features. The central and important goal of stimulating intuition of medical researchers and practitioners and of facilitating communication between them will not be addressed specifically since it appears more a principal topic of display oriented research and implementation. The paper concludes with a discussion of the present status and an attempt at identifying necessary and, hopefully, promising analysis strategies for the future.

2. Quantitating Multidimensional Measurements and Exploiting Bulk Data

The technical support offered by modern work-stations makes quantitative interactive measurements on 2-D images a trivial but generally tedious affair. Delineation of features in data-sets with three spatial dimensions requires realistic rendering schemes in conjunction with special interaction devices which are under technical development and should become widely available in a few years. In summary, completely interactive segmentation as with a surgical knife followed by readout and further analysis of the data within the defined region(s) may be considered as technically demanding tasks for which working solutions will be developed. Most efforts pertaining to surgical planning and its technical implementation can also be put into this context.

Automated and partly automated measurements require functionally correct and efficient segmentation procedures. The reliability of present segmentation methods must, in general, be treated with reserve as discussed in section 4. A notable exception is given by function studies on time series of 2-D images. In this case, segmentation can be made very reliable by automatically extracting discontinuities in the data and interactively selecting those features which delineate the desired anatomical structures. Elastic pattern matching techniques allow to retrieve these from one time frame to the next and to put them into one-to-one correspondence. Great savings in operator time result, in particular for extended time series, since interactive intervention is required on one time frame only. The approach and the steps involved are representative for recent advances in image analysis and are explained in greater detail.

2.1 Semiautomated Analysis of Extended 2-D Time Sequences

Time series of medical image data are used either when studying the movement of an anatomical structure (beating heart) or the dynamic behavior of the signal in specific areas within anatomical parts with the aim of extracting functional information. In Nuclear Medicine this is done by injecting tracers and assessing the temporal change of radioactivity within regions-of-interest (ROIs) from the recorded scintigrams.

In order to analyze localized regions a one-to-one correspondence between them must be given or must be constructed for the overall acquisition time. This prerequisite is uncritical in Nuclear Medicine because of poor spatial resolution but is hard to fulfill when high resolution methods such as MRI are to be exploited. In MRI, large scale function studies involving the use of contrast agents have been performed on kidneys to quantitate regional glomerular filtration. The target spatial resolution of ROIs was defined by the different functional parts of the renal anatomy. Due to the considerable movement and rotation of the kidneys between image frames the ROIs have to be realigned, either by redrawing them tediously in every frame or by developing and resorting to robust and efficient techniques for automatic recognition and geometric correction of kidney positions throughout the different sequences of more than 100 frames each.

The proposed image analysis processing scheme consists of a multi step feed-forward system containing essential elements of the classical SS paradigm and schematically described in Fig 1.

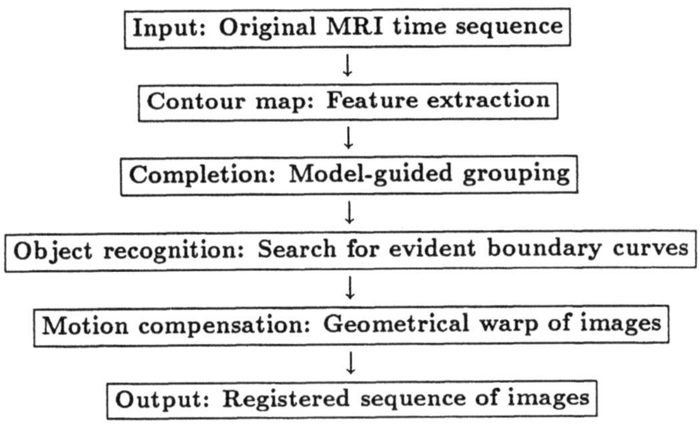

Fig 1: Processing scheme for high resolution renal function studies from extended time series of 2-D Magnetic Resonance Images

The processing scheme is motivated by the observations that the most important cues to image structures are normally given in the form of edges and contours but that edge detection techniques will supply fragmented and spurious contours. Grouping and selection thus becomes mandatory for object recognition. The geometrical distortion of the kidney structures represents a purely technical step to establish the desired one-to-one correspondence in the form of registered images making subsequent compilation of function curves over highly resolved ROIs a trivial task.

A look at a typical MR image (Fig 2a) reveals that large parts of the left and right kidneys are outlined by conspicuous valleys in image intensity. Well-known edge detectors [Haralick 1984, Canny 1986] in their standard line-detection variants will find all significant contour elements in the image meaning that far more than the kidney boundary shows up and that it is not delineated by a continuous line (Fig 2b). Although this lack of selectivity and completeness is a generally accepted part of the SS strategy which aims at first detecting all edges in a low-level process in order to group and do something useful with them in subsequent (intermediate-level) steps, it is exactly these non-elementary processes which are still largely missing in computer vision.

In the present case, grouping of edge primitives can be performed by a robust model-guided procedure. A valid model of the kidney contours is obtained by tracing it interactively or, alternatively, by selecting and linking contour-elements in one edge image (Fig 2c). Grouping of contour primitives in all edge images of the long time sequence is then performed by a generalized Hough strategy which links image-space and accumulator (=Hough) space [Gerig 1987].

The central idea of augmented Hough technique is to use the transform for finding elements which belong to a particular pattern (template matching) while retaining the exact spatial distribution of the contributing elements in image-space. Relative maxima in Hough space are regarded as representative of most evident groupings and are extracted without ambiguity by directed but exhaustive search driven by a second Hough transform which removes all irrelevant counts from the accumulator generated in the first transform. A pointer structure is generated which links the surviving nonzero accumulator cells with the elements of the various incarnations of partially represented patterns in image space. Discrimination between equally evident patterns with differently represented contour (e.g. 50% evident, long dashes-long gaps vs. hemi-contour or arrays of dots with different degrees of uniformity) thus becomes possible.

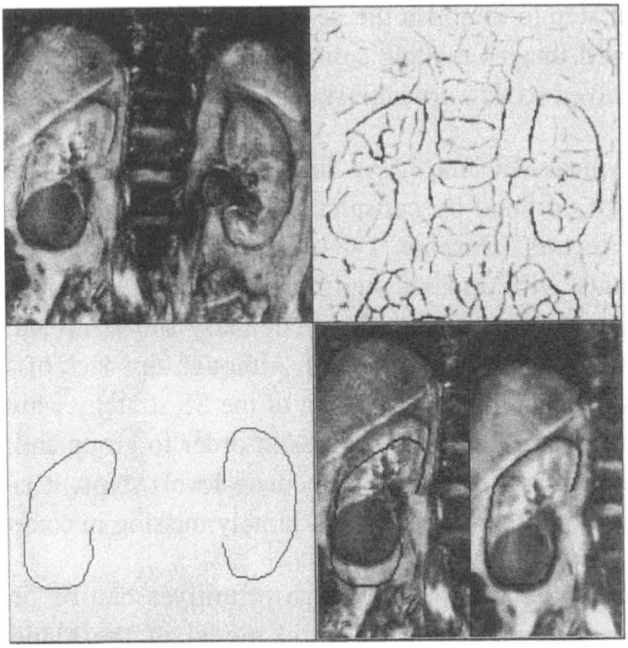

a | b
c | d

Fig 2: Detection of kidneys in MR images:
a : original, **b**: extracted edge-elements, **c**: left and right templates (= models),
d: original and corrected left kidney with overlayed template

The grouping and recognition strategy provided by the augmented Hough transform in its simplest form immediately yields the transform parameters for left and right kidneys in the sequence assuming rigid body motion. Differential contour deformation can also be treated by using a more elaborate kidney model than the initial rigid template [Morgue 1988]. The quality of the geometric correction is illustrated in Fig 2d and resulting function curves obtained on small areas of the renal cortex and renal medulla are shown in Fig 3.

The contour extraction, grouping, and organ recognition in the renal function studies represents an example of successful and complete segmentation and illustrates the spirit of the signals-to-symbols paradigm of computer-vision. Success is due to the efficient grouping strategy provided by the augmented Hough transform and the stratagem of interactively procuring a simple, explicit, and essentially rigid model of the structure under investigation. The grouping strategy will remain viable as long as similarly simple template-type model scan be used but it is not at all obvious how such models could be obtained if image sequences are not available. For reasons of computational complexity and memory requirements, the Hough transform is restricted to images with two spatial dimensions on present hardware while edge, or rather, surface extraction has been transcribed to 3-D (Section 4). The task of compiling functional information on renal glomeral filtration was solved completely. This should not conceal the fact that structural information had to be provided externally as ROIs were defined interactively. Other analysis tasks will demand structural answers in addition to segmentation, and again combining methods appears hold most promise.

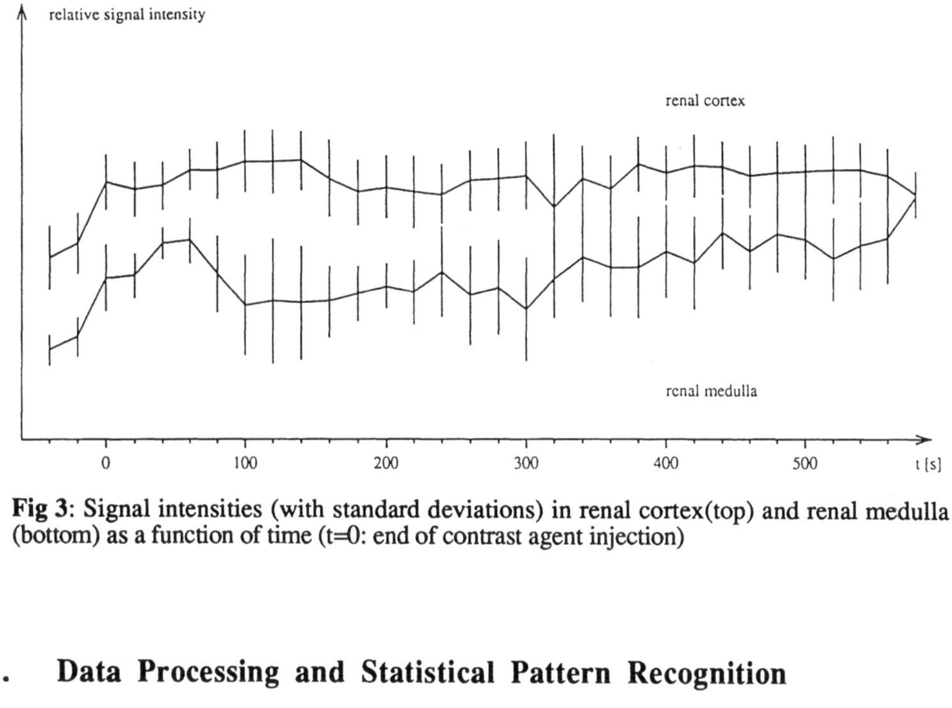

Fig 3: Signal intensities (with standard deviations) in renal cortex(top) and renal medulla (bottom) as a function of time (t=0: end of contrast agent injection)

3. Data Processing and Statistical Pattern Recognition

Statistical Pattern Recognition, although a standard method with an elaborate mathematical basis and of widespread use in image analysis applications such as remote sensing, has so far played a minor role in the analysis of multidimensional medical data. The notable exception of course is the simple two category classification problem of bone soft-tissue discrimination in CT images treated by elementary windowing or thresholding techniques.

With the advent of multi-echo MRI data interest in statistical methods has been revived. The pronounced noise-sensitivity of pixel- or voxel-based classification methods, however, leads to unattractive results on vector-valued MR measurements acquired under patient-compatible conditions. Since spatial relationships are disregarded in pointwise classification, additional methods have to be applied to derive a minimum of structural information.

Segmentation of MRI data into anatomically and functionally distinct regions calls for a multistage processing scheme. For dual-echo head images a scheme requiring minimal user interaction (Fig 4) has been defined [Gerig 1989] and applied to a substantial number of cases.

Input: multiecho MR volume data

↓

1: 3D edge preserving smoothing

↓

2: supervised multispectral classification

↓

3: 3D morphological postprocessing

↓

4: 3D connected component labeling

↓

5: interactive assignment of anatomical meanings

↓

Output: segmented data

Fig 4: Polystage segmentation of multi-echo multidimensional MR data

In the proposed segmentation scheme, human interaction is only necessary in stages 2 and 5. Training areas of distinct categories must be defined to apply supervised classification techniques and simple assignment have to be made as to the anatomical meaning of connected entities resulting from the labeling in stage 4. Thus, the brain is given by the structure containing the largest number of connected voxels while the ventricular system must be chosen from a number of liquid structures.

Basic anatomical knowledge yields simple spatial constraints on the 3-D distribution of gray and white brain matter and on the ventricular system, the structures which are to be determined in the analysis. The different categories form highly diverse but connected, directly adjacent and clearly distinguishable entities. The result of statistical classification should reflect this property in the best possible way. It will not do so, however, when applied to the im

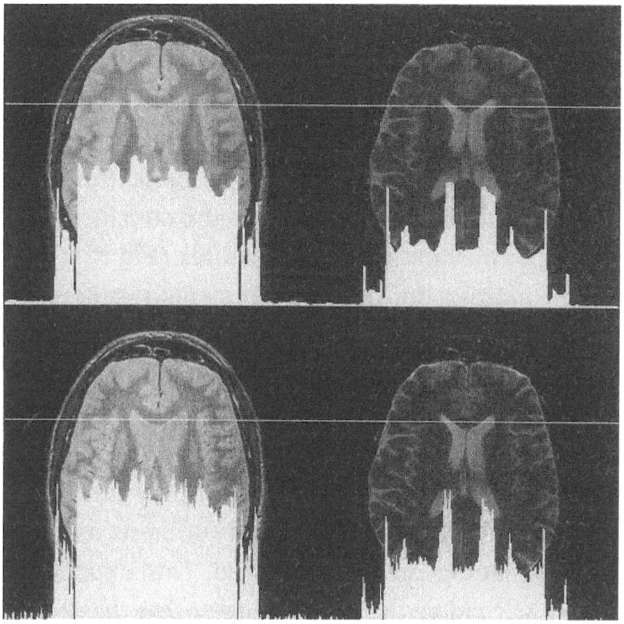

Fig 5: discontinuity preserving smoothing of dual-echo MR images, top: smoothed images, bottom: original images, Signal intensities along one scan-line (white) are shown as bar diagrams (white overlay)

mediate data with their high noise component, inevitable if patient-friendly short acquisition times are to be used.

Noise suppression has been a basic issue in image processing since its beginnings and a host of linear as well as non-linear techniques have been proposed. Linear smoothing as preparatory step to pointwise classification is of little value as edges and fine details are blurred and shifted from their true locations and accidental transition zones are generated disagreeing with the postulate of having distinct but directly abutting anatomical and functional regions. A number of discontinuity preserving non-linear smoothing methods have been proposed, the most effective of which appears to be an iterative scheme based on the multidimensional anisotropic diffusion equation where the diffusion coefficient depends on the spatial coordinates and the number of iterations (=time) [Perona 1987]. Every image element in a new iteration results from a weighted superposition of neighboring elements in the previous iteration. The weights are given by an exponential dependence with the negative (scaled) modulus of the image gradient forming the space (and time) dependent exponent. The rapid exponential decrease of the diffusion coefficient in function of increasing gradient leads to the superb behavior of anisotropic diffusion and distinguishes it from older and simpler gradient inverse smoothing schemes [Wang 1981]. The algorithm is easily implemented for any number of spatial dimensions and can be made to run very fast [Gerig 1990]. Anisotropically smoothed MR images show considerable noise suppression while retaining the relevant structural features (Fig 5).

The acquisition of multi-echo MR measurements results in enhanced sensitivity and discriminatory power, as different parameter combinations are available to characterize soft tissue and liquid containing structures. Applying multispectral pattern classification techniques becomes a natural choice under these circumstances, also with regard to the ready availability of such methods. The image data have to fulfill basic assumptions: they must be radiometrically homogeneous over their full spatial extent and different tissue categories must have characteristic signatures in feature space. Our experience with a large number of dual-echo MR studies clearly indicates that these conditions are satisfied. A typical feature plane and the result of classification are shown in Fig 6.

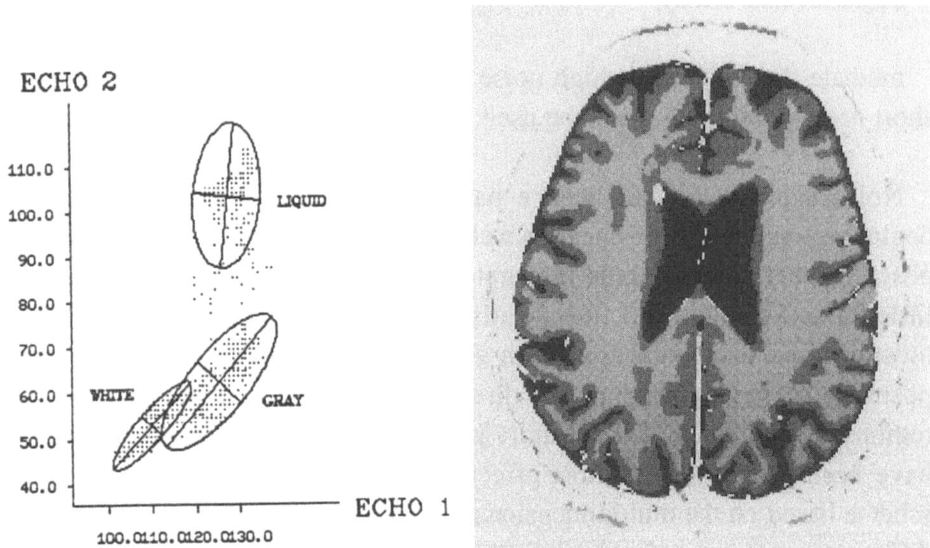

Fig 6: statistical classification: left: 2D feature space, right: classification result (white-, gray-matter, liquid (dark))

The display of classification results often detracts from the fact that each individual data element gets its category assigned without any regard to its neighbors. Human observers immediately reconstruct structural properties from the classified images which are, however, not represented in the data at this stage of processing. Important structural measures for segmentation should also be made available quantitatively such as topological and space filling properties of the classified data elements. An efficient means to procure this information is connected component labeling where we use the normal convention of defining object-connectivity in the 8-neigborhood sense in 2-D and in the 26-neighborhood sense in 3-D.

Once the connected components are known it becomes a trivial matter to assign anatomical meanings, either interactively or by simple arguments such as: ventricular system = liquid structure, completely surrounded by other material. The minimum on spatial information introduced by connected component labeling also allows to remove incorrectly classified spurious detail such as visible as fine disconnected outer rim encircling the brain tissue in Fig 6. The final segmentation results show all of the anatomical features expected from textbooks but they are still obviously deficient in terms of highly resolved structures, in particular on the ventricular system (Fig 7). Less conspicuous segmentation errors may exist but can only be inferred by a skilled observer with sufficient medical training.

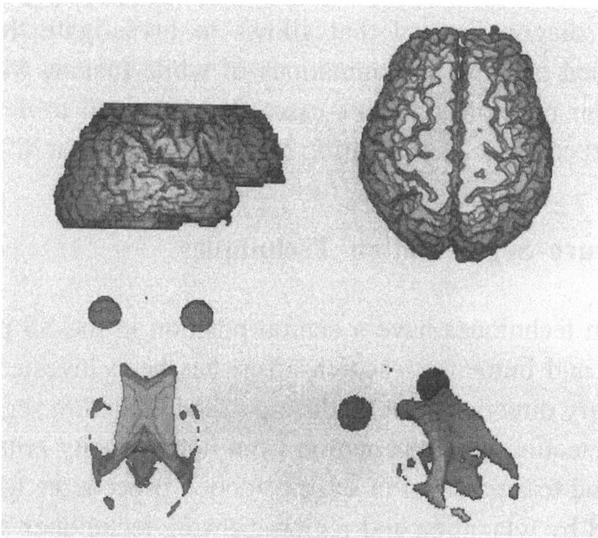

Fig 7: Visualization of segmentation results by statistical classification and morphological postprocessing. top: brain surface, bottom: ventricular system and eyes

Segmentation of dual-echo 3-D MR images of the human head by a combination of data-smoothing, supervised statistical classification, connected component labeling, and interactive assignment of anatomical meaning has yielded consistent results which are presently being compiled into an experimental data base in order to investigate the tissue characterization power of multispectral MRI. The choice of methods may be considered heuristic and tailored to a very special application. It illustrates, however, the strategy of exploiting the original measurements by extracting their information content with optimally suited procedures and presenting the results to the user in an easily understandable manner.

The processing chain can be regarded as a prototype of tight integration covering the whole range from data acquisition to final evaluation. It has delivered results which, although still in the research phase, appear of practical medical relevance. The extraction and visualization of the ventricular system of the brain is seen to give valuable contributions to the diagnosis of different types of hydrocephalus (e.g. occlusive vs. non-occlusive). Provided particular regions in the brain do reflect the early morphological neuropathologic changes of Alzheimer type dementia, then the characteristics of tissue atrophy within local regions of interest will have to be analyzed and automatic segmentation of the brain is a first step towards its detailed description. Finally, MRI represents a diagnostic tool that allows to investigate the morphological changes caused by local inflammations of white matter, which is of special importance for multiple sclerosis cases. It is planned to detect and describe such temporal changes by comparing MR data acquired at different dates.

4. Structure Segmentation Techniques

Segmentation techniques have a central position in the SS paradigm of computer vision and immense research effort has been invested into them. Two complementary directions can be distinguished: structure segmentation by discontinuity detection or segmentation from homogeneity criteria. The first approach has led to an arsenal of edge detection procedures while the second is characterized by relaxation and region growing techniques and texture analysis.

In reviewing structure segmentation techniques, we will restrict attention to edge and surface detection procedures since they far outnumber homogeneity based techniques and since they are increasingly used in general computer vision applications and in medical image analysis. In a comprehensive definition which we will adopt, segmentation is accompanied by a change of data format, i.e. a first symbolic description should be obtained in result of the segmentation process.

The common approach to localizing edges in images is to associate edges with the zero-crossings of the second derivative of a smoothed version of the image. A principal difference is given by electing to use rotationally invariant or directional derivatives and technical differences are defined by the choice of smoothing function. An excellent survey of the early days of edge detection and image understanding is given by Brady [Brady 1981] while a number of articles supplying detailed mathematical investigations and consolidating approaches have appeared over the last years [Haralick 1984, Canny 1986, Torre 1986, Yuille 1986, Bertero 1988, Clark 1989].

The edge detection scheme of particular prominence is the so called Laplacian-of-Gauss (LOG) or, in its popular approximation, Difference-of-Gauss (DOG) also known as Mexican hat operator. Edges are here defined by zeros in the (rotationally invariant) Laplacian derivative of the image smoothed in a regularizing manner by convolution with an isotropic Gaussian. The LOG edge detector is trivial to implement in 2-D and 3-D explaining part of its popularity. In addition, it has the great attraction of yielding zeros which are either closed or terminate on the boundary of the image. This property makes segmentation a seemingly trivial task since the image is directly partitioned into regions.

LOG surface extraction on 3-D MR images of a human head yields natural looking segmented brain structures [Kübler 1987] but mathematical investigations [Torre 1986, Clark 1989] and further experimental evidence have established that LOG edges are incorrectly localized and may be spurious. Comparative segmentation of a brain tumor (acousticus neurinoma) by LOG surface extraction and region filling methods reveals significant differences (color plate I).

A look at the kidney study illustrates how incorrect and misleading, although highly convenient, closed edges can be. The kidney outlines (=model curves) can impossibly be closed if anatomy is respected. The organ is more of an abstract than of a disjoint physical entity and segmentation methods without semantic information will not be able to describe it adequately.

Edge and surface detection methods with correct localization are available [Haralick 1984, Canny 1986]. They will produce edge elements where there is sufficient evidence in the data but the resulting fragments have to be grouped posing another - in general - unsolved problem. Considerable effort has gone into the definition of grouping strategies, with moderate success. The most recent proposal combines a particularly effective implementation of 3-D edge detection with a closing strategy and shows first results on MRI data [Monga 1990].

Structure segmentation techniques have been investigated in depth in the computer vision literature. Mathematical theorems have been supplied for conjectures made at earlier and many open issues have been cleared up. Practical applicability of these methods, however, is still very limited and becomes successful only in conjunction with grouping and interpretation strategies. Simple segmentation techniques have led to attractive visual results in the display of 3-D MRI data but, at present, quantitative analysis should make extremely careful and critical use of such methods.

5. Prospects of knowledge-based analysis

Representation of anatomical and physiological knowledge and means to incorporate it the interpretation process appear absolutely essential for the future of computer supported analysis of medical images and for the utilization and management of medical information. Image information must be condensed into a symbolic description in order to be combined with clinical information and to be organized in data bases for easy retrieval and use in comparative and long term studies.

Abstraction, representation and modelling of spatial structures has been introduced in the area of robotics and has resulted in considerable success. Robotics applications can profit from the great simplification of dealing with a structured environment consisting of technical objects many of which already exist in the form of CAD-models. Good strategies for general knowledge-based interpretation of multidimensional medical data sets appear much harder to define, given the enormous diversity of anatomical structures, in particular when soft tissue is involved, and the preliminary and deficient state of generally applicable structure segmentation methods. A reasonable approach to gain experience with knowledge based techniques for the interpretation of anatomical structures appears to start with structures of reduced complexity and segmentable images. In practice, this means dealing with skeletal anatomy in CT data-sets, defining and implementing tool, and endeavoring to expand them gradually and systematically into a comprehensive system.

We present first results obtained on identifying and selectively manipulating bones of the hip joint in the case a fractured pelvis and on determining the amount of dislocation in the case of a slipped capital femural epiphesis. The segmentation procedure adopted consisted in image binarization, connected component labeling, and simple model-guided object refinement. Information abstraction and representation of the geometry were used to analyse and quantitate the amount of slippage of the epiphesis.

The volume elements belonging to the bony objects are available from the segmentation process and their connectivity is known. Region filling algorithms allow to fill the interior parts of bones if required. In both cases under study, one hip joint had suffered abnormal changes such that pelvic and femur bones were not recognized as separate entities at normal CT resolution. By modeling the femur head as a hemisphere and fitting it to the data the bones could be separated (color plate II). Knowing the identity of the different bones facilitates visual assessment of the fractured pelvis as the crack running through the joint can now be seen in its full extension.

The dislocation of the capital femural epiphesis was quantified by determining the femur axis and putting it into relation with the center of the hemispherical epiphesis. The bone axis is derived by 3-D medial axis transformation and subsequent straight line fit. The axis together with the hemisphere yields the correct position of the femur and allow to determine the three components of the angular misorientation needed for a correcting wedge removal osteosynthesis.

The present cases must be regarded as extremely preliminary examples of a vast and complex area of research in which progress is mandatory if analysis of medical images with substantial computer support is to have a bright future. We advocate working on the skeletal anatomy and temporarily using CT data to define initial tools for model-guided analysis since the complexity and geometry of the scene appears realistic but manageable and segmentation of bony structures is successful. There is an abundance of suggestions for the representation of natural shape but implementations are preliminary or seem to be lacking and no cogent evidence has been produced to prefer one over the other. Interpretation strategies and AI tools for multidimensional data in medicine clearly are still in their infancy and need all the support and effort they can get.

6. Conclusion

Advanced imaging methods increasingly enable researchers in medicine, biology, and other areas of science and technology to acquire true 3-D information in the form of densely sampled contiguous volume elements (voxels). Understanding voxel data necessitates effective tools to display, manipulate, enhance, and analyse the underlying structures while retaining their full dimensionality.

Manipulating 3-D structures on the display-station is an area of very active technical development requiring, however, that the 3-D data sets and the individual structures be segmented into their constituent parts. Part of this task can be done automatically but a certain amount of interactivity remains which can be facilitated by preprocessing and image enhancement techniques virtually all of which are direct transcriptions of well known 2-D procedures.

Analysis and automated segmentation of 3-D data sets can proceed only in part according to traditional signal processing and computer vision paradigms. The objective of extracting edges, surfaces, and texture properties and of combining them for a structural and functional description is made difficult by non-unique contrast assignments, different scales of useful resolution and complicated geometries. Automated scene interpretation is still in its infancy due to the general lack of knowledge-bases for medical and biological objects and the absence of explicit rules governing their composition, appearance, and functional behavior. Substantial efforts aim at filling this gap and to apply symbolic reasoning and structural pattern recognition methods.

Succes has been obtained for simple structures where a priori-knowledge could be incorporated into the segmentation process in the form of geometrical constraints and data structures could be defined and used. The scope of these preliminary methods is limited, however, and immense research effort is needed to produce generally useful analysis methods.

7. References

Bertero, M., T.A. Poggio and V. Torre: Ill-posed problems in early vision, Proc. IEEE 76(8), 869-889, 1988

Canny, J.F.: A computational approach to edge detection, IEEE PAMI 8(6), 679-698 (1986)

Clark, J.J.: Authenticating edges produced by zero-crossing algorithms, IEEE PAMI 11(1), 43-57, 1989

Fischler, M. and O. Firshein (eds): Readings in Computer Vision: Issues, Problems, Principles, and Paradigms, Los Altos CA: Morgan Kaufmann 1987

Haralick, R.M.: The digital step edge from zero crossings of second directional derivatives, IEEE PAMI 6(1), 58-68 (1984)

Gerig, G.: Linking image-space and accumulator-space: a new approach for object recognition, Proc. 1st Internat. Conf. on Computer-Vision (ICCV '87), 112-117, London 1987

Gerig, G., W. Kuoni, R. Kikinis and O. Kübler: Medical Imaging and Computer Vision: an integrated approach for diagnosis and planning, In: Proc. 11. DAGM Symposium (H. Burkhardt, K.H. Höhne and B. Neumann, eds.), Informatik Fachberichte, Vol. 219, pp. 425-432, Berlin-Heidelberg: Springer 1989

Gerig, G., O. Kübler and R. Kikinis, Significant improvement of MR image data quality using anisotropic diffusion filtering, Technical Report 112, Institut für Kommunikationstechnik, ETH-Zentrum, CH-8092 Zurich, March 1990

Kübler, O., J. Ylä-Jääski and E. Hiltebrand: 3-D segmentation and real-time display of medical volume images, In: Proc. Internat. Symposium on Computer Assisted Radiology (CAR '87) (H.U. Lemke, M.L. Rhodes, C.C. Jaffee, R. Felix (eds.)), pp. 637-641, Berlin-Heidelberg: Springer 1987

Monga, O., R. Deriche, G. Malandain and J.-P. Cocquerez: Recursive filtering and edge closing: two primary tools for 3d edge detection, Proc. 1st European Conference on Computer Vision, Antibes 1990

Morgue, D. and G. Gerig: Recognition of nonrigid objects using the generalized Hough transform, In: Proc. 10. DAGM Symposium (H. Bunke, O. Kuebler and P. Stucki, eds.), Informatik Fachberichte, Vol. 180, pp. 151-157, Berlin-Heidelberg: Springer 1988

Perona, P. and J. Malik: Scale space and edge detection using anisotropic diffusion, Proc. IEEE Workshop on Computer Vision, 16-22, Miami 1987

Torre, V. and T.A. Poggio: On edge detection, IEEE PAMI 8(2), 147-163, 1986

Wang, D.C.C., A.H. Vagnucci and C.C. Li, Gradient inverse weighted smoothing scheme and evaluation of its performance, Computer Graphics and Image Processing (CGIP) 15, 167-181, 1981

Yuille, A.L. and T.A. Poggio: Scaling theorems for zero crossings, IEEE PAMI 8(1), 15-25, 1986

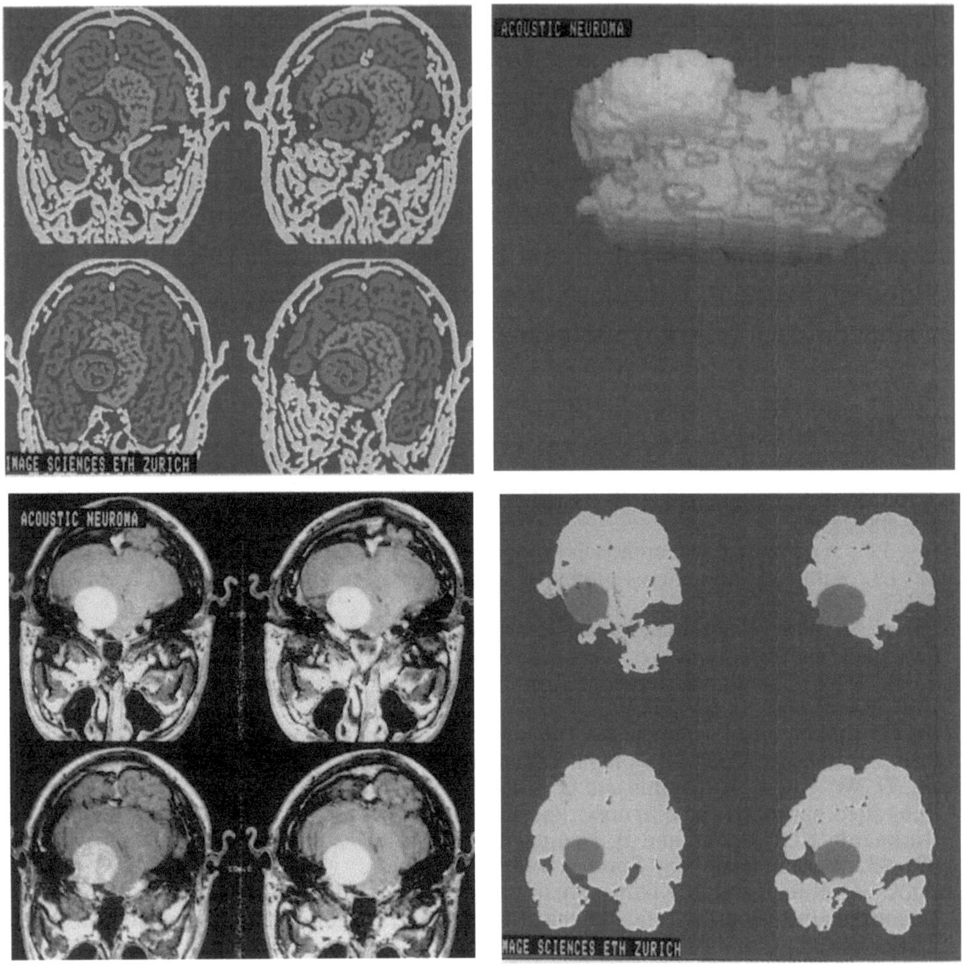

a | b
c | d

Color Plate I: 3-D segmentation of acousticus neurinoma: **a.**original MR image, **b** 3-D structure of tumor, **c**: LOG segmentation, **d**: region growing segmentation,

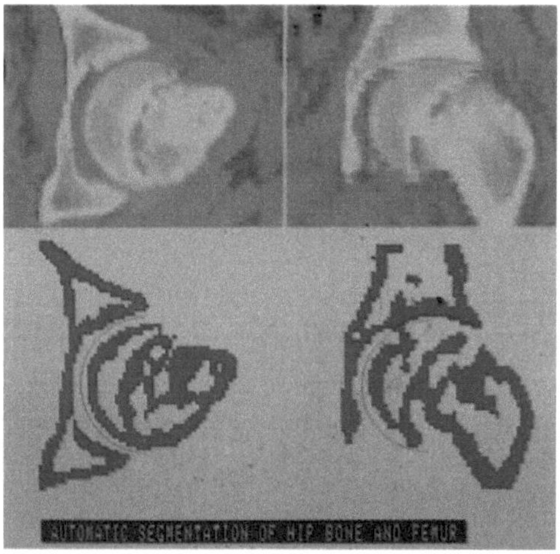

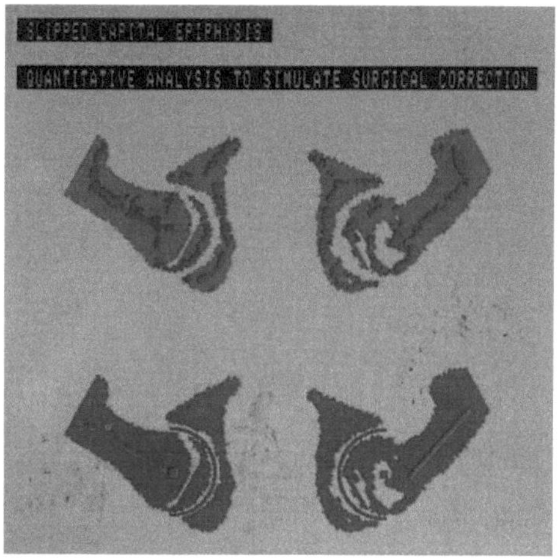

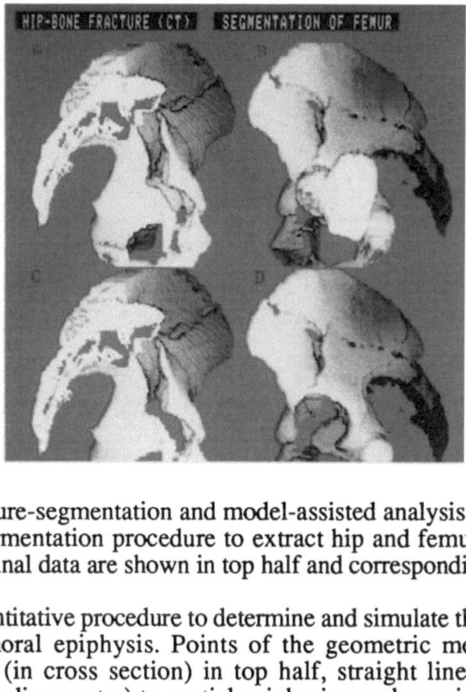

a | b
c

Color Plate II: Structure-segmentation and model-assisted analysis of CT images
a: Example of the segmentation procedure to extract hip and femur in a set of CT slices. Planar sections of original data are shown in top half and corresponding segmentation results in bottom half.
b: Illustration of a quantitative procedure to determine and simulate the surgical correction of a slipped capital femoral epiphysis. Points of the geometric medial axis of the bony structures are shown (in cross section) in top half, straight line fit to femur axis and hemispherical fit (including center) to captial epiphesis are shown in lower half. Notice the good and poor alignment in the healthy and afflicted joints.
c: Shaded surface views of a hip fracture. Hip with femur in its original position are shown in top half; femur is extracted after segmentation and pelvic bone is shown in isolation in bottom half

Toward Interactive Object Definition in 3D Scalar Images

Stephen M. Pizer, Timothy J. Cullip, Robin E. Fredericksen

Medical Image Display Research Group
University of North Carolina
Chapel Hill, NC 22599-3175, USA

Abstract

We present a method of object definition designed to allow fast interactive definition of object regions in 2D and 3D image data by a human user based on an automatically computed image description of sensible image regions. The image description is a quasi-hierarchy of ridges (or courses) and subridges (or subcourses) in the intensity surface corresponding to the image. Two methods of ridge computation are presented, one based on the intensity axis of symmetry and another based on flow lines in the intensity surface. A system for interactive object definition using this approach is described, and the use of this approach on a variety of medical images is evaluated. Generalizations of these descriptions and the interactive object definition tool to 3D are discussed.

Keywords: Segmentation, 3D display, object definition, multiscale, geometry.

1. Introduction

The most time-consuming step of methods involving general volume visualization is that of selecting regions within the 3D image data that correspond to the objects to be visualized. For example, in 3D scalar ("grey-scale") medical images the display of all but objects appearing at the highest contrast typically requires hours of hand selection, slice by slice, of regions enclosing the objects before the elegant rendering methods can be applied to the resulting regions to produce good visualizations of the organs or other anatomic objects of interest. For the so-called volume rendering methods the region selected does not need to fit closely to the surface of the object of interest but may include the object and part of its background, as long as that background does not include other objects that after rendering will occlude the

NATO ASI Series, Vol. F 60
3D Imaging in Medicine
Edited by K. H. Höhne et al.

object of interest. For surface rendering, and also for the purposes of object measurement, the surface of the region indeed needs to follow the surface of the object. However, given the surface of the former type, the active contours method of Kass et al [1987] can be used to find the actual object surface with little user interaction. As a result this paper focuses on methods to determine the more loosely fitting region just described.

The definition of an object depends on both properties inherent in the image data and properties of the scene. The properties of the image data give geometric information about region structure (syntax) but do not use our knowledge of possible scenes (semantics). We suggest that both the early human visual system and a computer can derive the syntactic region structure by using geometric analysis, but while the human cognitive system also knows how to bring knowledge of possible scenes to bear, given the regions produced from the syntactic analysis, it is not very well known how to make computers do this. However, for the regions to be useful for rendering, they must be known not simply by the human but by the computer. We are thus led to design a method whereby we attempt to have the computer derive the syntactically defined regions and let these serve as a means of communication between the human and the computer such that the user interactively and quickly can specify to the computer semantically correct regions from the syntactically defined ones.

This paper is therefore divided into section 2 on how to produce an image description in the form of a graph of sensible regions and section 3 on an interactive method for producing a meaningful region based on the image description.

2. Image Description by Multiscale Geometry

We suggest that if the image is viewed as an intensity surface, where height corresponds to image intensity (see figure 1), then shape properties of this surface will determine the region definitions. Humans seem to use ridges of intensities as organizing features for light objects on dark backgrounds, or courses for dark objects on light backgrounds. In keeping with its being a good model for human vision, ridge/course analysis is largely insensitive to rotations of the co-ordinate system, spatial scaling, and monotonic transformations of intensity and involves a focus on curvature, orientation, and two-sidedness. In this paper we will focus on regions lighter than their background, but all of the ideas will apply to the problem of finding regions darker than their background. The relation between these two sets of regions is also of interest if we are trying to imitate human visual behavior but is beyond the scope of this paper.

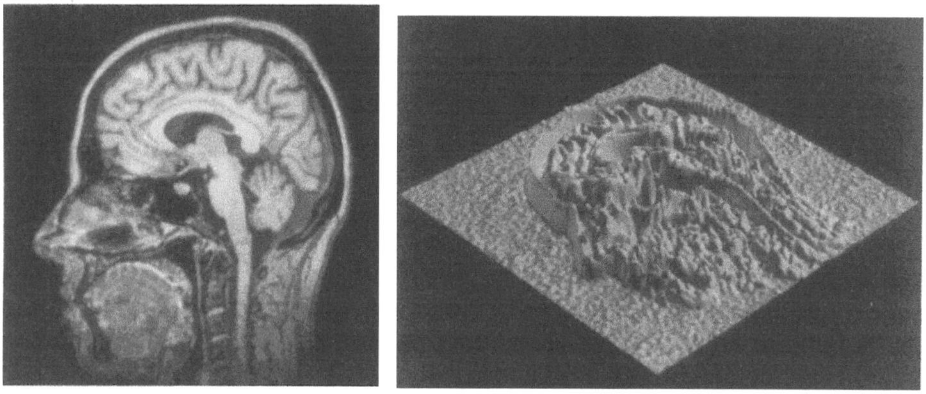

Fig. 1. MRI brain image and corresponding intensity surface

The fact that analysis into ridges may provide a reasonable model of human visual segmentation is relevant for 2D images but not necessarily for 3D images. Nevertheless, the effectiveness of this analysis for 2D images has led us to investigate its usefulness for 3D images. For each form of ridge analysis we will discuss first the ideas as applied to 2D images and then their generalization to 3D.

Ridges are structures with a ridge top and flanks falling to a course separating this ridge's flank from that of the next. The ridge top is defined by curvature properties of the surface there. Ridges have a multiscale structure, in the sense that small ridges may fall on the flank of or branch from larger ridges. For 2D (see figure 2), a) a ridge may fork into two ridges, and b) these two may rejoin, so that the pair surround an indentation with a course and even possibly a pit at its bottom. Furthermore, c,d) one ridge may appear on the flank of another. Finally, e) a ridge can begin on the flank of one ridge and end on the flank of another. The result is that the intensity surface forms a graph of ridges, with the directed arcs deriving from the child ridge being on the flank of or branching from the parent ridge. The fact that a ridge may fork and rejoin or connect two or more other ridge flanks mean that a single ridge can have more than one parent.

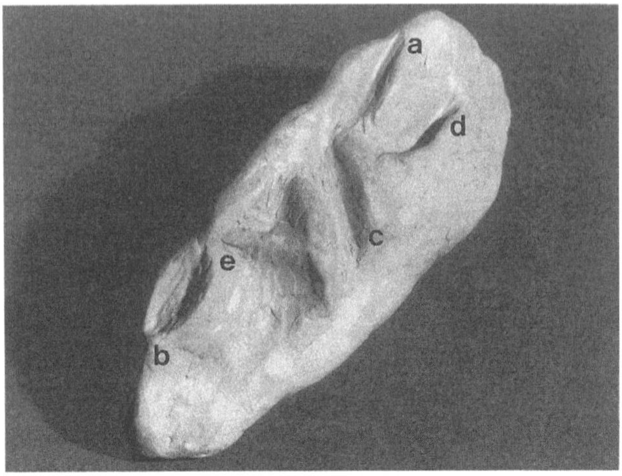

Fig. 2. Ridge types: a) ridge fork, b) ridge fork and rejoining, c) monotonic ridge on flank of another ridge, d) nonmonotonic ridge on flank of another ridge, e) ridge connecting two other ridges.

It follows that an analysis of an intensity surface into ridges must have a multiscale, geometric flavor such that ridge regions are identified and parent/child relations are generated among these ridge regions. Two different strategies suggest themselves. In the first, the quasi-hierarchical structure of ridges and subridges is captured by a quasi-tree of connected geometric elements with each element corresponding to a ridge, and the elements are followed to annihilation into other elements under increase in scale (blurring) of the image. The ridge corresponding to an annihilating element is then taken as the child of the ridge corresponding to the element into which the annihilation took place. In the second strategy the ridges (with their flanks) are analyzed directly via geometric properties, and the ridge/subridge relationship follows from the geometric properties of one ridge's flank containing the other ridge. Sections 2.1 and 2.2 explore these respective strategies.

2.1. Connected Structures with Annihilation

2.1.1. The Intensity Axis of Symmetry

We suggest that a basic property of "ridgeness" is symmetry, so that for each point on one flank of a ridge there is another point on the other flank to which it has some sort of symmetry relation. Symmetry is an important visual property -- regions are seen as two sided, i.e., ridge flanks are visually grouped to form the objects that we see. An analysis in terms of this symmetry should thus generate regions and hierarchical relationships that provide a natural medium for human and computer communication for segmentation.

As a result we define an intensity axis of symmetry (IAS). This axis, made of a forest of branching or looping sheets, fits under the ridges midway between the two symmetric flanks of the ridge. In order to avoid sensitivity to any monotonic transformation of intensity and also to avoid the incommensurability of intensity and space, the image (intensity surface) must be considered as a one-parameter family of slices in the intensity dimension, with the IAS made by stacking axes of symmetry defined for each slice. Thus, the IAS is the one-parameter family of medial axes of the intersections of the intensity surface with a series of slicing surfaces (see figure 3).

For now we have been slicing at isointensity levels, so that each axis of symmetry in the family is the medial axis of a level curve of intensity, even though this slicing focuses too greatly on intensity levels and too little on local image structure. The present IAS has the advantage that there is a 1-1 relation between each branch and a ridge top as defined by the locus of maxima of positive curvature of intensity level curves (these loci are called *vertex curves*) [Gauch, 1988a, 1989]. However, it has the disadvantage that it is too directly tied to absolute intensity and ridge flanks go down only to the higher of the two courses bounding a ridge, so future consideration of slicing strategy is indicated.

As illustrated in figure 3, associated with each point on the IAS are (normally) two points in the image where the maximal disk centered at the IAS point is tangent to a level curve at its intensity. The basic sheets that are the leaves in the IAS quasi-tree thus have associated with them a set of image points that are taken as the primitive regions of the image.

2.1.2. Generating the Ridge/Subridge Relationships

Associated with a monotonic ridge on the flank of another ridge (its parent), then, is a branch of the IAS that grows from the IAS sheet lying under the parent ridge. If the child ridgetop branches from the parent ridgetop, the IAS child branch also begins at the top of the IAS sheet of the parent (see figure 4a); otherwise from the middle of the parent sheet (see figure 4b). A splitting

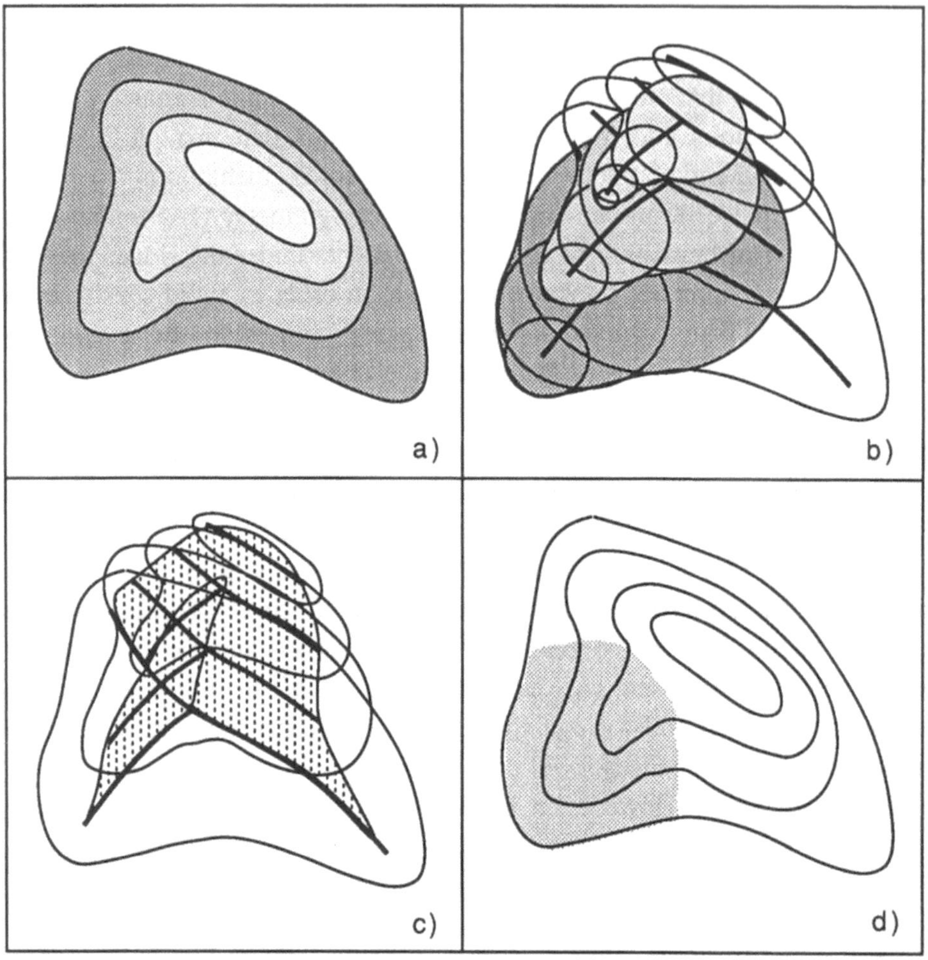

Fig. 3. The IAS as a one parameter family of slice axes of symmetry.
a) Level curves of a simple image; b) Level curves on intensity surface, their medial axes, and selected maximal disks; c) Level curves on intensity surface, and IAS; d) Image region associated with an IAS branch.

and merging of the ridgetop results in a scoop in the IAS sheettop (see figure 4c), and a ridge connecting two other ridges corresponds to an IAS sheet connecting two other sheets (see figure 4d). The IAS under a nonmonotonic subridge forms a loop on the parent IAS sheet; at the intensity of the loop bottom the parent sheet tears (see figure 4e). More complicated forms arise if pits appear in inter-ridge valleys. In any case, we have the important relationship that although ridgetops may not form a connected structure, the corresponding IAS sheets do form a connected structure reflecting the subridge on ridge flank or ridge branching relationships. Because a given subridge may be on the flank of or branch from more than one other ridge, a given ridge may be the child of more than one parent.

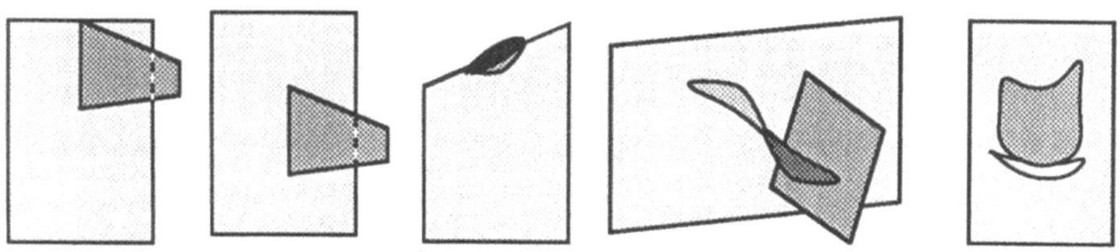

a) branch at sheet-top   b) branch below sheet-top   c) scoop in sheet-top   d) connecting sheets   e) loop & tear below sheet-top

Fig. 4. IAS branchings

Region analysis thus involves computing a graph in which each node corresponds to a ridge, i.e. what we are taking to define sensible, coherent regions. The arcs in the graph are directed to ridges from subridges that describe sensible, coherent subregions of the parent ridge in question. In addition the parent ridge contains pixels that are on the parent's flank but not in any subridge.

Our problem is to decide which ridge is a subridge of which and to form the graph which describes these relationships. Our strategy is to compute the region containment relations induced by the connectivity of the IAS branches and by the annihilation of one branch into another as scale is increased (see figure 5). That is, one ridge is taken as a subregion of another ridge if the IAS branch corresponding to the former disappears into the IAS branch corresponding to the latter as the scale at which the image is considered is successively increased.

In the process of increasing spatial scale, it can be shown that a simple IAS branch (obtained by either ridge branching or by one ridge being on the flank of another ridge) will begin as a sheet but shrink from the top and bottom until it disappears smoothly into the sheet to which it is attached. This disappearance can be used to establish the child/parent relationship. Looping subsheets will normally disappear by narrowing down in their middle and breaking into two simple IAS sheets. An IAS sheet connecting two other

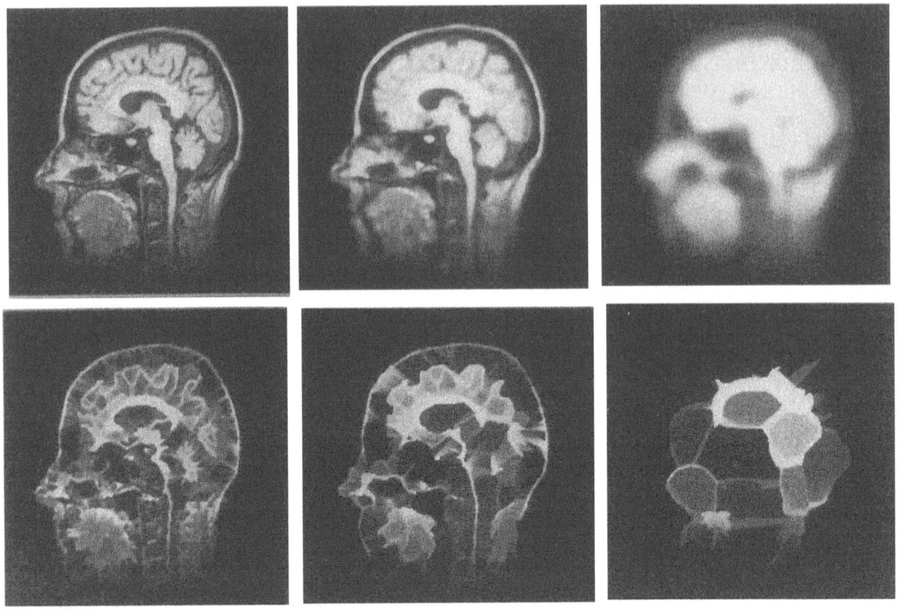

Fig. 5. An MRI brain image at various scales and corresponding IAS's. Scale change was obtained by edge-affected anisotropic diffusion.

sheets can detach first from one of its parents and then from the other, the intensity levels at which it is attached to one not being the same as those at which it is attached to the other.

Scale increase is performed by diffusion (Gaussian blurring), since this form of scale increase avoids most strongly the creation of structure [Yuille, 1983]. Simplicity suggests isotropic and stationary Gaussian blurring, but it seems preferable that the diffusion rate be matched to the scale of local objects. That is, diffusion across the interior of a large object should be faster than across a similar region made up of small objects or across the edge of the large object -- the scale change should be nonstationary. Moreover, diffusion along an edge should be faster than diffusion across it. To achieve this adaptively nonisotropic diffusion, we have begun using a program written by De Moliner [1989] at E.T.H., Zürich, in which a negative exponential function of intensity gradient magnitude (edge strength) is used as the local conductivity in a variable-conduction form of the diffusion equation [Perona, 1987].

2.1.3. Computing the 2D IAS-Based Image Description

IAS Computation. To compute the IAS [Gauch, 1988b], we first view the 2D image as an intensity surface in 3-space. We define a parameterization of the 2D image as f(u,v) = (x(u,v),y(u,v),z(u,v)) with initial values x(u,v)=u, y(u,v)=v, and z(u,v)=I(u,v). The surface thus defined is allowed to move in the spatial directions (x(u,v) and y(u,v) are allowed to change values) but held fixed in intensity as it is moved toward the IAS. Think of this as a collapsing of the original surface onto the IAS skeleton. By not allowing motion in the intensity direction, we insure that the basic structure of the image, as defined by the ridges, is not lost.

The surface is moved toward the IAS by an iterative relaxation algorithm. During each iteration each point on the surface is examined to see if moving it to a neighboring location will increase its distance, inside the intensity surface and at its intensity, from that surface, as measured by the image symmetry function, g(x,y,I). If so, the point is moved to that new location.

This method breaks up surface continuity. In order to re-establish this continuity, at least at places where the surface should be continuous (see figure 6), we extend the concept of active contours [Kass, 1987] to a deformable surface, starting from the broken surface resulting from the unconstrained collapse. This active surface model iteratively minimizes an energy functional defined on the surface with terms that capture the smoothness constraints of the surface while also keeping the surface near the IAS. The smoothness constraints aid in the identification of individual IAS branches. The equation we are currently using is given by

$$\text{Energy} = \iint [w_1(f_u(u,v)^2+f_v(u,v)^2) + w_2(f_{uu}(u,v)^2+f_{vv}(u,v)^2+f_{uv}(u,v)^2) - w_3(g(f(u,v)))]\, du\, dv,$$

where the u and v subscripts indicate the partial differentiation of the surface equation. The weighting factors determine the degree to which the surface behaves like a smooth membrane (w_1), or a spline (w_2). To allow desired discontinuities of the type illustrated in figure 6, these weights are reduced in those areas of the image that have strongly moved away from each other during the original collapse. We are also currently investigating other surface energy equations that better match the surface geometry of the IAS.

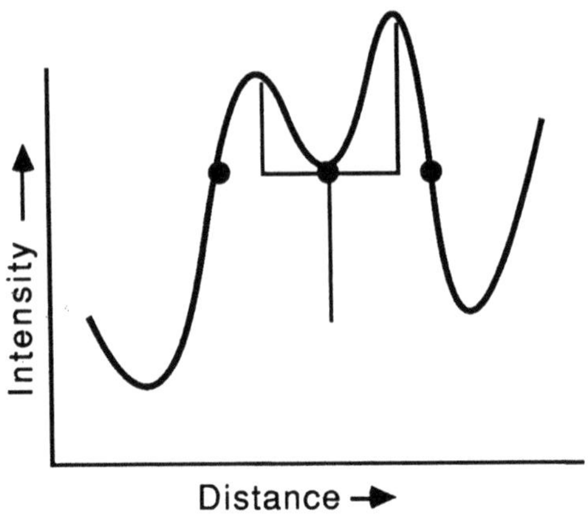

Fig. 6. A cross-sectional slice of a double ridge. Areas of the image where the continuity constraints need to be relaxed are marked with dots.

Since the image symmetry function g is used extensively during the iterative collapse routine, its values are precomputed for every point in the image at each intensity level. To limit the size of this 3D data array the image is first scaled to a reasonable number of discrete intensity levels, typically 50. The calculation can take advantage of the fact that at a given pixel location, the distance value at a lower intensity must be no less than the distance value at a higher intensity ($g(x,y,I+1) \leq g(x,y,I)$).

Identifying Branches. Once the surface has been collapsed onto the IAS, we need to identify the individual branches within this structure. We do this for each branch by identifying the two patches of the active surface which have been drawn to the opposite sides of the IAS branch. Before we do this, we must first identify for each point in the image its involute: the point on the other side of the IAS that mapped to the same point on the IAS (see figure 7). The collapsing of the surface onto the IAS has given us a mapping for each point from the original surface to the IAS. Note that each point on the surface maps to a single point on the IAS, but each point on the IAS may have several image points mapped to it. Determining involute pairs is achieved by first allocating a bucket for each IAS point, then filling the buckets with the original image points that map to each bucket location. Each point in a proper involute pair should satisfy the following conditions:

1) moved the same distance in getting to the IAS,
2) be distant from its involute to prevent linking points on the same side of the axis sheet,
3) moved to the same IAS point (bucket).

Due to the discrete nature of the data and the approximation of the IAS resulting from the surface minimization, these conditions may only be approximately satisfied. By using a weighting of the first two conditions, and an outward search based on the third condition (start with the current bucket and then search neighboring buckets), we have achieved a reasonable linking of involutes. We now have for each point in the image two mappings, the first maps the point to its IAS point, and the second maps the point to its involute.

The identification of individual branches of the IAS now becomes a simple region growing process. The process starts with an arbitrary point, placed in set A, and its involute in set B. You then examine all neighboring points of points in set A, adding them to A (and their involutes to B) if their involutes are in set B or neighbors of points in B. You do likewise for points in B with their involutes in A. This process is repeated until neither set A nor B increase. The growing stops when the involute of a neighbor is not a neighbor of the other set, that is, at the branch points of the IAS (see figure 7).

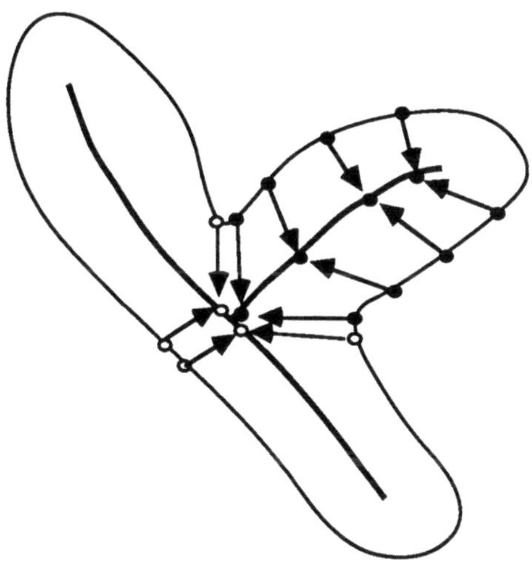

Fig. 7. Involute pairs and their relationship at a branch point in the IAS. The outer curve represents a single isointensity contour.

Region/Subregion Identification. By identifying the individual branches of the IAS, we have partitioned the image into primitive regions, one region per branch. In order to group these regions in a hierarchical manner, we follow the simplification of the branching structure through scale space. Since the IAS branching structure simplifies through successive blurring levels, the goal is to track the branches from level to level, recording when individual branches annihilate.

At present we are associating each branch at a given blurring level (scale) with the branch at the successive scale with whose region its region has the greatest overlap. If more than one branch at the lower scale are associated with a single branch at the higher scale, a connection in the hierarchy is recorded. We are currently investigating more robust methods of tracking the IAS branches themselves.

2.1.4. Toward 3D IAS-Based Image Descriptions

All of the 2D definitions for IAS generalize to 3D. An intensity surface over three spatial variables has an IAS based on an intensity family of 3D medial axes based on maximal spheres. The resulting structure can be formed into a graph of connected elements, and these can be followed to annihilation into each other as the 3D image scale is increased by diffusion.

The extension of the algorithms to 3D images is straightforward. The intensity surface now becomes a 3D surface in 4-space: f(u,v,w) = (x(u,v,w), y(u,v,w), z(u,v,w), I(u,v,w)). The IAS is now a branching structure in 4D. The image symmetry function is now defined over a 4D data set.

We have had some success with the 3D algorithms on small, simple test images, but are still investigating the process. We have found that the proper identification of involutes is both a more difficult task due to the added spatial dimension, and more sensitive to the proper termination of the branch growing algorithm. We are currently modifying the algorithms to take advantage of a newly acquired MasPar SIMD parallel processing computer which should reduce the hindrances to our research caused by the large amounts of time and space the current sequential algorithms require.

2.2. Surface Regions with Containment

2.2.1. Reverse Gravity Watersheds

The goal of image segmentation is to produce visually meaningful primitive regions and a hierarchy on those regions that allows a sensible grouping of those regions into larger scale objects. The IAS achieves the first goal by associating regions of the image with individual branches of the IAS. It achieves the second goal by following the annihilations of the branching structures through scale space. Several aspects of the current implementation of IAS have led us to consider alternative approaches. One disadvantage of the IAS is that its dependence on isointensity slicing results in ridge flanks that do not necessarily reach down to valleys on both sides (see figure 8). Also the complex nature of the IAS branching and connectivity makes it difficult to follow the structure through scale space (consider that a loop at one scale can break into two branches at a larger scale). Finally, the computational cost of the IAS is significant, especially when considering the extension to 3D.

Instead of focusing on the medial axis of ridges, one can focus on their surface properties [Koenderink, 1990]. Consistency with the medial axis approach would suggest the use of vertex curves to define ridge tops, but this does not associate flanks with these tops. Another, not equivalent, means of characterizing a ridge is via properties of integral curves of principal curvature. Yet another, not equivalent, means of characterizing a ridge is via the properties of flow lines on the surface, that is curves of maximum slope on the surface. A weakness of this approach is its sensitivity to monotonic transformations of intensity. Nevertheless, the computational and theoretical simplicity of this approach has led us to work on its development and implementation.

Perhaps the simplest way of using flow lines to partition the surface into ridge/flank structures that span valley to valley is to group them into watersheds (with reverse gravity) (see figure 8). A reverse gravity watershed is a region of the image where if water were to fall on the region, the water would flow uphill and collect into a single maximum of that region. Once the surface has been partitioned into watersheds, you can produce a hierarchy on those regions by successively blurring the image and determining the order in which watersheds blur together. Unfortunately since a single long ridge may be composed of several watersheds (several local maxima along the ridge) we have found that they do not always blur together before the ridge blurs into a neighboring ridge. This results in a hierarchy that does not adequately reflect the ridge structure of the image.

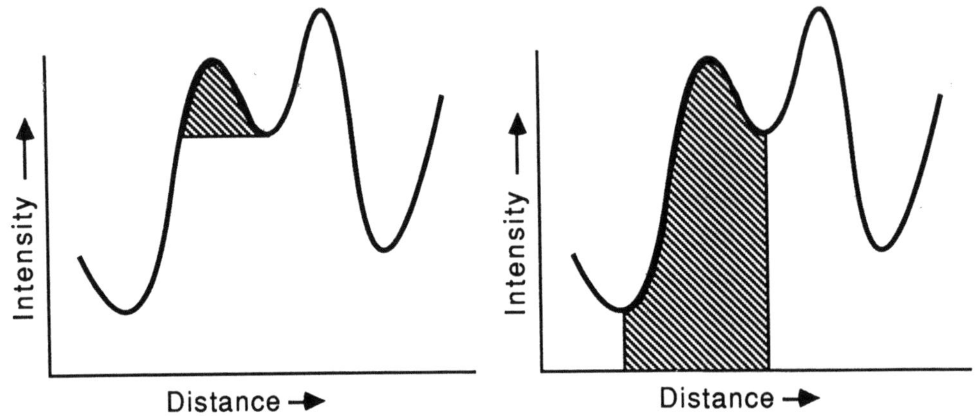

Fig. 8. IAS region (left) vs. reverse gravity watershed (right)

An alternative approach for producing a hierarchy on watershed regions has been tried with promising results. It examines surface properties rather than behavior under blurring. The concept is to hierarchically group together primitive regions based on a measure of the cost of moving from one patch to another. When you move from the maximum of one watershed, through the saddle point on the border of the two regions, to the maximum of the neighboring watershed, a reasonable cost measure is the effort spent going downhill and back up again. This would reflect a relatively low cost when moving along a ridge line (allowing watersheds along a ridge to be connected lower in the hierarchy) while moving from one ridge to another would be relatively more expensive (linking separate ridges higher in the hierarchy).

Partitioning the surface into watersheds is a relatively simple task. The image is scanned once initially to build a list of pixels for each discrete intensity. To grow each region from its highest intensity point downward these lists are processed one at a time. Starting with no primitive regions and the highest intensity list, you choose any pixel in that list as the seed for the first region. The region starts growing by repeatedly scanning the list, each time removing any pixels from the list and adding them to the region if they are neighbors of the region. If the region stops growing, but there are still pixels in the list, then you choose one of these remaining pixels as a seed for a new region. Continue the process until the list is exhausted. Then start processing the next lower intensity list, adding neighboring pixels to the existing regions and creating new regions each time a list still has pixels that aren't neighbors to any existing region.

To compute the cost function between each pair of neighboring regions you need to determine the intensities at the maximum of each region and at the saddle point between the regions (cost equals the average of the two maximum intensities minus the saddle point intensity). The intensities at each maximum is easily determined since the maximum point is the seed point mentioned in the previous paragraph. The saddle point intensity can be found with two scans of the image, the first in row major order, the second in column major order, checking the intensity each time you cross from one region to another and retaining the largest of those intensities for each pair of regions. Once the cost function has been determined, the hierarchy is built simply by ordering the costs from lowest to highest. Typical results can be seen in figure 10.

Unfortunately, watersheds have significant disadvantages as the primitive region generator. It has already been mentioned that a single long ridge may be composed of several watersheds. Even worse, one watershed may contain more than one ridge structure. For example, a child ridge may branch off a parent ridge where the child ridge does not have its own local maximum. Therefore we are currently investigating other surface properties involving flow lines that may be used to partition the surface into ridge/flank primitive regions. For example, the separation and reconvergence of nearby flow lines may be utilized in defining a subridge. Also, the cost function used so far has been a very simple one and we plan to investigate more robust alternatives.

2.2.2. Towards 3D Watershed-Based Image Descriptions

Flow lines and their computation generalize directly to 3D. The definition of reverse gravity watershed regions in 3D, however, requires new concepts because in 3D there are more types of critical points and ridges/courses.

In a 2D image a critical point has either two negative principal curvatures (a peak, where the surface peaks along both principal directions), one negative and one positive principal curvature (a saddle, where the surface peaks along one direction but bottoms out along the other), or two positive principal curvatures (a pit, where the surface bottoms out along both directions). In 3D there are four types of critical points, where the signs of the principal curvatures are respectively --- (peak), --+ (saddlepoint of type 1), -++ (saddlepoint of type 2), and +++ (pit). A reverse gravity watershed ridge in a 2D image goes from maximum to a saddlepoint to maximum. A reverse gravity watershed ridge in a 3D image goes from a maximum to a saddlepoint of the first type to a maximum. The role of the connections between saddlepoints of the two types needs to be investigated.

The 2D reverse gravity watershed algorithm extends naturally to 3D where each region is again grown from a maximum intensity seed point until it collides with a neighboring region. The cost function between pairs of regions is still the average of the maxima minus the largest intensity point along the border between the two regions (a saddlepoint).

2.3. Image Description Quality

The speed of the computation of the image description is not critical because it is totally automatic. Nevertheless, this speed affects the computing cost and the extendability of the method to 3D. We have implemented both 2D image description methods on a DEC3100 workstation (12 MIPS). Our programming has taken some care for speed, but we have not searched for optimal algorithms. Using 12 levels of blurring, the IAS-based method requires approximately six hours on a 256 x 256 image. Approximately 30% of this was taken by the distance calculations, approximately 20% by the surface collapses, and approximately 35% by the branch identifications. The reverse gravity watershed-based method is two orders of magnitude more efficient: the region growing and the hierarchy computation take less than a minute. The 3D generalization of this algorithm has been tested on a 128 x 128 x 32 MRI brain image. The computation time was approximately 15 minutes. We are in the process of transforming these to operate on a MasPar MP-1 4K-processor SIMD parallel computer (8000 MIPS). Speedups of a factor of approximately 100 are achieved by using the MasPar MP-1.

We have applied the algorithms for each of the two approaches to a number of 2D MR and CT images. These include objects that are elongated and possibly branching, such as the brain stem and the bronchial tree, and objects that are more compact, such as the cerebellum and kidneys. They also include objects with sharp edges and others with indistinct edges or portions of the organ boundary that have not been imaged as an edge. Examples of regions found on an MRI image of a brain are given in figures 9 and 10.

Most of the regions, both primitive ones and ones appearing higher in the quasi-hierarchy, from both approaches seem reasonable and provide a useful basis for a user to build meaningful object regions using the interactive tool described in section 3. For example, bronchial branches and their connections within the bronchial tree were well found in a CT chest image. The brain stem and the cerebellum in an MR brain image were each made up from just a few regions.

The IAS regions cover the full range of scale, and it appears that the regions as mathematically defined generate a reasonable hierarchy. The reverse gravity watershed primitive regions capture large sensible object regions more frequently than the IAS regions do. Our limited trial of the 3D generalization of the reverse gravity watershed calculation on a portion of a 3D MR image surrounding the cerebellum captures a number of reasonable 3D regions such as the scalp and most of the cerebellum, but the method sometimes failed to isolate visually sensible regions from neighboring regions.

Besides its inefficiency the main weaknesses of the IAS-based method have been that the resulting regions

1) sometimes are uncomfortably small,
2) connect image regions in an unnatural way, or
3) have boundaries that did not come up to the natural object boundary.

The small regions were frequently correct, but of a scale smaller than one desired to build a meaningful object. This is essentially a desirable situation because these regions are real ones that could for some small objects be just the regions needed to build these objects. However, for larger objects, some means of selecting only regions of a satisfactorily large scale may be usefully built into our interactive object region definition tool.

As for the unnaturally connected IAS regions and some of the small regions, it appears that these were largely the result of the computation not calculating the regions that the mathematics would define. Largely, this followed from improperly identifying involutes due to an imperfect surface contraction onto the IAS. An example of an area of disappointment was in trying to follow the convolutions of the cortex in an MR image. The primitive regions corresponded well to the individual folds, but moving up the hierarchy through scale space quickly brought in regions outside of the brain. We expect that improvements in the hierarchy and the primitive regions will result from using non-stationary, anisotropic edge sensitive diffusion as opposed to the Gaussian blurring we have used so far and from using a surface energy function that better captures the geometry of the desired IAS. Also, improvements can be made in our method for tracking the IAS branches through scale space.

Finally, the unnatural IAS region boundaries seem to have two causes: problems with the computational algorithm and the fact, illustrated in figure 8, that region edges are tied to absolute intensity levels of one of the courses bounding the region.

The main weaknesses of the reverse gravity watershed method have been that sometimes small regions were simply incorporated into larger regions that were taken to be primitive and that some unnaturally connected regions were computed. We attribute both of these failings to two facts. First, only reverse gravity watershed regions and not regions reflecting flow line separations have been computed to date. Second, the cost function used to connect ridge pieces together requires improvement. Our short term research goals are to attack these problems.

Fig. 9. Brain MRI regions defined via IAS
Left: Primitive regions. Right: The scalp and cerebellum selection each from a single point plus moving up the hierarchy. The blue region shows what would be added by moving once more up the hierarchy.

Fig. 10. Brain MRI regions defined from reverse gravity watersheds
Left: Primitive regions. Right: The scalp and cerebellum each from a single point plus moving up the hierarchy.

Fig. 11. A complex selected object region: the cortex

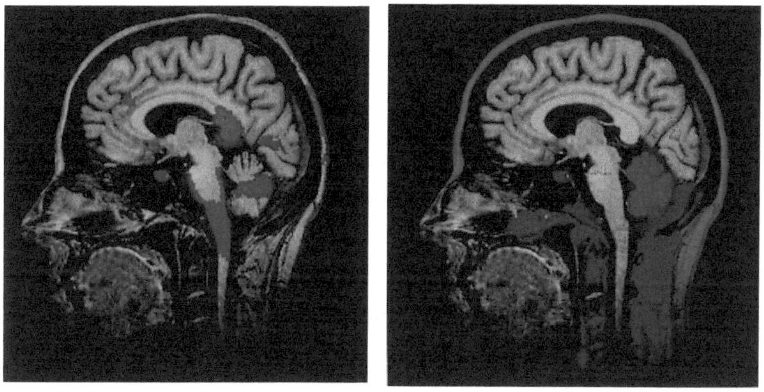

Fig. 9.

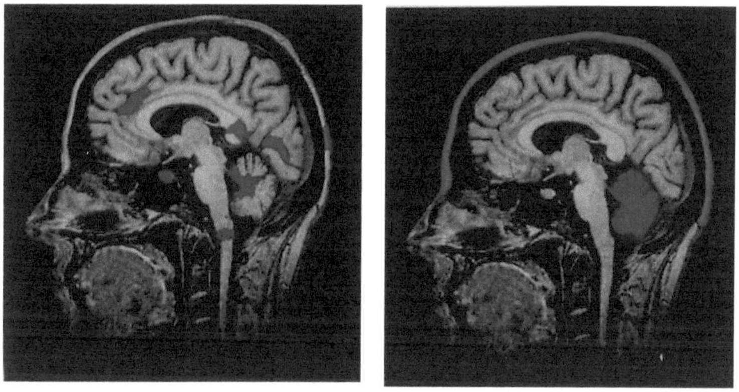

Fig. 10.

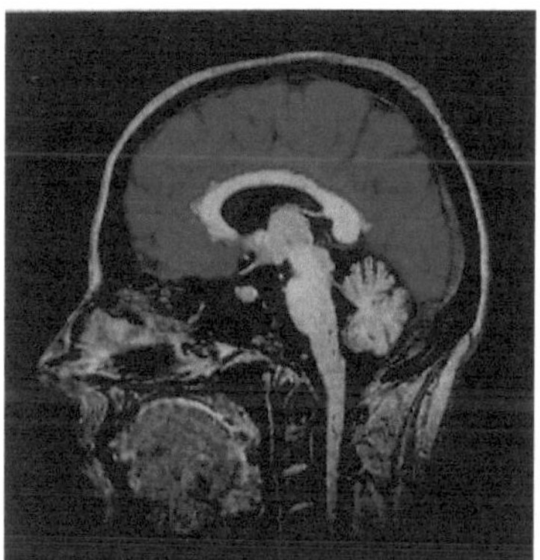

Fig. 11.

3. Interactive Object Definition from Descriptions

3.1. 2D Interactions

In our approach to interactive object definition the user defines an object region using simple mouse-based interactions. The user causes a set of regions from the image description to be selected as an object region and displayed with a color wash to indicate their selection. While the option of editing an object region by pixel painting is provided, it is infrequently necessary, except when the object being defined really is not fully captured in the grey-scale differences in the original image data.

The simplest interaction of the user is that of pointing to and clicking on a pixel, which causes the smallest region containing the selected pixel to be chosen. If the user is in the mode of adding regions to the object set, the chosen region is added to the set. If the user is in difference mode, the chosen region is removed from the set. 'Region painting' is accomplished by holding the selection button down and sweeping the cursor over the region of interest. The union of the regions corresponding to all the selected pixels is selected for addition into or removal from the the object set, depending on the mode. A "parent" button causes the presently selected description region to move to its parent in the description graph. This higher level region is then added to or removed from the object set, depending on the mode. It has proven quite useful to avoid the frequent use of the 'undo' operation by color washing, in a special color, additional pixels in the parent of the presently selected description region.

This method of object definition has proven to be very intuitive for physicians and to allow the definition of complicated anatomic objects in CT and MR slice images to take place in under a half-minute. The system implementing this method runs in an X environment on color workstations.

3.2. Speed of 2D Object Definition

The following examples of object definition speeds are typical using our interactive system. In our example MR brain image, using pointing and hierarchy traversals, the scalp could be identified in 10 seconds by selecting one primitive region followed by seven parent operations for IAS regions and eight parent operations for reverse gravity watersheds. The cerebellum required seven atomic actions over 8 seconds in each case, but the reverse gravity watershed method used one primitive region followed by six parent operations, whereas the IAS method required one primitive region selection followed by five parent operations followed by an additional primitive region selection. Selecting the cortex by groups consisting of the selection of a primitive region and a number of hierarchy traversal operations required six groups of 3-7 operations over 60 seconds for reverse gravity watersheds and 12 groups of 3-4 operations over 120 seconds for IAS regions. Thus the timings were about the same for the IAS regions and the reverse gravity watersheds except for the cortex, where the reverse gravity watersheds provided faster definition. In all these cases, with both methods, the object definition was rather speedy.

Region painting (see figure 11 for the selected regions) provided a considerable speedup for the more complex objects. For example, to define the cortex, a speedup by a factor of 3 was obtained for IAS and a factor of 2 for the reverse gravity watersheds. To define the cerebellum, no speedup was obtained.

Applying this object definition approach slice by slice to 69 MRI slices covering the cerebellum allowed us to specify a 3D region including the cerebellum and a bit of its background but no occluding organs. Only two of

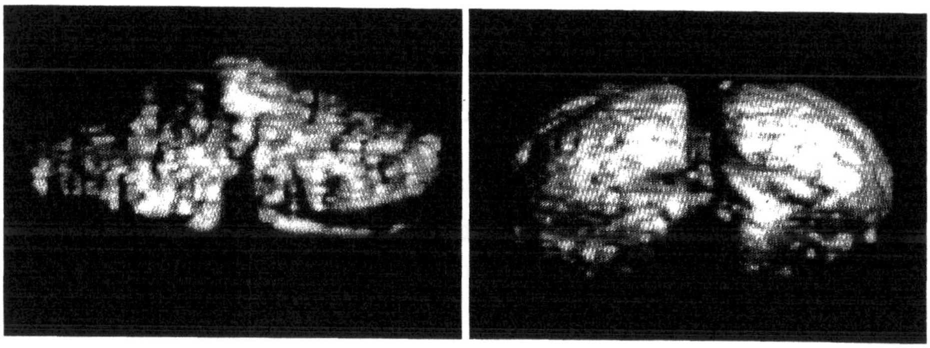

Fig. 12. Volume rendering of cerebellum from region selected from MRI study

these slices required editing by pixel painting. The application of Levoy's volume rendering technique [Levoy, 1988] to the image intensities in this 3D region produced the images in figure 12. Thus we have demonstrated the capability of this approach to select 3D regions of interest for the volume rendering of internal organs in 3D medical images.

3.3. Toward 3D Interactive Object Definition

We are working toward a system for interactive specification of 3D object regions based on quasi-hierarchical 3D image descriptions. We anticipate interactions essentially the same as the ones provided in our 2D system. In such an interactive system the user must be able to specify locations in 3-space, visualize the regions selected and unselected relative to the grey-scale values in the original 3D image data, and visualize the 3D objects in the selected region. The latter capability, with a speed of approximately one second from a 256^3 data set, will be provided by Pixel-planes 5, described in another paper in this volume [Fuchs, 1990]. The possibility of modification of viewpoint, rendering parameters, and clipping planes with optional display of the original grey-scale data on these planes, with a response via progressive refinement of about 0.1 seconds, is anticipated.

A major component of the system displays the slices making up the original image data as an array of small images ("postage stamps") in an X window on a CRT. Any postage stamp can be selected for display at full sampling in another X window. Pixel (voxel) pointing will be possible either on any postage stamp or on the fully sampled slice window. The presently selected 3D object region will appear as a color wash on the collection of slices, such that the selection of any pixel would cause a 3D region including pixels on many slices to be colored.

We are also considering region painting interactions using a 3D mouse controlling a 3D cursor that would appear on the volume rendered image as well as on the 2D postage stamp display. It remains to be seen whether a way can be developed to communicate in 3D the relation of the cursor to the 3D grey-scale data and whether such a means of communication is necessary.

In any case we are aiming to have such a system allow the volume rendered presentation of any anatomic region appearing in any 3D data set with a selection time of under 2 minutes, after a considerably longer image acquisition and automatic image description computation stage.

Acknowledgements

We are indebted to John Gauch for his work on IAS and watershed algorithms and theory, on which much of this paper is based; to Drs. Fred Bookstein, Robert Gardner, and Jan Koenderink and Mr. Michael Tuchman for useful discussions on geometry; to James Coggins for his contribution in designing our object region definition system; Graham Gash for providing our image access, processing, and display environment; to Bo Strain for photography; and to Siemens, Inc. for providing the MRI image data. The research reported herein was carried out with the partial support of NIH grant number P01 CA47982.

References

Fuchs, H, "Systems for Display of Three-Dimensional Medical Image Data", in this volume, 1990.

Gauch, J, Oliver, W, Pizer, S, "Multiresolution Shape Descriptions and Their Applications in Medical Imaging." *Information Processing in Medical Imaging* (IPMI X, June 1987): 131-150, Plenum, New York,1988a.

Gauch, JM and Pizer SM, "Image Description via the Multiresolution Intensity Axis of Symmetry." *Proc. 2nd Int. Conf. on Comp. Vis.*, IEEE Catalog #88CH2664-1:269-274, 1988b.

Gauch, JM, *The Multiresolution Intensity Axis of Symmetry and Its Application to Image Segmentation*, PhD Dissertation, University of North Carolina, 1989.

Kass, M, Witkin, A, and Terzopoulos, D, "Snakes: Active Contour Models." *Proc 1st Int. Conf. on Comp. Vis.*, IEEE Catalog #87CH2465-3:259-268, 1987.

Koenderink, JJ., *Solid Shape*. MIT Press, Cambridge, 1990.

Levoy, M, "Display of Surfaces from Volume Data." *IEEE Computer Graphics and Applications* **8**(3):29-37, 1988.

De Moliner, R., *Kantenerhaltende Glättung*, Internal Report, Institut für Kommunikationstechnik, Fachgruppe Bildwissenschaft, Eidgenössiche Technische Hochschule, Zürich, Switzerland, 1989.

Perona, P and Malik, J., "Scale Space and Edge Detection Using Anisotropic Diffusion", *Proc. Workshop on Computer Vision*: 16-22, IEEE, 1987.

Yuille, AL and Poggio, T, "Scaling Theorems for Zero-Crossings." A.I. Memo 722, MIT, 1983.

Steps Toward the Automatic Interpretation of 3D Images

N. Ayache [1], *J.D. Boissonnat* [2], *L. Cohen* [1], *B. Geiger* [2], *J. Levy-Vehel* [1], *O. Monga* [1], *P. Sander* [1]

[1] INRIA - Domaine de Voluceau - Rocquencourt
BP. 105, F-78153 Le Chesnay Cédex - FRANCE
[2] INRIA Sophia Antipolis - 2004, route des Lucioles
F-06565 Valbonne Cédex - FRANCE

Abstract

We describe in this paper part of the research being performed at Inria on the automatic interpretation of three-dimensional images. We identify three common key problems which we call **segmentation, representation**, and **matching** of 3D regions. We describe our approach for solving these problems, our current results on 3D medical images, and give the trends of our future work.

Keywords : 3D Images, Interpretation, Segmentation, Representation, Matching.

1 Introduction

We describe in this paper part of the research being performed at Inria on the automatic interpretation of three-dimensional images.

An interpretation task usually includes operations such as **display, quantification, registration** and **labelling** of 3D regions coming from one or several 3D images. *Display* consists in producing a useful image of a 3D object; *quantification* consists in extracting some significant numerical features characterizing its shape and intensity, *registration* consists in superposing several 3D images of similar 3D objects; and *labelling* consists in attaching a name to specific 3D regions.

Automating these operations is very useful for improving the diagnosis, and also for improving the *planning, simulation* and *control* of a therapeutic action, such as a medical gesture or a radiotherapy [LCD*89].

To solve most of the interpretation tasks, it appears necessary to solve three common key problems which we call **segmentation, representation**, and **matching** of 3D regions.

1. **Segmentation** consists in a partitioning the original image into regions of interest. This is a very crucial stage, necessary for concentrating the following operations towards a more compact and pertinent amount of data. Regions can be extracted by regrouping voxels which form homogeneous areas with respect to a given homogeneity criterion, or by detecting edges, i.e. points of transition between two homogeneous areas. In our approach, we have chosen to detect

NATO ASI Series, Vol. F 60
3D Imaging in Medicine
Edited by K. H. Höhne et al.

regions by combining both a local edge-extraction and global region-extraction. We describe in section 2 the use of linear and non-linear filtering methods to extract *local* edge chains, and in section 3 the use of deformable models to extract *global* regions, taking advantage of the previous local analysis and enforcing some important regularity properties.

2. **Representation** of extracted regions is a dual task. On one hand, one must describe the *shape* of the extracted regions, i.e. produce a geometric representation, and on the other hand, one must describe the distribution of intensity within each region, i.e. produce a *textural* representation. We describe in section 4 the representation we have chosen to describe shapes, which is based on the Delaunay triangulation, and we discuss its advantages, compared to the more widely used voxel-based representations. In section 5, we describe our current work on the representation of textures using fractal and integral geometry.

3. **Matching** of regions is based on their representations, and can be performed between the representations of two distinct regions or between the representations of a region and a *model*. In the first case, matching solves the registration problem, and allows the superposition and comparison of two 3D images. In the second case, matching solves the labelling problem, by attaching the names and properties of the model description to the corresponding regions in the 3D image. We describe in section 6 our preliminary work on this topic, which consists in extracting first some surface singularities which will later guide the matching task.

Much work has already been done in these areas, and the interested reader can refer to the following short and necessarily incomplete list of references: on display and representations, [SH88], [BGL*88], [FLP89], [RB89], [SS89a], [PFL*89], [Mon87]; on registration and labelling [STH89], [LCD*89], [BK89a], [SS89b], [SKM89].

2 Edge Extraction

Edges are defined as being points of transition between homogeneous regions. Our model of an ideal edge is a step between 2 regions of constant intensity, corrupted by additive Gaussian noise. Following the approach of Canny, Deriche and Shen [Can83,Der89,Der87,SC86], we generalized in 3D an *optimal* linear filter with respect to objective quality criterions: accuracy and robustness [MD89,MD86]. For the sake of efficiency, we chose a *recursive* and *separable* implementation of the filter in 3D. Therefore, the computational complexity no longer depends on the width of the impulse response of the filter, and is of the order of a single convolution with a $3 \times 3 \times 3$ mask. Our edge detector compares very favorably with previous 3D edge detectors [Lui77,ZH81,MR81].

We also improved our results *locally* by developing a new method to fill up the small gaps between two disconnected edge chains; this is done by searching for a path minimizing a criterion which combines distance, curvature and gradient [MDMC89,DC88].

An edge extraction example in a 3D MR image is shown in figure 1. Despite the high level of noise and the presence of artefacts due to blood motion, one can see that the important boundaries are correctly recovered. This is mainly due to a better elimination of the noise with 3D filtering. Moreover, this approach insures *3D coherence* among the detected edges, which is not the case if one applies 2D operators on successive cross-sections. This important result is examplified on the right bottom pictures of figure 1.

3 Applications of deformable models

The previous approach only applies *local* analysis to detect object boundaries. But in many situations, one might have some *global a priori* knowledge on the shape of the desired edges. For instance, one might look for a *closed* contour, whose curvature varies *smoothly* with abscissa, in a *given region* of the image.

As in [KWT87,Ter88], we consider the application of deformable models [Coh89,CC90]. We introduce an elastic curve in the image and let it evolve from an initial position under the action of both internal (elasticity of the curve) and external forces (attraction towards the local edges).

We have improved the original method on several points:

1. We take into account the previously extracted local edges.

2. We introduce an internal pressure force by considering our curve as a balloon that we blow out. From an initial oriented curve we add to the previous forces a pressure force pushing outside as if we were introducing air inside. The curve expands and is attracted and stopped by edges as before. But if the edge is too weak, since there is a pressure force, the curve can pass through the edge if it is a singularity with regard to the rest of the curve being blown out.

   Beginning with a small curve, we inflate a balloon which then sticks to the edges. This permits taking an initial curve far away from the solution, and makes the result much more insensitive to the initial conditions (see figure 2).

3. We use a finite element technique instead of finite differences, since in finite differences we only see what happens at points of the subdivision. The forces are applied to a grid of points, and the distance between successive points has to be small enough so as to not miss too much information. This has effects on the size of the linear system being solved. In finite elements, we are really working on curves independent of the size of the grid, so that the whole curve between two points of the grid is considered in the equilibrium of forces.

Once an object is extracted in a 2D image, we follow its contour from one slice to the other, and rebuild a 3D surface from the contours [ABB*89]. We use the balloon model to initialize one slice, and then we take as initial value for each succeeding slice the final result of the previous slice. Figure 3 shows the reconstruction of the left ventricle. This reconstruction is almost automatic. Indeed, when the contour undergoes a big change from one slice to the next, the initial curve in that slice may have to be redefined in order to obtain a good contour. This problem can be avoided by adding interpolated slices when necessary.

The next step in our research is to follow the deformation of this surface in time. This can be done either slice by slice or globally by generalizing this approach to a 3D surface model which would be a real balloon since the active contour model is a particular case of the deformable models of [Ter88].

We can add internal forces to control the deformation for following the contours. This is the case if we have a physical model of the desired object (for example, following the deformation of a ventricle during a cardiac cycle), or for making the curve expand or contract from the initial data using some knowledge of the deformation properties.

Another orientation of our research is the elastic matching of extracted features to an atlas, which is related to [BK89b]. A deformation between the pattern and the object is allowed. Thus we deform a curve to best match the pattern using some of the measure distorsion to achieve the deformation, such as the area between the two curves. This was also studied in [YCH89] with simple geometric shapes as templates which are deformed to match the image.

4 Shape representation

In the recent past, several concepts of 3-dimensional reconstruction and rendering of volume data have been presented. They basically belong to two different classes: voxel based methods and triangulation based methods. The common aim is to visualize the surface of the objects of interest, and to use the resulting data for FEM or volume calculation.

In voxel based methods, after a segmentation of the volume data, the voxels are treated as opaque cubes. A different approach uses a kind of ray tracing, deducing the orientation of the surface normals from the gray scale gradient along the ray [HB86] [Lev88]. The main disadvantage of these methods is the use of the entire data for visualisation yielding high computational cost. Real time display is only possible on dedicated machines. With standard visualisation algorithms and machines, it is necessary to fit geometric primitives (typically triangles) to the object surface.

In triangulation based methods, edge-tracking is performed by means of image processing (c.f. previous sections) or manually or both resulting in a polygonal approximation of the object contours. Then a strip of triangles can be constructed that links two contours in successive cross-sections.

The resulting triangles are smoothened and a conventional surface rendering algorithm (Z-buffer, Phong/Gouraud shading) is performed. On a graphics workstation like HP, IRIS, real time motion is possible, but there is already hardware available for running the visualisation on personal computers.

The advantages of the triangulation are the following ones:

1. it provides directly a facet representation,
2. there is a small number of surface triangles (linearly related to the number of points per contour),
3. it provides an interpolation between the cross sections avoiding any anti–aliasing step.

We propose to use a specific triangulation, the Delaunay triangulation. We can show that the reconstruction based on the Delaunay triangulation introduced in [Boi88] has some properties, that allow to compare the assignment pixel–voxel with the assignment triangle–tetrahedron: indeed, the Delaunay triangulation links a point to its nearest neighbours in the adjacent sections. Thus this triangulation method joins the advantage of the voxel method – accurate connections of the cross sections – with those of the triangulation – small data amount and the possibility of applying widely distributed shading software and hardware.

The algorithm has been implemented in C, is fully automatic and has been successfully tested on various medical data: heart, brain, lungs, liver, aneurism of aorta, hip joints. The results shown in Figures 4 to 6 have been obtained from 256×256 resolution MR images, provided by the Universitätsklinik Heidelberg, Germany. It is to point out, that the number of cross-sections is small and corresponds to the normal rate of scans taken at a clinical examination. The reconstruction time is less than 70ms per point on a Sun3 Workstation.

5 Texture Segmentation Using Fractal and Integral Geometry

Besides the description of the shape of extracted regions, it is necessary to describe their intensity distribution, i.e. to produce a *textural* representation.

Both fractal geometry [Man82,Man77,Vos85] and integral geometry [San76] are new approaches that we are investigating for characterizing the texture of very complex medical images [LV89,LVB90].

The main interest of Fractal Geometry is that it provides a way to quantify irregularity. Also, fractal techniques differ from the classical texture analysis methods [Gag83,HS85], and better account for the infinite degree of correlation found in many natural objects. This is because classical techniques are based on the assumption that textures can be modelized by Markov fields and so have the property that sufficiently distant points are nearly independent, which is not always true.

With Integral Geometry, classical measures such as length, perimeter and surface are made on sets of the space. The interest is that the computation of these parameters is much more robust to noise, because it is done with global knowledge and properties of the sets.

We only have room for one example in this paper. Figure 7 shows the global value of *lacunarity* (a fractal parameter to characterize second order irregularities) for three types of lung diseases, and figure 8 shows the visual representation of a *local computation* of the same parameter, again for three classes of lung diseases, exemplifying its discrimination power for texture classification. The reader will find a detailed description of these methods and applications in [LV89,LVB90].

6 Matching

6.1 Extracting Surface Features

An increasingly important notion in computer vision is that of *qualitative description.* For many tasks, including object identification, computation of qualitative features is more important than precisely estimating structure everywhere. In this spirit, we develop methods for the computation of surface structure permitting the reliable determination of second derivative singularities [San89].

Qualitative features can only be useful, however, when they are *stable*, that is *generic* (insensitive to small perturbations of the form of the object), *intrinsic* (not dependent on the situation of the object in space), and *robust* to a certain amount of noise. Two such stable second derivative singularities of surfaces are: parabolic points, those zeros of Gaussian curvature which appear on smooth surfaces between regions of elliptic and hyperbolic points; and umbilic points, the singularities of the principal direction fields. Generic parabolic points on object surfaces form curves which can be used to effect a decomposition into stable parts. Umbilic points are generally isolated, with their number and type forming an intimate link between local differential surface structure and global topological properties.

Shape characterization and decomposition provide the overall context in which to place this research. We strongly believe that the sorts of descriptions that will underlie successful theories of shape must satisfy the attributes of stability mentioned above — and principal curvature and direction fields on object surfaces permit the calculation of generic singularities which do so. They thus provide a qualitative foundation of precisely the sort on which third paradigm computational vision stands [Zuc87].

6.2 Applications

6.2.1 Registration and matching

We are investigating two applications of our methods to biomedical imagery.

The first one involves registration between images taken from different viewpoints, e.g., transverse and sagittal MR images, or images from different modalities, e.g., MR, positron emmission or X-ray tomography. As an illustration, we show in figure 9 the detection and localization of 3 umbilic points from the MR image of a human face.

The second application involves matching structure in clinical images to structure in a standardized atlas, with the eventual goal of automatic structure identification. As an illustration, we shall start with the analysis of the surface of the brain ventricles (see figure 10), which were reconstructed from the cross-sections of the anatomical atlas of Talairach [TT88].

Singularities also permit the decomposition of object surfaces into stable parts useful for the registration and matching tasks.

6.2.2 Shape deformations

In addition to applications to static 3-D images, singularities provide insights for the analysis of image sequences. Changes in surfaces produce characteristic changes in the surface singularities, and a longer term theoretical goal is the analysis of the evolution of singularities with time to yield information about dynamic surface deformation as, for example, in a sequence of images of a beating heart. This research will be connected to the one on deformable models described in section 3.

7 Conclusion

We presented an integrated approach to the difficult problem of 3D image interpretation. This approach focuses on three main topics which we called segmentation, representation and matching. Segmentation transforms the raw data into regions of interest, combining edge-detection and deformable models. The shape of these regions can be described by geometric models involving the Delaunay triangulation, while their texture can described by Fractal and Integral Geometry.

Finally, characteristic geometric features derived from the Differential Geometry of surfaces can be extracted to guide a spatial registration between two images or with a model, providing the semantic labeling of the extracted regions.

We presented several promising experimental results on each topic, but much research remains to be done to complete a fully integrated system.

Acknowledgments

Part of this research is supported by Digital Equipment Corporation. This research has benefited of fruitful discussions with J.M. Rocchisani. MR images of the heart come from Kremlin-Bicêtre Hospital, Paris (Dr J. Bittoun). Digitalization of the brain atlas was done in collaboration with H. Dunant Hospital (Dr Thiébaut & Uhl). We thank Dr Wolfgang Wrazidlo at the Universitätsklinik Heidelberg for providing the brain images and Christoph Fischer at the University of Karlsruhe for his help to read them.

INRIA is the French National Research Institute in Computer Science and Automatic Control.

References

[ABB*89] N. Ayache, J.D. Boissonnat, E. Brunet, L. Cohen, J.P. Chièze, B. Geiger, O. Monga, J.M. Rocchisani, and P. Sander. Building highly structured volume representations in 3-D medical images. In *Computer Aided Radiology*, 1989. Berlin, West-Germany.

[BGL*88] Ch. Barillot, B. Gibaud, O. Lis, L.L. Min, A. Bouliou, G. Le Certen, R. Collorec, and J.L. Coatrieux. Computer graphics in medicine: a survey. *CRC Critical Reviews in Biomedical Engineering*, 15:269–307, 1988.

[BK89a] R. Bajcsy and S. Kovacic. Multiresolution elastic matching. *Computer Vision, Graphics, and Image Processing*, 46:1–21, 1989.

[BK89b] Ruzena Bajcsy and Stane Kovacic. Multiresolution elastic matching. *Computer Vision, Graphics, and Image Processing*, 46:1–21, 1989.

[Boi88] J.D. Boissonnat. Shape reconstruction from planar cross-sections. *Computer Graphics and Image Processing*, 44:1–29, 1988.

[Can83] J.F. Canny. *Finding edges and lines in images.* Technical Report TR. 720, MIT, June 1983.

[CC90] Laurent D. Cohen and Isaac Cohen. Using a finite elements method for active contour models and its application to 3-D reconstruction from cross sections. In *submitted to ECCV*, 1990.

[Coh89] Laurent D. Cohen. On active contours models. In *Proceedings of NATO ASI Active perception and Robot vision*, Maratea, July 1989. rapport INRIA 1075.

[DC88] R. Deriche and J.P. Cocquerez. An efficient method to build early image description. In *International Conference on Pattern Recognition*, Rome, 1988.

[Der87] R. Deriche. Using Canny's criteria to derive a recursively implemented optimal edge detector. *International Journal of Computer Vision*, 1(2), May 1987.

[Der89] R. Deriche. Fast algorithms for low level vision. *IEEE Transactions on Pattern Analysis and machine Intelligence*, 1989.

[FLP89] H. Fuchs, M. Levoy, and S.M. Pizer. Interactive visualization of 3-D medical data. *Computer*, 46–51, 1989.

[Gag83] A. Gagalowicz. *Vers un modèle de textures.* PhD thesis, Université de Paris VI, Mai 1983. Thèse de Doctorat d'Université.

[HB86] K.H. Hoehne and R. Bernstein. Shading 3-D images from CT using gray-level gradients. In *IEEE Trans. Medical Imaging*, pages 45–47, 1986.

[HS85] R.M. Haralick and L.G. Shapiro. Image clustering techniques. *Computer Vision Graphics and Image Processing*, 29:100–132, 1985.

[KWT87] Michael Kass, Andrew Witkin, and Demetri Terzopoulos. Snakes: active contour models. *International Journal of Computer Vision*, 1:321–331, 1987.

[LCD*89] S. Lavallée, Ph. Cinquin, J. Demongeot, A.L. Benabid, I. Marque, and M. Djaid. Computer assisted driving of a needle into the brain. In *Proc. of Computer Assisted Radiology*, pages 416–420, 1989.

[Lev88] M. Levoy. Display of surfaces from volume data. *IEEE Computer Graphics & Application*, 29–37, 1988.

[Lui77] H.K. Lui. Two and three dimensional boundary detection. In *Comput. Graphics Image Process.*, pages 123–134, 1977. Vol. 6.

[LV89] J. Lévy Véhel. Using fractal and morphological criteria for automatic classification of lung diseases. In *Visual Communications and Image Processing IV*, Philadelphia, Pennsylvania, November 8–10 1989. SPIE, The International Society for Optical Engineering.

[LVB90] J. Lévy Véhel and J.-P. Berroir. 3-D fractal model for lung morphogenesis. In *Medical Imaging IV*, New Port Beach, Los Angeles, February 1990.

[Man77] B.B Mandelbrot. Fractals: form, chance and dimension. In *CA: Freeman*, San Fransisco, 1977.

[Man82] B.B Mandelbrot. The fractal geometry of nature. In *CA: Freeman*, San Fransisco, 1982.

[MD86] O. Monga and R. Deriche. A new three dimensional boundary detection. In *International Conference on Pattern Recognition*, Paris, 1986.

[MD89] O. Monga and R. Deriche. 3-D edge detection using recursive filtering. In *Computer Vision and Pattern Recognition*, IEEE, San Diego, Juin 1989.

[MDMC89] O. Monga, R. Deriche, G. Malandain, and J.P Cocquerez. Recursive filtering and edge closing: two primary tools for 3-D edge detection. In *submitted to IEEE Transactions on Pattern Analysis and Machine Intelligence*, 1989. also Research Report INRIA 1103.

[Mon87] O. Monga. An optimal region growing algorithm for image segmentation. In *International Journal of Pattern Recognition and Artificial Intelligence*, pages 351–376, December 1987.

[MR81] M. Morgenthaler and A. Rosenfeld. Multidimensional edge detection by hypersurface fitting. *PAMI-3*, 4, July 1981.

[PFL*89] S.M. Pizer, H. Fuchs, M. Levoy, J.G. Rosenman, R.E. Davis, and J.B. Renner. 3-D display with minimal predefinition. In *Proc. of Computer Assisted Radiology*, pages 723–736, 1989.

[RB89] R.A. Robb and C. Barillot. Interactive display and analysis of 3-D medical images. *IEEE Transactions on Medical Imaging*, 8(3):217–226, 1989.

[San76] L.A. Santalo. *Integral Geometry and Geometric Probability.* Addison-Wesley Publishing Company, 1976.

[San89] P.T. Sander. Generic curvature features from 3-D images. In *IEEE Transactions on Systems, Man, and Cybernetics*, November 1989. Special Issue on Computer Vision.

[SC86] J. Shen and S. Castan. An optimal linear operator for edge detection. In *Conference on Vision and Pattern Recognition*, IEEE, USA, Juin 1986.

[SH88] H.H. Shen and T.S. Huang. A survey of construction and manipulation of octrees. *Computer Vision, and Image Processing*, 43:409–431, 1988.

[SKM89] T. Schutz, H.-J. Kretschmann, and D. Muller. Automated identification of neuroanatomical structures in CT- and mr-images. In *Proc. of Computer Assisted Radiology*, pages 287–291, 1989.

[SS89a] H.S. Shen and M.R. Sontag. Representation, display, and manipulation of 3-D digital scenes and their medical applications. *Computer Vision, Graphics, and Image Processing*, 48:190–216, 1989.

[SS89b] K. Spitzer and H.S. Stiehl. The paradigm of computer assisted radiology and computer assisted neurology. In *Proc. of Computer Assisted Radiology*, pages 292–295, 1989.

[STH89] C. Schiers, U. Tiede, and K.H. Hohne. Interactive 3-D registration of image volumes from different sources. In *Proc. of Computer Assisted Radiology*, pages 666–670, 1989.

[Ter88] Demetri Terzopoulos. The computation of visible-surface representations. *IEEE Transactions on Pattern Analysis and Machine Intelligence*, PAMI-10(4):417–438, July 1988.

[TT88] J. Talairach and P. Tournoux. *Co-Planar Stereotaxic Atlas of the Human Brain. 3-Dimensional Proportional System: An Approach to Cerebral Imaging.* Thieme Medical Publisher, Inc., Georg Thieme Verlag, Stuttgart, New York, 1988.

[Vos85] R.F. Voss. Random fractal forgeries. In Springer-verlag R.A. Earnshaw, editor, *Computer Graphics*, 1985.

[YCH89] A.L. Yuille, D.S. Cohen, and P.W. Hallinan. Feature extraction from faces using deformable templates. In *Proceedings of Computer Vision and Pattern Recognition*, San Diego, June 1989.

[ZH81] S.W. Zucker and R.A. Hummel. A three dimensional edge operator. *IEEE Transactions on Pattern Analysis and Machine Intelligence*, 3(PAMI-3), May 1981.

[Zuc87] S.W. Zucker. The emerging paradigm of computational vision. In *Ann. Rev. Comput. Sci.*, pages 69–89, 1987.

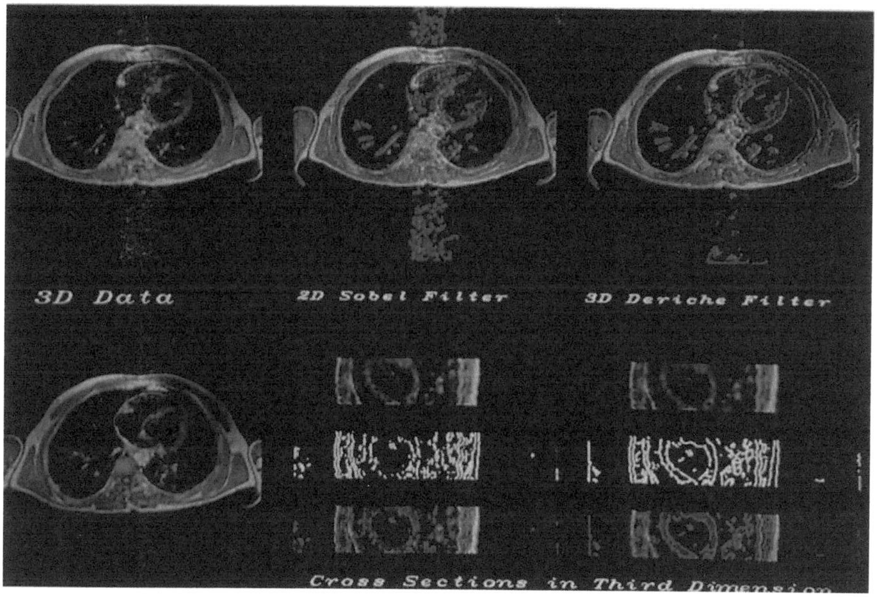

Figure 1: Comparison between 2D and 3D edge extraction: better robustness to noise, and spatial coherence of the results.

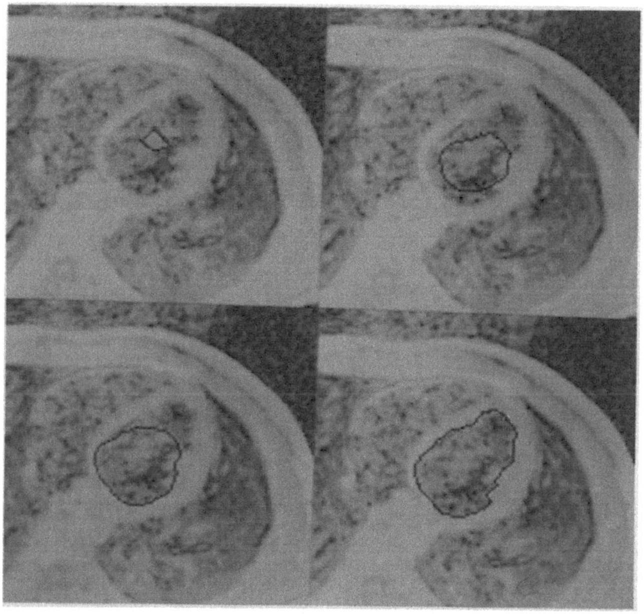

Figure 2: Automatic segmentation of the left ventricle using a deformable balloon model.

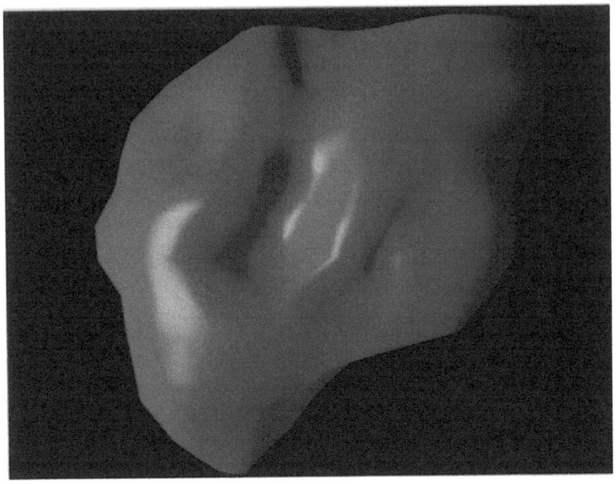

Figure 3: 3-D reconstruction of the inside cavity of the left ventricle of the heart.

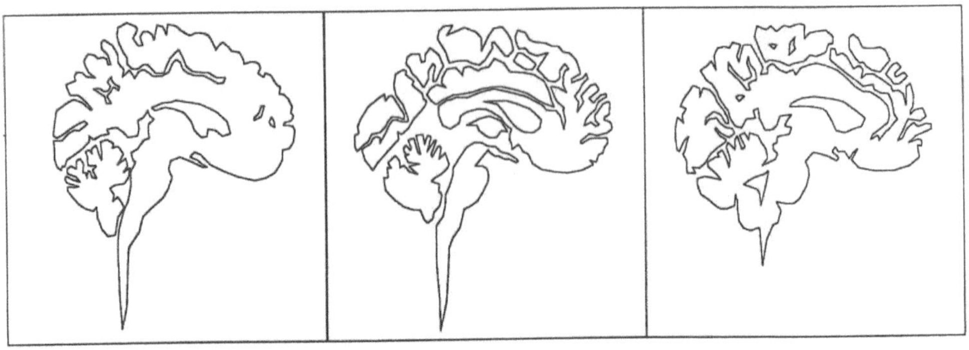

Figure 4: The edges extracted in three successive cross-sections of the brain. Note the complexity of the these contours.

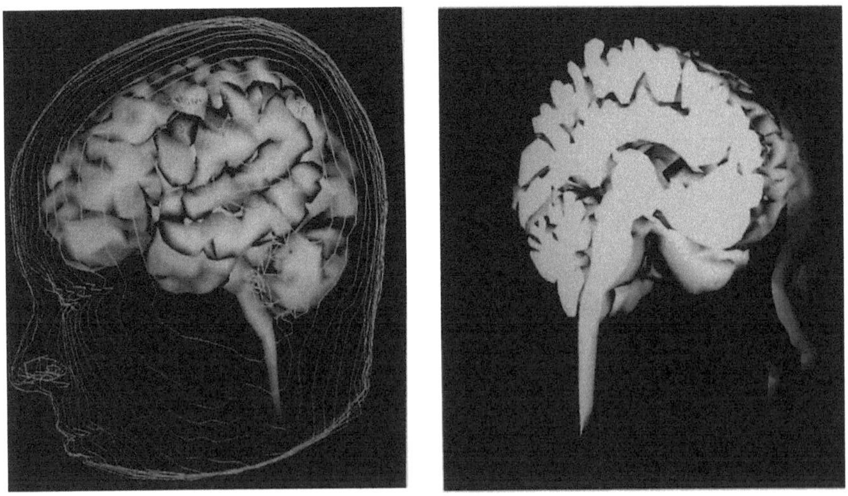

Figure 5: Human brain: 15 cross-sections, 104 contours, 3300 points, 7900 triangles.

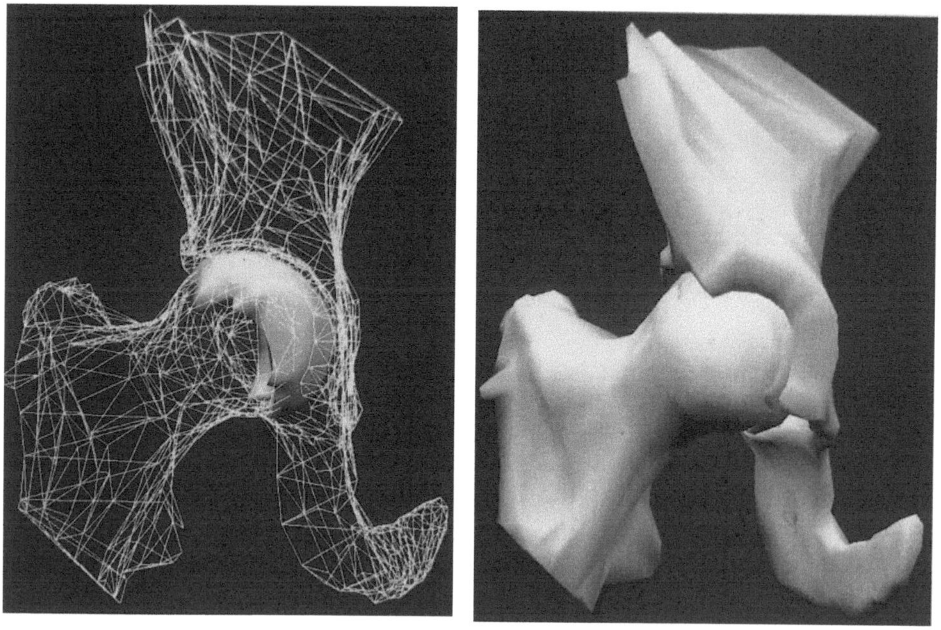

Figure 6: Hip joint: 14 cross-sections, 38 contours, 1103 points, 2362 triangles. The aim was to visualize the cartilage of the femoral head.

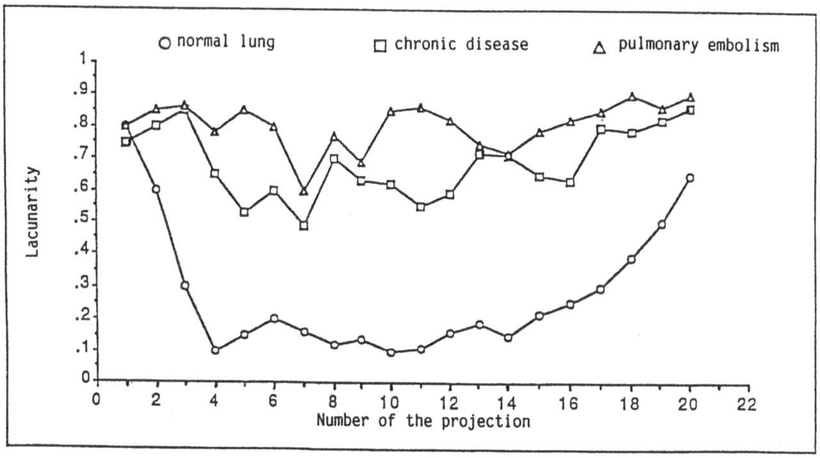

Figure 7: Typical values of fractal lacunarity for normal lung, chronic disease, pulmonary embolism.

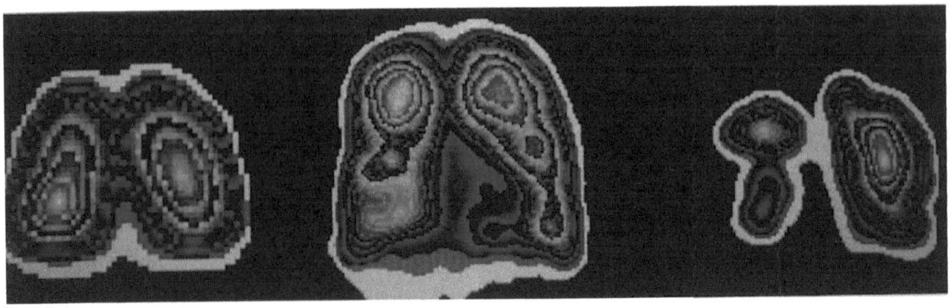

Figure 8: Local computation of fractal lacunarity on SPECT images of normal lung, chronic disease and pulmonary embolism.

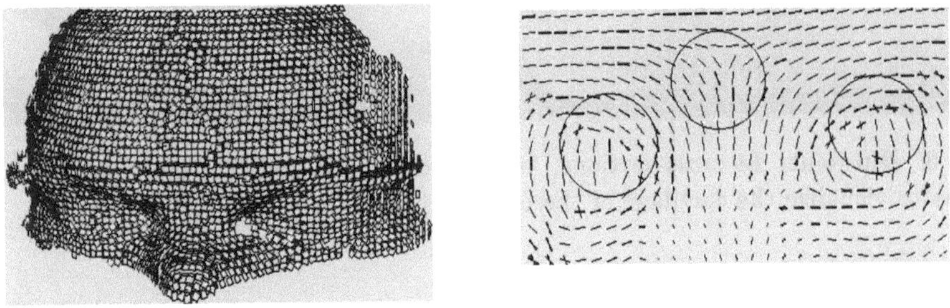

Figure 9: Computation of umbilic points on the MR image of a human face. (a) Tangent planes. (b) Principal direction field showing three typical singularities at umbilic points.

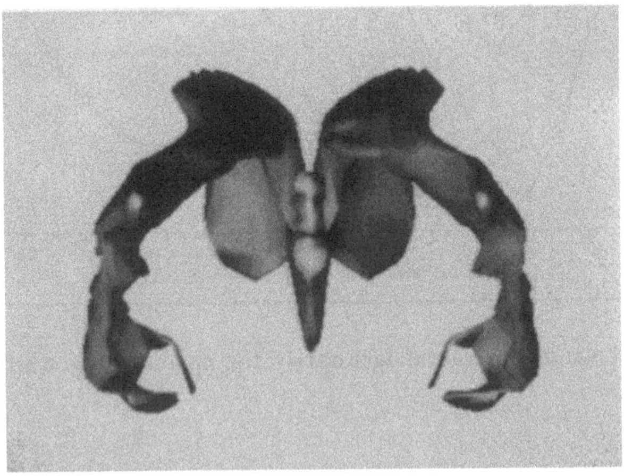

Figure 10: Surface of the brain ventricles, reconstructed from the Talairach Atlas.

Image Processing of Routine Spin-Echo MR Images to Enhance Vascular Structures: Comparison with MR Angiography

Guido Gerig[1], *Ron Kikinis*[2] *and Ferenc A. Jolesz*[2]

[1]Institute for Communication Technology, Image Science Division, ETH-Zürich, CH-8092 Zürich, Switzerland
[2]Department of Radiology, Harvard Medical School and Brigham and Women's Hospital Boston MA 02115, USA

Abstract

Magnetic Resonance Imaging (MRI) is a highly flexible diagnostic imaging technique providing complex information about the morphology of various normal and abnormal tissues. Besides this clinically valuable anatomic and pathologic information, there are some unique functional features which can be extracted from MR images. The influence of macroscopic and microscopic motion on MR images as revealed by more or less specialized pulse sequences may demonstrate the presence of physiologically important processes such as blood or cerebrospinal fluid (CSF) flow, tissue perfusion and diffusion [Axel84, Demoulin87, Demoulin89, Haacke89, Wehrli87]. Recent progress in the implementation of these MRI methods shows that while these applications have not yet been fully exploited, their use in clinical practice is not far in the future.

The use of image processing techniques can definitely improve the visualization, display and interpretation of MR images. The computerized post-processing techniques may also have a very significant role in accessing the physiologic information available from MRI. In this work we applied image processing techniques to both routine, standard spin-echo MR acquisitions and to MR angiograms (MRA) in order to extract and display vascular structures with intraluminal flow. We demonstrated that it is possible to segment a standard spin-echo data set into brain parenchyma (white and gray matter), cerebrospinal fluid (CSF) and vascular structures. The segmented images can be displayed using 3D surface rendering and selective clipping. The extracted vascular structures obtained with the MRI and MRA methods were compared .

NATO ASI Series, Vol. F 60
3D Imaging in Medicine
Edited by K. H. Höhne et al.

While the vessels were better delineated by the MRA method, especially after image processing, the major morphologic features were also accessible after segmentation of the standard spin-echo images.

Keywords: Medical Image Processing, Magnetic Resonance Imaging, MR-Angiography, Vascular Imaging, 3D Segmentation, 3D Display

1. Introduction

The basic purpose of various MR angiography (MRA) techniques is to suppress or eliminate signal from stationary tissue and to enhance the appearance of flow, and by doing so to provide optimal contrast between flowing blood and surrounding soft tissue. MRA methods can be used efficiently as a complement to standard MRI. The simultaneous display of intraluminal and extraluminal information can demonstrate not only details about the vessel walls and perivascular structures but can also delineate the spatial relationship of the vessels with other non-vascular structures. Usually, the MRA is acquired as an additional pulse sequence along with conventional MRI, and therefore prolongs the imaging session significantly.

The enhancement and segmentation of specific structures is the fundamental role of digital image processing. While the performance of most image processing techniques depends on the quantitative and qualitative information directly available from the digital data, it is possible to utilize image processing techniques even when the information available from the images is quite limited. While vascular flow data are more easily accessible from specialized MRA images we investigated the possibility of obtaining similar information from standard MR images by applying image processing tools. This approach is justified since spin echo images, although limited in some respects, are sensitive to blood flow and at the same time are optimal for the visualization of the stationary tissue elements. In spin-echo images, the signal intensity of rapidly flowing blood is low or absent and the vascular structures are visible due to their relative contrast to the surrounding tissues. The fact that our visual system clearly distinguishes blood vessels from surrounding non-vascular structures even on these standard non-angiographic images motivated the work presented herein.

This report is divided into three parts. In the first section, we describe a preprocessing method which preserves the finely detailed information of the vessels and significantly increases the SNR. In the second part, a 3D line (tube) detection algorithm is presented which recognizes linear structures in 3D volume data. The segmentation algorithm is applied to a routine clinical spin-echo data set to demonstrate its ability to enhance tubular structures. Once extracted, the vascular system can be displayed using standard 3D volume rendering or surface rendering procedures. We finally demonstrate the segmentation of spin-echo volume data into brain parenchyma (white and gray matter), CSF, and vessels. A combined 3D display of the segmented surfaces and the vascular morphology is also demonstrated.

2. MRI methods

Spin echo images of the brain were obtained as part of routine clinical studies. Imaging was performed at 1.5 Tesla using a General Electric SIGNA MRI system. Data acquisition used a double echo sequence (TR 3000ms, TE 30/80ms) with 3mm axial slice distance and thickness. Contiguous slices were obtained by combining two interleaved sequences. Imaging time was reduced by using half-Fourier sampling (.5 NEX). We used flow compensation (gradient moment nulling) and presaturation of a slab inferior to the head to reduce flow related artifacts and to obtain low intra-arterial signal intensity. Forty to 50 double echo slices were used to cover the brain volume. The imaging time was under 8 minutes.

The method used to obtain additional MRA was a multiple thin-slab 3D gradient echo technique implemented in the General Electric SIGNA system [Parker89]. The 15 mm 3D slabs were further divided with 16 phase encoding steps into nominally 1 mm thin sections. The slabs were acquired in a sequential manner and the reconstructed one half slabs were spaced in a contiguous fashion. The gradient echo (TR=50 msec, TE=5 msec flip angle=80 degrees) were acquired within 15 minutes.

3. Image Processing Methods

3.1 Three-dimensional Preprocessing

The MRIs are corrupted by a considerable noise component, which interfered with the recognition of fine vessels. In order to smooth this noise without losing the fine and detailed vascular structures, we have developed a processing scheme based on anisotropic smoothing [Gerig89, Gerig90]. The new filtering is based on a locally nonlinear diffusion process, which has a high smoothing capacity while preserving and even increasing the sharpness of object boundaries and fine detailed structures. These properties are best suited to preprocess MR images, which can be assumed to be approximately piecewise constant. The discrete implementation of the diffusion process is formulated as an iterative adaptive smoothing, where the filter coefficients (c_i) are nonlinear functions of the local gradients (∇I_i):

$$I(t+\Delta t) \approx I(t) + \Delta t * I_t = I(t) + \Delta t * \sum_{i=1}^{n} (c_i * \nabla I_i) \qquad | \sum_i \text{: local neighborhood}$$

$$c_i = \left(1 + \left(\frac{\nabla I_i}{\kappa}\right)^2 \right)^{-1} \qquad | \ \kappa\text{: conduction parameter}$$

The iterative filtering results in image structures of homogeneous intensity without blurring of edges or fine lines, and without altering the appearance of their exact positions. A great advantage of the filtering process applied to isotropic 3D volume data is full 3D implementation. Local estimations about image structures and averaging are performed within 3x3x3 voxel neighborhoods, resulting in a significant decrease of noise even after one iteration (adaptive averaging over 27 neighbors in homogeneous areas increases SNR by more than factor of four [Gerig90]) and in a sharpening of 3D discontinuities.

3.2 Segmentation of 3D Tubular Structures

Common solutions for optimal filtering to enhance structural features can be found in digital signal processing, the most appropriate filters often match the characteristics of the structures themselves. A typical analysis by Canny [Canny83] proposes the use of twodimensional directive Gaussian derivatives in a multiresolution simultaneous convolution scheme. Extended to three dimensions, a filter for the detection of linear features is designed as an isotropic 3D Gaussian filter with second derivatives in two orthogonal directions (Laplacian of Gaussian). Instead of convolving the volume data with a set of directive filters the differentiation can be performed on the convolution result of the image data with the isotropic Gaussian filter, resulting in a maximum computational efficiency:

$$\left(\frac{\partial^2}{\partial x^2}+\frac{\partial^2}{\partial y^2}\right)\left[G(x,y,z;\sigma)\otimes I(x,y,z)\right] \qquad \mid \sigma\text{: smoothing parameter}$$

Differentiation is approximated by discrete operations within 3x3x3 voxel neighborhoods, we used 13 filters spatially oriented in the three immediate, six plane-diagonal and the four volume-diagonal directions. The maximal response among the multiple filter output is chosen to match optimally the local intensity structure. It determines the local direction and the significance of the curvilinear three-dimensional structure.The segmentation scheme consists of the following steps:

- summation of the two echoes (if applied to multivariate spin-echo data)
- multiple convolution with directive Gaussian derivatives
- Half-wave rectification (discrimination between dark and bright features)
- connected component labeling starting from interactively selected seeds

4. Results

We applied the discontinuity-preserving filtering to both MRA and routine MR images. In the case of MRAs the goal of preprocessing is to give maximal small vessel enhancement. Equally important is the possibility of displaying these fine vascular structures. Maximum intensity projection favors the demonstration of high intensities along projection rays over finer structures even when they are located nearer to the observer. Furthermore, a 3D impression is obtained only in moving image sequences or in stereoscopic views. MRA takes the role of an acquisition based segmentation, the vessel identification process is therefore carried out by thresholding and connected component labeling. Such a labeling serves as a prerequisite for a subsequent volume- or surface-rendering of structures of interest. Figures 1a,b and 2a,b illustrate maximum intensity projections and 3D surface renderings of a MRA data set.

The segmentation method to enhance tubular structures was applied to the contiguous multi-slice spin-echo MR images, as well. Again, the iterative preprocessing algorithm was applied to remove noise while preserving and even enhancing edges and fine structures. A special extension to perform simultaneous multiecho-smoothing has been applied [Gerig90]. Our segmentation concept is based upon the observation that on spin-echo images parts of the blood vessels are visible as low signal intensity curvilinear structures in contrast to higher signal intensity and mostly non-linear surrounding tissues. This information may obscured or misrepresented by several flow-related artifacts. Along the phase-encoding direction, a parallel bright line is visible representing a kind of "shadow" caused by misregistration of the flowing blood. The distance of the dark and bright lines as well as their intensities depend on the flow velocity and the phase encoding direction [Wehrli87]. Flowing blood causes local extinction of signal; by localizing the curvilinear dark structures we can characterize the true positions of high flowing blood. Since signal loss occurs in both echoes, the two echoes are averaged, resulting in a further increase of the SNR and in a less pronounced occurrence of the bright "shadow" lines, which vary in position on the two echoes. Figure 3 shows an original spin-echo slice (left) and the result after the segmentation of tubular structures (right). Corresponding views of 3D reconstructions of the arteries of the MRA (left) and the spin-echo images (right) are shown in figure 4. A pure visual comparison of the two views shows a more complete vascular system obtained from the MRA data. The original spin-echo input data are acquired with an anisotropy of 1:3, which does not allow the extraction of vessels with isotropic resolution. With thinner slices a better segmentation of vascular structures will undoubtedly be possible. It is obvious that the best performance of the segmentation scheme will be obtained using isotropic voxels.

Nevertheless, the first results encourage us to continue the development of image processing methods to detect structural features. Further refinements are necessary to increase the resolution of the algorithm to detect even finer vessels. The primarily local processing must be extended to a more complex segmentation scheme. Furthermore it is planned to apply the processing to nearly isotropic volume data.

5. Complete Segmentation of Two-echo Spin-echo Volume Data

A great advantage of the image processing approach lies in the fact, that we can fully explore the volume data by applying different segmentation schemes to obtain information as much as possible.

An earlier work proposed a multistage segmentation scheme [Gerig89] which performed a segmentation of the brain (white and gray matter) and the ventricular system in multispectral MR volume data with only minimal user interaction. In that method, the most important factor was the fully 3D formulation of the segmentation steps. Processing stages included discontinuity-preserving filtering, multispectral user-supervised classification, and morphological postprocessing using spatial context information. Human interaction is needed only in the second stage by interactively selecting meaningful subregions of tissue categories to train the classifier and to calculate multidimensional discrimination functions. Our experience clearly shows that the proposed segmentation scheme does not require tedious interactive corrections of the segmented areas, which is very important for use in daily clinical routine.

The multispectral segmentation scheme is applied to carry out a segmentation of multispectral MR volume data. The two volume data sets specified by the two different echoes are preprocessed using the iterative anisotropic smoothing procedure. Noise can be reduced significantly after three iterations while finely detailed structures are preserved. Our analysis focused on the three categories CSF, white matter and gray matter. The segmentation defines labeled objects in 3D space and their surfaces can be displayed directly using standard rendering techniques.

Many attempts to visualize multimodality, multivariable medical image data demonstrate its importance to clinical applications (e.g. [Hu89]). In the approach presented here, the combination of tissue segmentation and extraction of vascular structures is a straightforward process, as both sets of information are obtained using the same input data. No matching or reformatting problems arise.

Marching cube soft surface calculation and selective clipping facilities [Cline88] were applied to generate 3D views. Figure 5 illustrates the brain surface and the ventricular system in combination with the main arteries extracted by the segmentation scheme proposed herein. The anatomical surfaces shown in figure 5 are extracted from a routine spine-echo data set, demonstrating the variety of information contained in the original data.

6. Conclusions

The MRA is typical of a class of "goal directed MRI techniques", which select specific MR scan parameters based on distinctive imaging goals. In this case the segmentation, or the enhancement of specific anatomical structures is performed using a special acquisition technique. An alternative approach to segmentation of MR volume data into meaningful anatomical structures is the application of image processing methods. When fully evaluated routine clinical data reveal sufficient information for gross morphological evaluation. Application of image processing methods can provide contrast- or feature-enhancement of anatomical structures or functions in the postprocessing stage.

The segmentation and enhancement of specific information using image processing has several advantages over specialized, goal-directed MR-acquisition techniques. Furthermore it offers a great potential in giving access to otherwise hidden quantitative information. Postprocessing can be obtained retrospectively if necessary. If flow-related and stationary data are available from the same data set, no geometric mapping of image data from different sources is required.

In summary, the use of computer assisted advanced image processing for the delineation of vascular structures from routine MRI data sets yields several advantages, which in some cases may favor this approach over special acquisition techniques especially when imaging time is restricted and when the information obtained is otherwise sufficient. The possibility of retrospectively deriving vascular data from routine images is very promising, and this method may have unique clinical applications. Additional to segmentation methods based on the spectral properties of single voxels, structural segmentation methods have been developed and applied in which 2D line-extraction methods were extended to 3D to extract blood vessels from volume data. We have shown that, despite the large amount of data, the presence of noise, and the complexity of imaged structures, it is possible to apply efficient procedures to segment and visualize anatomical structures. The 3D display of segmented objects allows the study and qualification of complex shapes and interrelations of different structures, and provides specific reference to the position of abnormalities with respect to the surrounding anatomical structures.

An additional important role of the image processing techniques presented here is to improve the visibility of fine vascular structures from highly specialized MRA data sets in which this information is present but not necessarily displayed in an optimal way. We have demonstrated that the SNR of MRA can be improved significantly using a discontinuity-preserving 3D-filter. A 3D rendering of the segmented vascular system improves the visual impression of structural relationships.

The quality of the segmentation results and 3D renderings is evaluated by radiologists and surgeons specifically considering its potential use in surgical planning, where the access to major morphological features is required. Although the vascular system obtained by the processing of MRA data turns out to be more complete and finer detailed, the segmentation of spin-echo images yields promising results and support a further development of three-dimensional structural segmentation methods.

We believe this research is a step towards the realization of a system that allows maximum information to be obtained from a single noninvasive procedure. The extraction of specific information, which may vary from case to case, is done by the computer. The possibility of retrospectively answering of multiple questions, and the excellent cost-effectiveness of using only one brief patient encounter and minimal computational expense favors such an approach over specific goal-directed MR-acquisitions.

Acknowledgments: The research reported in this article is supported by the Swiss National Science Foundation grant number 4018-11082 and by the NIH PO1 CA41167 (USA).

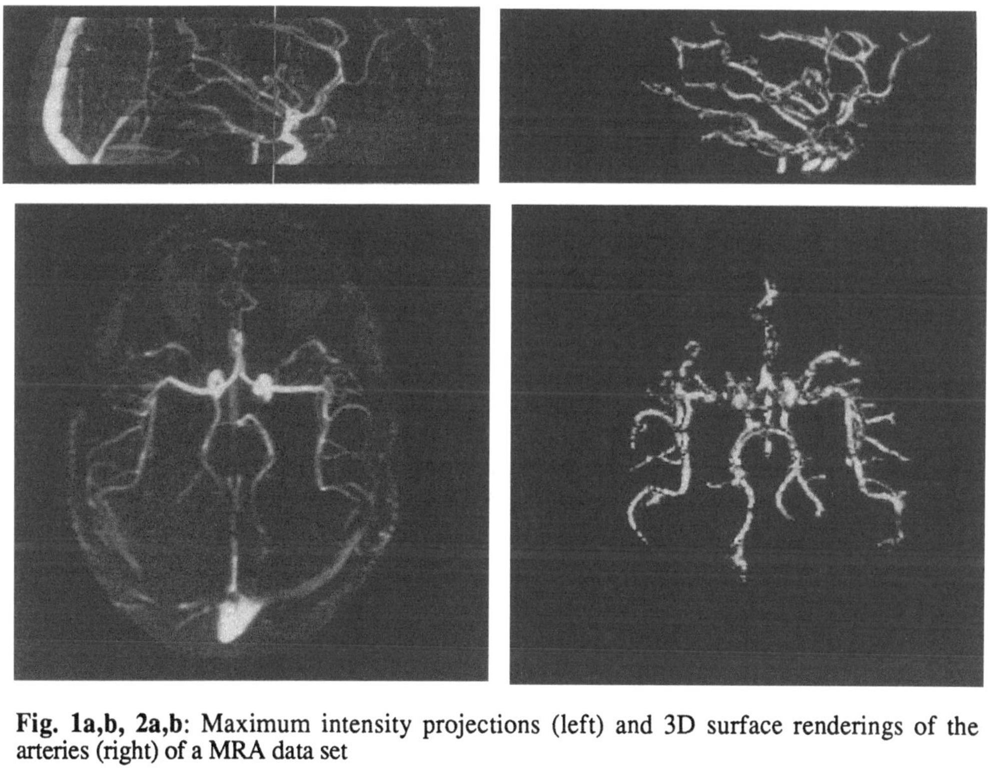

Fig. 1a,b, 2a,b: Maximum intensity projections (left) and 3D surface renderings of the arteries (right) of a MRA data set

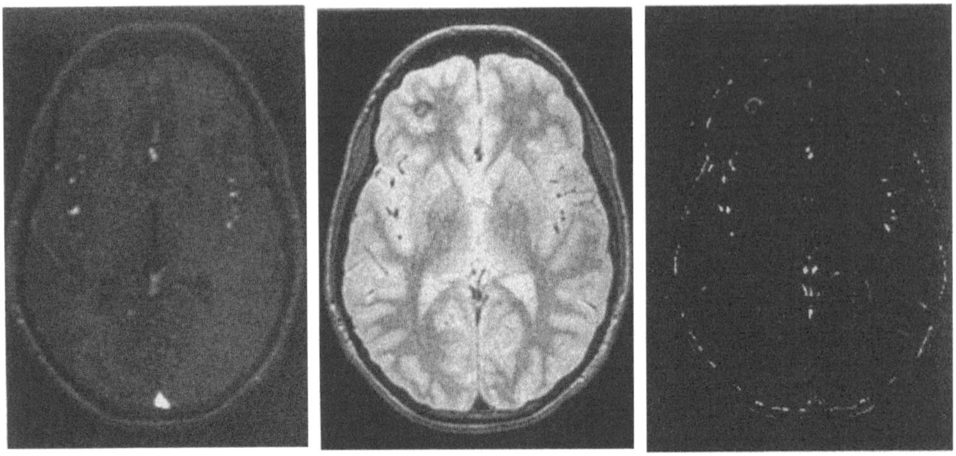

Fig. 3: Corresponding slices obtained by Magnetic Resonance Angiography (left), Spin-echo acquisition (center) and result after the enhancement of tubular structures (right)

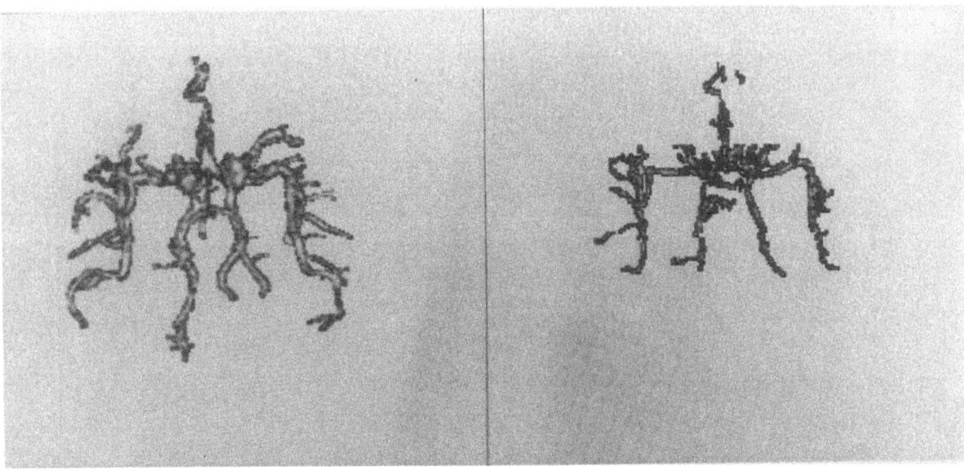

Fig. 4: 3D surface rendering of the arteries segmented from the MR-angiogram (left) and from the spin-echo image data (right). The circular structure represents a hematoma.

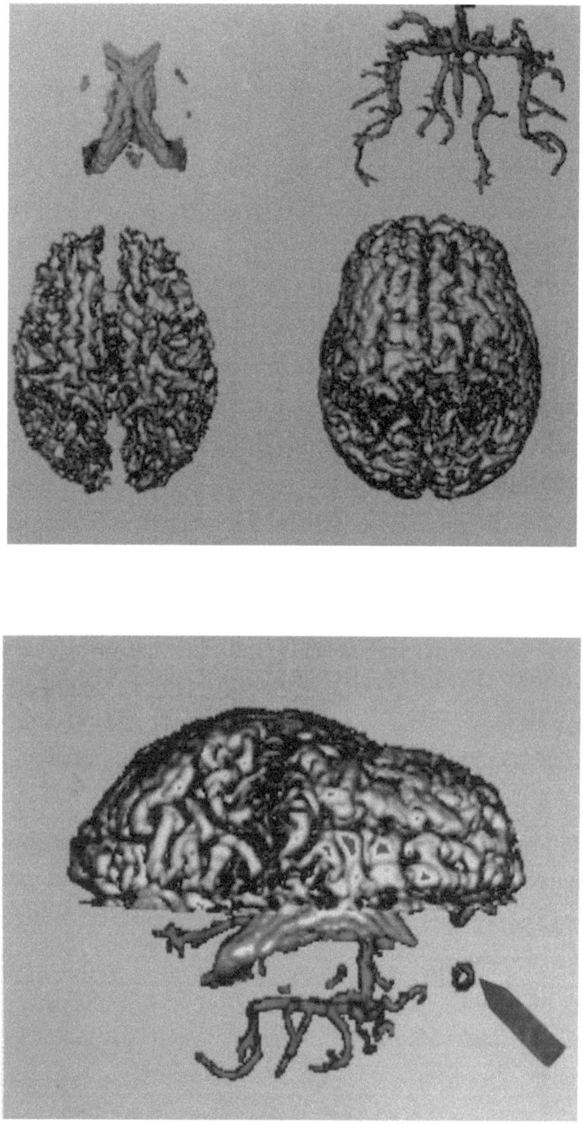

Fig. 5: Brain surface, surface of white matter and the ventricular system in combination with the main arteries extracted by the proposed segmentation scheme. The complete information is obtained by the processing of a routine spin-echo acquisition. The arrow points to the hematoma.

7. References

Axel L, Blood Flow Effects in Magnetic Resonance Imaging, Review, American Journal of Radiology, vol.143:1157-1166, 1984

Canny, Finding Edges and Lines in Images, Technical Report 720, MIT Artificial Intelligence Laboratory, Department of Dept. of Electrical Engineering and Computer Science, M.I.T., Cambridge Mass., 1983

Cline HE, Lorensen WE, Ludke S, Crawford CR, Teeter BC, Two algorithms for the three-dimensional reconstruction of tomograms, Med. Phys. 15(3), May/June 1988

Dumoulin CL, Cline HE, Souza SP, Wegh MA, Walker MF. Three-dimensional time-of-flight magnetic resonance angiography using spin saturation. Magn Reson Med 1989;11;35-46

Dumoulin CL, Souza SP, Feng H, Multiecho magnetic resonance angiography. Magn. Reson Med 1987; 149:1097-1109

Gerig G, Kuoni W, Kikinis R and Kuebler O, Medical Imaging and Computer Vision: An integrated approach for diagnosis and planning, 11th DAGM-Symposium Mustererkennung '89, Hamburg Germany, Informatik Fachberichte IFB 219, Springer Berlin, pp. 425-432

Gerig G, Kikinis R and Kuebler O, Significant Improvement of MR Image Data Quality using Anisotropic Diffusion Filtering, Technical Report TR112, Institute for Communication Technology, Image Science Division, Zurich, Switzerland, March 1990

Haacke EM, Masaryk TJ, The Salient Features of MR Angiography , Radiology 1989; 173:611-612

Hu X, Tan KK, Levin DN, Galhotra S, Pellizzari CA, Chen GTY, Beck RN, Chen C-T, and Cooper MD, Volumetric rendering of multimodality, multivariable medical imaging data, Upson C, ed. Proc of Chapel Hill Workshop on Volume Visualization, Chapel Hill, University of North Carolina, 1989; 45-49

Parker, Blatter, Pelc, Radiology 1989, Volume 173(P), p.413 (abstract)

Wehrli F, Field ShA, Operation Manual GE Signa, Revision 0, page 6, chapter: Displacement Flow Artefact, 1987

A Multispectral Pattern Recognition System for the Noninvasive Evaluation of Atherosclerosis Utilizing MRI

Michael B. Merickel[1], *Theodore Jackson*[1], *Charles Carman*[1], *James R. Brookeman*[1,2] *and Carlos R. Ayers*[3]

[1]Biomedical Engineering, University of Virginia
Charlottesville, VA 22908, USA
[2]Radiology, University of Virginia
Charlottesville, VA 22908, USA
[3]Cardiology, University of Virginia
Charlottesville, VA 22908, USA

Abstract

This paper describes an image processing, pattern recognition and computer graphics system for the noninvasive identification and evaluation of atherosclerosis using multidimensional Magnetic Resonance Imaging (MRI). Particular emphasis has been placed on the problem of developing a pattern recognition system for noninvasively identifying the different plaque classes involved in atherosclerosis using minimal a priori information. This pattern recognition technique involves an extension of the ISODATA clustering algorithm to include an information theoretic criterion (Consistent Akaike Information Criterion) to provide a measure of the fit of the cluster composition at a particular iteration to the actual data. A rapid 3-D display system is also described for the simultaneous display of multiple data classes resulting from the tissue identification process. This work demonstrates the feasibility of developing a "high information content" display which will aid in the diagnosis and analysis of the atherosclerotic disease process. Such capability will permit detailed and quantitative studies to assess the effectiveness of therapies, such as drug, exercise and dietary regimens.

Key Words: Magnetic Resonance Imaging, multispectral pattern recognition, atherosclerosis, classification, and 3-D display.

NATO ASI Series, Vol. F 60
3D Imaging in Medicine
Edited by K. H. Höhne et al.

1. Introduction

Magnetic Resonance Imaging (MRI) is rapidly becoming accepted as a valuable diagnostic tool in modern day radiology. One of the most valuable aspects of MRI is that it provides multidimensional (or multispectral) imagery which emphasizes different soft tissue characteristics. However, techniques are just beginning to be developed to combine information from different pulse sequences for the purpose of characterizing and identifying different tissue types. This paper describes our work on the development and integration of multidimensional (i.e., multispectral) image processing, pattern recognition and 3-D display techniques which are capable of effectively utilizing multispectral MR imagery. This work demonstrates the feasibility of utilizing these techniques for the noninvasive evaluation of a disease such as atherosclerosis.

We have focused the development of these techniques on atherosclerosis because it represents a major health risk in the United States and other developed countries. The diagnosis and management of the consequences of atherosclerosis is a major contributing factor to health care costs. The current gold standard for the diagnosis and evaluation of atherosclerosis is x-ray angiography, an invasive procedure only justified after severe cardiovascular problems have already developed, such as stroke, myocardial infarction or peripheral vascular ischemia. Additionally, angiography carries the risks of high x-ray exposure, reaction to the contrast medium, as well as potential embolism. Most importantly, detection of the disease at an advanced stage means that much of the damage has been done, and the disease process is not generally reversible. More recently, techniques for producing an MRI correlate of x-ray angiography, called MRI angiography, have been developed which noninvasively show the lumen of the vessel. All such angiographic techniques suffer from the problem of not being able to characterize the tissues in the wall of the vessel, which is essential when evaluating the atherosclerotic lesion and stage of the disease.

Work by other investigators has demonstrated that different MRI pulse sequences are able to enhance the contrast between certain atherosclerotic tissue types [Kaufman et al 1982; Herfkens et al 1983 and Wesbey et al 1986]. Most work to date has concentrated on visualization of protrusional atheromatous tissue. We have been specifically interested in developing techniques for automatically identifying the different atherosclerotic tissue class types involved in the disease process for both protrusional as well as non-protrusional lesions. We have concentrated on recognizing the tissue classes of normal muscle wall (i.e., intima and media) and the major atheromatous plaque constituents, which include: (1) fatty, lipid containing plaque; (2) fibrous, dense connective tissue; (3) complex mixed fatty and fibrous tissue; (4) calcification; and (5) thrombus. Previous work of ours has concentrated on application of these techniques to fresh excised human aorta specimens [Merickel et al 1986 and Merickel et al 1988]. This paper describes new techniques developed to extend this work to the noninvasive evaluation of atherosclerosis in patients. The ability to reliably identify different tissue types is an important goal because it will permit the stage of the disease process to be quantitatively evaluated and followed over time.

2. The General Approach

2.1 Soft Tissue Characterization with Multispectral MRI

MRI provides multispectral or multidimensional information regarding soft tissue types due to the ability of the machine operator to control pulse sequence parameters. Different spin-echo pulse sequences were employed which enable different physicochemical aspects of the tissue to be emphasized. Contrast between different tissue constituents in MRI is due to the distribution of water and lipid (fat) protons, as well as inherent differences in their nuclear spin relaxation times. Multiple images emphasizing different tissue characteristics can be obtained at the same slice position to produce a multispectral data set of overlying image planes.

The standard spin-warp imaging method was employed with spin-echo pulse sequences chosen to emphasize the MRI parameters of T_1, the spin lattice relaxation time, T_2, the spin-spin relaxation time, and proton density. In MRI, the time between the excitation pulse and the spin-echo maximum is denoted as TE, and the repetition time between serial excitation pulses is referred to as TR [Bottomley et al 1984]. Images emphasizing differences in proton density were obtained using relatively short values of TE (< 32 ms) and long values of TR (> 1.7 s) to obtain maximum signal intensity. Images emphasizing differences in T_1 were obtained using relatively short values of TE (< 20 ms) and TR (< 0.9 s), while images emphasizing differences in T_2 were obtained with relatively long values of TE (60 ms) and long TR times (> 1.7 s) to allow equilibration between scans. Such T_2 weighted images were usually obtained as the second echo of a dual echo sequence. Specific pulse sequence parameters employed are given in the legend for Figure 2.

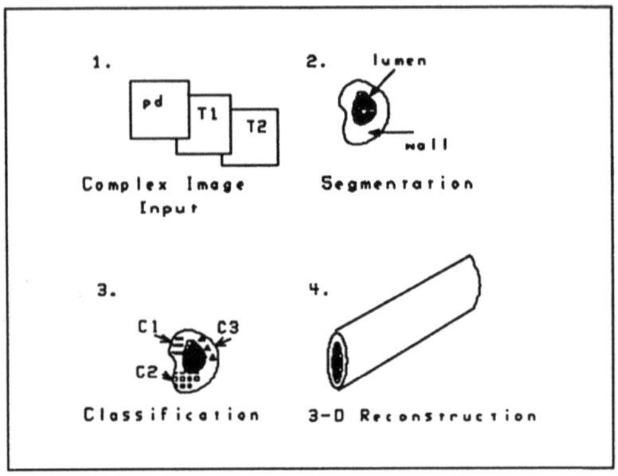

Fig. 1. Image processing, pattern recognition and graphical display system for creating 3-D display of aorta as described in text

2.2 Pattern Recognition Protocol

The pattern recognition system which has been developed to classify the major tissue types involved in atherosclerosis based on the above multispectral MR imagery is shown in block diagram form in Figure 1. The major steps in the process are listed below. 1) The input to the system is multiple pulse sequence images (usually proton density (pd), T_1 and T_2) of a major vessel taken in cross section. 2) The entire scene is segmented to extract the major vessel(s) of interest from the rest of the scene. 3) The inner and outer wall boundaries of the major vessel are then identified using contour following methods to identify the vessel wall; 4) Classification techniques are utilized to identify the different soft tissue types of interest in the vessel wall identified in (3) above. 5) A 3-D data base for the different soft tissue types is then created which is displayed using 3-D graphical techniques. Steps 2 and 3 involve determination of the vessel wall boundaries. Steps 4 and 5 involve classification of the tissues in the vessel wall and creation of a 3-D data base of the resulting class types (eg., "normal wall" and the major atheromatous plaque constituents of fatty plaque, fibrous plaque, and complex plaque with possible calcification). The identification of these soft tissue types can be directly utilized for evaluation of the physiological stage of the disease. This is done by examination of the ratios of the various plaque constituent volumes to the total wall volume.

3. MR Imaging of the In Vivo Aorta

The work reported in this paper has focused on imaging the lower abdominal aorta due to its large size as well as the relatively small movement artifacts due to respiration. Images were generated using a Siemens spine coil on a 1 Tesla Magnetom Imager. The present study employed transverse T_1, proton density (Pd) and T_2 weighted images obtained from approximately 10 different slice positions. The work reported in this paper was performed on a patient with advanced atherosclerosis (confirmed by prior angiography) as well as a "normal" subject with no significant disease.

3.1 Artifact Reduction

The raw MR images contained several artifacts which had to be removed before segmentation and classification could be performed. Most artifacts were caused by the use of the spine coil and are typical of surface coils currently employed in MR imaging to increase spatial resolution and the signal to noise ratio. The sensitivity profile of the surface coil, such as the spine coil employed in this investigation, causes the average intensity of the image to fall off with distance from the center of the coil. This intensity fall off results in an intensity gradient across the image of an individual slice as well as a reduction in average intensity of slices imaged further from the coil.

The intensity gradient within a slice was reduced by employing the technique of dividing each image by a spatially smoothed copy of itself on a pixel by pixel basis [Haselgrove & Prammer 1986]. This technique significantly reduces the low frequency surface coil intensity gradient and produces some high frequency emphasis which enhances the contrast of edges [Bottomley 1984].

The average intensity of slices fell off with distance from the surface coil which required all of the slices belonging to each of the 3 pulse sequences to be normalized to some standard. This was accomplished by normalizing each of the sequences to the mean intensity value of a small set of easily identifiable tissues common to all slices. For this data set we used skeletal muscle and interstitial tissue as the readily identifiable tissues for normalization. The mean values for skeletal and interstitial tissue for each pulse sequence were then used to define the set of standard intensity values for these reference tissues, which then allowed computation of the T_1, Pd and T_2 normalization scale factors for each slice.

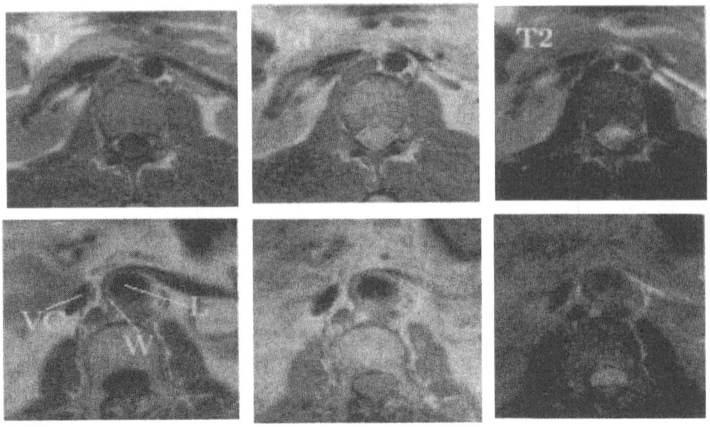

Fig. 2. Representative cross sectional T_1, Pd and T_2 images through the lower abdominal aorta in "normal" subject (Norm, top row) and patient with advanced atherosclerosis (Ath, bottom row) confirmed with angiography. The labels in the T_1 weighted image from the diseased aorta are: W = aorta wall; and L = aorta lumen. The pulse sequence parameters for the "normal" subject are: T_1 = SE 500/20, pd = SE 1700/32 and T_2 = SE 1700/64; x-y resolution = 0.59 mm/pixel; distance between slices = 15 mm; and slice thickness = 10 mm. The parameters for the patient with advanced disease are: T_1 = SE 900/20, pd = SE 3600/20 and T_2 = 3600/60; x-y resolution 0.59 mm/pixel; distance between slices = 10 mm; and slice thickness = 5 mm. White scale bars in bottom right hand corner are 4.5 cm.

A third major artifact was within plane misregistration between the T_1, Pd and T_2 pulse sequences for an individual slice. This misregistration is due to a combination of patient movement and configuration of the current Siemens reconstruction software. The shifts were corrected by spatially shifting the T_1, Pd and T_2 images until they lined up to one another by visual inspection.

3.2 Basic Characteristics of the Aorta Viewed with MRI

Figure 2 shows a set of T_1, Pd and T_2 weighted transverse images through a representative section of the "normal" subject and the patient with advanced disease. The T_1 and T_2 weighted images through the diseased vessel clearly show the wall of the aorta to be thickened, particularly on the dorsal side next to the vertebral column. The wall thickening and distribution of plaque tissues can be quite complex in patients with advanced disease.

In general, the lumen of a vessel appears dark with all pulse sequences due to spin washout, except in cases exhibiting significant flow artifact associated with the movement of the blood and the long T_1 of blood (approx. 1 sec, see [Bradley 1987] for review). Flow artifacts which can cause the lumen to appear bright are entry phenomenon and second echo rephasing. The lumen of most of the images in Figures 2 and 3 are dark compared to the surrounding vessel wall. However, the lumen appears gray in the T_1 weighted pulse sequence of the normal subject in Figure 2 because it is the first slice in the sequence (i.e., entry phenomenon).

4. Pattern Recognition and Classification of In Vivo Tissue

Previously, we have demonstrated the ability to identify the plaque constituent classes of normal wall (primarily smooth muscle), fatty lipid containing plaque, fibrous plaque (primarily connective tissue) and mixed complex plaque with regions of calcification in an excised aorta preparation [Merickel et al 1986 and Merickel et al 1988]. We have found that simple pattern recognition classifiers, such as the minimum distance to the means and Fisher linear classifiers, are sufficient for the classification of excised aorta specimens using multispectral image input consisting of T_1, Pd and T_2 weighted image sequences. However, these supervised classification techniques cannot be utilized in vivo due to the lack of appropriate a priori training data, such as histopathology [Tou & Gonzalez 1974]. We have therefore developed an unsupervised classification technique, based on the ISODATA clustering algorithm, which does not require training data.

4.1 Segmentation of Aorta

Following acquisition of transverse images of the abdominal aorta, the T_1 weighted images are used for aorta segmentation. The T_1 weighted images are utilized for image segmentation because they have been found to emphasize the structural information involving the vessel wall boundaries. The segmentation is accomplished by convolving the images with a Laplacian of a Gaussian operator. Peaks in the resulting image correspond to the position of the aorta, the vena cava and various other structures. The spot detected sequence of images is thresholded and an average of this sequence is calculated. Since the aorta and vena cava are the only circular structures with a consistent spatial location between slices, the peaks in the average of the sequence correspond to only the aorta and vena cava. Since the positions of the aorta and vena cava relative to each other are known, it is a simple matter to select the peak corresponding to the aorta. Extractions are performed about a peak in each convolved slice nearest the aorta's peak in the average of the sequence, and the resulting extracted regions are passed to the aorta segmentation procedure.

The goal of the aorta segmentation procedure is to locate a better approximation to the center of the aorta than that obtained through spot detection of the images and to distinguish the outer wall of the aorta from the surrounding tissue. An automatic threshold selection is performed on each of the extracted regions. The first goal is achieved through an erosion process to find circular structures in the resulting binary images. The spatial average of the remaining points are taken as the center of the aorta. A smaller sub-image is extracted about this center, further reducing the quantity of tissue outside the aorta.

The aorta segmentation procedure uses these sub-images as a starting point to accomplish it's second goal. A threshold for each of these sub-images is automatically determined and thresholding is performed. A radius filter is used that eliminates the regions in the thresholded sub-image which lie beyond the aorta wall. The result is a region with size, shape, and position corresponding to that of the aorta, as shown in Figure 3. A one pixel border is added to this region to ensure inclusion of all of the aorta wall.

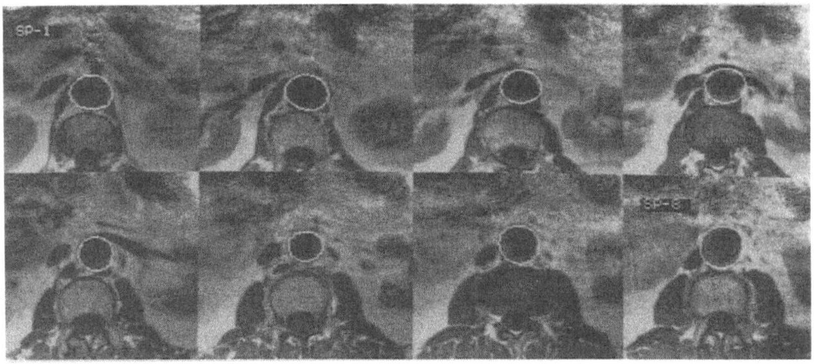

Fig. 3. Resulting boundaries (white curves) of the outer wall of the aorta for the first eight successive sections of the diseased aorta <u>in vivo</u> data set which are superimposed on the T_1 weighted MR images.

4.2 Classifier Description

Most clustering techniques utilize task dependent heuristics or analyst intervention to supervise the progress of the clustering algorithm. We have been interested in developing a clustering procedure which utilizes as little a priori information as possible in order to determine the intrinsic number of tissue classes in the MRI data. Carman and Merickel [1990] have developed an internally supervised clustering technique for classification of the in vivo aorta images which requires little a priori information. The ISODATA clustering algorithm was utilized as the basic clustering strategy due its well known robustness and previous success in similar multispectral problems, as well as its relative ease of implementation [Anderberg 1973 and Ball & Hall 1965]. However, one of the major problems with ISODATA, as well as other unsupervised classifiers, is their sensitivity to initial parameter selection. Our clustering technique is based upon a modification of the ISODATA clustering algorithm and utilizes a statistical measure, the Consistent Akaike Information Criterion (CAIC), to provide supervision of the splitting and merging operations as described below.

ISODATA is a variation of the K-means clustering algorithm, in which data points are assigned to the nearest cluster (i.e., class) [Tou & Gonzalez 1974]. Many variations of ISODATA have been developed over recent years in which the definition of distance measure as well as the split and merge heuristics may differ. ISODATA provides more flexibility than the K-means method because the number of clusters can be modified to fit the data more accurately. The algorithm consists of iterative splitting and merging of the clusters until the number of clusters remains constant. In the original algorithm, a cluster is split when it has a single dimensional variance larger than a user selected threshold. The number of clusters is decreased by the merging of two clusters which are closer (measured by Mahalanobis distance) than a user selected threshold. There is also the provision for ignoring the data points which group into clusters with fewer than a user selected number of points. This last feature was used to remove wild points from the data as they were identified.

In the internally supervised ISODATA algorithm, an information theoretic criterion was utilized to provide a measure of the global fit between the current cluster composition and the actual data. The Consistent Akaike Information Criterion as derived by Bozdogan [1986 and 1987], is based on a data model of multivariate Gaussian distributions with varying mean vector and covariance matrices. The CAIC essentially provides a measure of negative entropy of the model and thereby provides a global measure of fit between the cluster results and data. In our procedure the CAIC measure was calculated during each iteration of the ISODATA algorithm. The clustering algorithm was terminated when a minimum CAIC value was detected, indicating the best global fit.

The CAIC is an asymptotically unbiased and consistent estimate of the negative expected log likelihood of the true data probability distribution. In other words, as the number of data points goes to infinity the mean of the CAIC equals the negative expected log likelihood (unbiasedness) and the variance of the CAIC goes to zero (consistency). The CAIC is evaluated using the

Maximum Likelihood (ML) estimates for the probability distribution parameters. In addition, the CAIC contains a term that corrects for the estimation bias of the ML estimates. The general form of the CAIC measure [Bozdogan 1986], is:

$$\mathrm{CAIC} = -2 \ln[L(\Theta_k)] + m(k) \ln(n) \qquad \text{(Eqn. 1)}$$

where ln is the natural logarithm, L[] is the likelihood function, Θ_k is the ML estimate parameter vector for k clusters, and m(k) is the number of parameters estimated for k clusters. For our specific implementation we chose a data model consisting of the normal probability distribution with unique mean vectors and diagonal covariance matrices (variances only) for each cluster. The form of the CAIC for this case is:

$$\mathrm{CAIC} = np\,[1 + \ln(2\pi)] + \Sigma\,[n_j \ln |S_j|] + 2kp \ln(n) \qquad \text{(Eqn. 2)}$$

where n is the number of data points, p is the number of data dimensions, k is the number of clusters, n_j is the number of data points in cluster j, and S_j is the ML estimate of the covariance matrix for cluster j. The modified ISODATA procedure starts with all data points in one cluster and splits out one new cluster on each iteration until at least two iterations in a row have CAIC values larger than the running minimum. It then merges cluster pairs (one pair per iteration) until a CAIC value is found to be close to a previous minimum for that number of clusters. At this point, the cluster model and data for the smallest CAIC value are output to a file and the procedure ends.

4.3 Classification Results

The aorta data set from the normal subject produced 6 clusters while the data set from the patient with advanced atherosclerosis produced 7 clusters using the internally supervised ISODATA algorithm. Some of the clusters in each of these data sets were expected to belong to the same tissue type and were therefore merged using a hierarchical (complete link) clustering procedure. The hierarchical clustering procedure produced 3 clusters for the normal subject aorta and 4 clusters for the diseased aorta, which are shown in Figure 4.

The cluster mean values from the clustering results were also used to classify the entire preprocessed images to produce a theme map as shown in Figure 5. The clusters were labeled with a specific tissue type by comparing the clusters in the aorta wall with surrounding homogeneous tissues that were readily identifiable. The process of identifying different soft tissue types in the slices was significantly aided by the expert advice from radiologists familiar with the anatomy of the region. The section of normal aorta wall contained the 3 main classes of lumen, normal wall (i.e., smooth muscle), and fatty tissue (primarily adventitial fat). The aorta wall from the patient with advanced atherosclerosis contained the 4 classes of lumen, normal wall, mixed fibrous plaque, and fatty plaque mixed with adventitial fat. No complex plaque or calcification were found in any of the slices from the diseased aorta studied in this project. Previous studies have demonstrated that complex plaque and particularly calcification (due to lack of signal), are readily visible in MR images [Merickel et al 1986 and Merickel et al 1988].

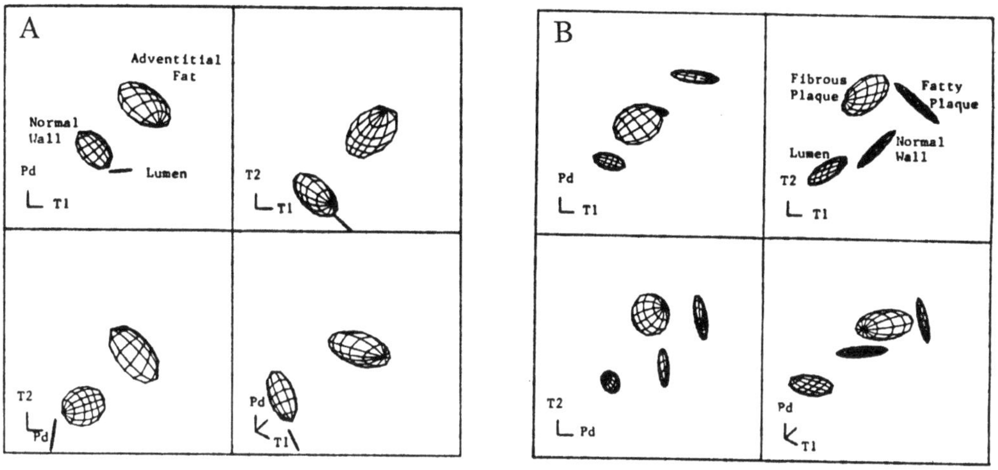

Fig. 4. Cluster plots for normal subject aorta (A) and diseased aorta (B). Each cluster has been labeled as described in text.

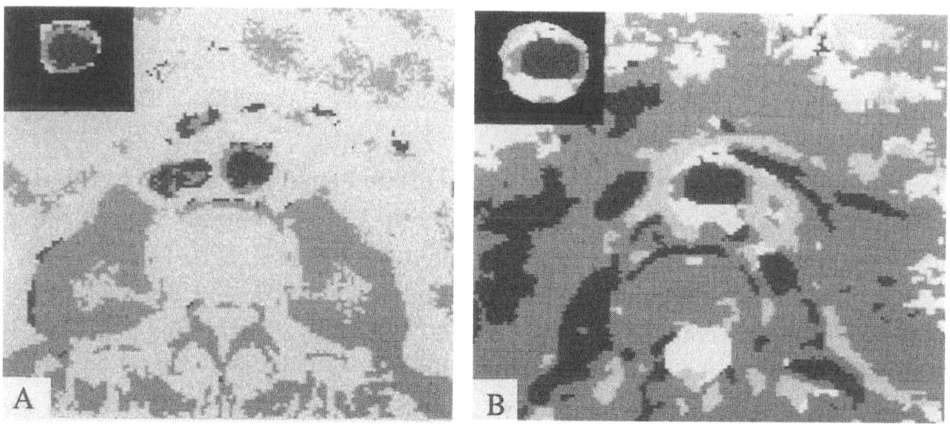

Fig. 5. Theme maps for normal subject aorta (A) and diseased aorta (B) generated for the same slice positions shown in Figure 2. The classes are represented as different gray levels for the normal and diseased aorta as follows. Normal aorta: black - lumen of aorta; normal aorta wall - medium gray; and white - fatty plaque and adventitial fat. Diseased aorta: black - lumen; dark gray - normal aorta wall; light gray - mixed fibrous plaque; and white - fatty plaque and adventitial fat.

5. 3-D Display of Pattern Recognition Results

In order to quantify the plaque volume and visualize the spatial extent of the different plaque constituents, a 3-D data base is constructed from the 2-D pattern classes obtained in Section 3. Various methods have been utilized to display the spatial relationship between the plaque 3-D pattern classes which include systems we have designed [Merickel et al 1990] as well as commercially available systems such as Voxvu on the Sun Microsystems TAAC system as well as ChapVolumes on the PIXAR.

The volumes of plaque tissue types are evaluated by counting the number of voxels in the different plaque tissue classes and expressing them as a ratio with respect to the total wall volume. Table 1 shows the volume of the aorta wall, fatty plaque and fibrous plaque expressed as in percentage of the total wall volume. Lumenal narrowing was quantified by measuring the lumen cross sectional area at each slice position.

Tissue Type	Volume (ml)	Volume (%)
Fatty Plaque	6.9	37.7
Fibrous Plaque	6.5	32.7
Normal Vessel Wall	6.4	32.6

Table 1. Volume of each plaque constituent identified in diseased aorta. The volume of each component is also expressed as a percentage of the total wall volume.

6. Conclusion

This paper describes an image processing, pattern recognition and 3-D graphical display system which demonstrates the feasibility of noninvasively evaluating atherosclerosis using MRI imagery. This approach has been specifically applied to the evaluation of atherosclerosis in the abdominal aorta of a normal subject and a patient with advanced atherosclerosis. A key concept to this approach is the combination of information from the multispectral MR imagery which is not available with a single pulse sequence. It is believed that these techniques are essential to exploiting the tremendous amount of information available in new biomedical imaging modalities, such as MRI.

The application of this approach to imagery from patients requires that several problems unique to the in vivo environment be addressed. A particularly difficult problem is relating the relative image intensity values observed in one patient to other patients without the aid of "absolute" information regarding the signature of different tissues. We have approached this problem by developing an internally supervised classifier based on ISODATA. Such an internally supervised classifier is able to effectively determine the number of clusters in the data, but still requires that the cluster be labeled as a specific tissue type using a priori information, such as provided by a radiologist. Other, more consistent methods of labelling the clusters need to be developed which require less operator intervention.

One of the most important aspects of the unsupervised classifier described in this paper is that it provides a mechanism for determination of the number of tissue types which can be differentiated with multiple pulse sequences without the bias imposed by a supervising operator. Both supervised classifiers (eg., minimum distance to the means and Fisher discriminant classifier) as well as unsupervised classifiers (eg., clustering algorithms) require important input from the operator which controls the number of classes to be determined. The unsupervised classifier described in this paper permits the number of tissue types to be compared between different data sets. The results have demonstrated that the diseased aorta from a patient with advanced atherosclerosis has more tissue classes than an aorta from a normal subject. Most importantly, the number of classes in the normal and diseased aorta correlates well with the expected number of classes based upon examination by radiologists.

The ability to noninvasively quantify and display the different tissue components associated with atherosclerosis is an important goal for the early detection of the disease in high risk, asymptomatic patients. This capability provides a means of directly assessing the effectiveness of various therapies, such as drug, exercise and dietary regimens, by measuring changes in the volume of different plaque constituents over time. Additionally, the ability to measure the volume of these plaque constituents in vivo can be expected to provide a means of staging the disease process to more accurately measure its time course and response to intervention with various therapies, such as cholesterol lowering drugs, exercise and nutrition.

This work has been supported by the Whitaker Foundation (MBM, JRB and CRA), the Virginia Center for Innovative Technology, E.I. DuPont, Dynatech Computer Systems, AMVEST Corp. and the University of Virginia School of Medicine Pratt Foundation. The authors would like to express their appreciation to the Radiology Department of the University of Virginia for their support of this project. The authors would particularly like to acknowledge the expert help of Dr. John Mugler and the expert advice of Dr. Wayne Cail.

6. References

Akaike, H.: A Bayesian analysis of the minimum AIC procedure. Annals of the Inst. of Stat. Math. 30, 9-14 (1979)

Anderberg, M.R.: Cluster Analysis for Applications, San Francisco, Calif.: Academic Press 1973

Ball, G.H. and Hall, J.D.: ISODATA, A Novel Method of Data Analysis and Pattern Classification, Tech. Rep., Menlo Park, Calif.: Stanford Research Institute 1965.

Bottomley, P.A. Foster,T.H., Argersinger, R.E. and Pfeifer, L.M.: A review of normal tissue hydrogen NMR relaxation times and relaxation mechanisms from 1-100 MHz. Med. Phys. 11, 425-449 (1984)

Bozdogan, H.: Multi-sample cluster analysis as an alternative to multiple comparison procedures. Bull. Info. Cyber. 22, 95-130 (1986)

Bozdogan, H.: Model selection and Akaike's information criterion (AIC): The general theory and its analytical extensions. Psychometrika. 52(3), 345-370 (1987)

Bradley, W.G.: Sorting out the Meaning of MRI Flow Phenomena. Diagnostic Imaging. 9(5), 102-111 (1987)

Carman, C.S. and M.B. Merickel, M.B.: Supervising ISODATA with an Information Theoretic Stopping Rule. In Press, Pattern Rec. 1990

Gonzalez, R.C. and Wintz, P.: Digital Image Processing, Reading, Mass: Addison-Wesley Publishing Co. 1977

Haselgrove, J. and M. Prammer, M.: An Algorithm for Compensation of Surface-Coil Images for Sensitivity of the Surface Coil. Mag. Res. Imaging. 4, 469-472 (1986)

Herfkens, R.J., Higgins, C.B. et al.: Nuclear Magnetic Resonance Imaging of Atherosclerotic Disease. Radiology. 48, 161-166 (1983)

Jiang, T. and Merickel, M.B.: Identification and Boundary Extraction of Blobs in Complex Imagery. Computerized Medical Imaging and Graphics. 13, 369-382 (1989).

Kaufman, L., Crooks, L.E. et al: Evaluation of NMR Imaging for Detection and Quantification of Obstructions in Vessels: Invest. Radiol. 17, 554-560 (1982)

Merickel, M.B., Carman, et al: Multidimensional MRI Pattern Recognition of Atherosclerosis. Proc. of the 8th Ann. Conf. of the E.M.B.S. Vol 2., pp. 1142-1145, Nov 1986.

Merickel, M.B. Carman, C.S. et al: Identification and 3-D Quantification of Atherosclerosis Using Magnetic Resonance Imaging. Comput. in Bio. and Med. 18(2), 89-102 (1988)

Merickel, M.B., Carman, C.S. Watterson, W.K. Brookeman, J.R. and Ayers, C.R.: Multispectral Pattern Recognition of MR Imagery for the Noninvasive Evaluation of Atherosclerosis. In Press, Computerized Medical Imaging and Graphics.

Tou, J.T., and Gonzalez, R.C.: Pattern Recognition Principles, Reading, Mass: Addison-Wesley Publishing Co. 1974

Wesbey, G.E. Higgins, C.B. et al: Magnetic Resonance Applications in Atherosclerotic Vascular Disease. Cardiovascular and Intervent. Rad. 8, 342-350 (1986)

3D Reconstruction Of High Contrast Objects Using a Multi-scale Detection / Estimation Scheme

Didier Saint-Félix, Yves Trousset, Catherine Picard and Anne Rougée

GENERAL ELECTRIC CGR - Engineering and New Products Division
283 Rue de la Minière - 78530 BUC - FRANCE

Abstract

Reconstructing a three-dimensional (3D) volume from a set of two-dimensional X-ray projections raises theoretical, instrumental and computational difficulties. Focused on high contrast objects, solutions are proposed for the successive steps of a 3D reconstruction procedure, from the raw measurements on an image intensifier up to the reconstruction algorithm based on a multi-scale detection / estimation scheme.

Keywords: 3D reconstruction, computed tomography, prior information, volume segmentation, multi-scale reconstruction algorithm, dual energy

1. Introduction

Reconstructing a three-dimensional (3D) volume from a set of two dimensional (2D) X-ray projections raises difficulties on different grounds: reconstruction theory, instrumentation, computation and display. This paper presents possible solutions to those various difficulties that enable one to go from the raw data measured on an image intensifier (II) up to the visualization. After a general presentation of the 3D reconstruction in the medical context, the basic principles of our approach are stated. Then a quick overview is given of the calibration procedures and of the measurement pre-processing that provide corrected data suitable to the reconstruction algorithms described subsequently. Results obtained on actual data are presented at the end.

1.1 3D Reconstruction: an *Ill-posed* Problem

The reconstruction of a 3D object from its 2D X- ray projections cannot be

NATO ASI Series, Vol. F 60
3D Imaging in Medicine
Edited by K. H. Höhne et al.

reduced to a trivial extension of the 2D reconstruction case. First, in practical situations where a 2D projection is obtained from a single position of the X-ray source assumed to be a point source placed at a finite distance, the projection geometry is conic. A given slice in the volume cannot be reconstructed independently of the surrounding slices and the 3D volume thus cannot be obtained by just stacking 2D reconstructions of parallel cross-sections. Second, the direct inversion of the integral equation describing the cone-beam projection, using the Radon transform for instance, requires certain conditions on the source trajectory [Kirillov 1961, Tuy 1983, Smith 1985] that often cannot be met for practical reasons such as gantry size and limited flux of X photons. Then the reconstruction problem has no solution in the usual sense. Extra difficulties are also added by the limited area of 2D detectors which introduces a truncature of the projections, at least in the direction of the body axis. For these reasons, the 3D problem should be considered, in medical applications, as a reconstruction problem with missing data. This point of view is reinforced when one is interested in physiological systems where dynamic phenomena occur, such as a vascular tree opacified by transient injection of a contrast agent. The number of available projections can then be dramatically low because of the time constraint.

Considering 3D reconstruction as a reconstruction problem with missing data has major consequences. The integral equation of the first kind modelling the cone-beam X-ray projection makes the reconstruction problem an *ill-posed* one and its intrinsic difficulties are increased by the data incompleteness [Rangayyan 1985]. The usage of regularization techniques is then a must. Since they explicitly introduce prior information on the solution to stabilize it [Nashed 1981, Saint-Félix 1988], one must therefore admit that a 3D reconstruction method cannot be universal, and that its efficiency increases with its specificity. This point of view is reinforced when the characteristics of today's 2D detector are considered.

1.2 3D Reconstruction of Highly Contrasted Objects

The II is the only existing device suitable for fast digital radiography. It exhibits good spatial resolution and good sensitivity, but it suffers from geometrical and densitometric distorsions and, above all, from a low dynamic range. We thus focus on highly contrasted objects, such as opacified vasculatures and bone structures, where the morphological information is more relevant than the densitometric information. The a priori informations that may characterize both classes are the non-negativity, their sparseness (except for the skull) and their high contrast with respect to the background [Trousset 1989]. The last two points are even more significant if subtraction techniques are used to remove the background in the projections: logarithmic subtraction for vascular trees, dual energy techniques for bone structures.

The ability of reconstruction techniques to incorporate prior information largely differs from one method to another. Let us focus on 3D reconstruction of sparse objects, such as vascular trees. Besides the techniques based on computer vision tools and using two or three projections only, the others can be classified according to the kind of object model they use. A first group use parametric models where the number of unknown parameters is reduced by decomposing the solution on a small set of "basis functions" carefully chosen. The advantage is double: the solution is well regularized and the amount of computation required by the estimation of few parameters can be dramatically low, specially if fast algorithms can be used when the model enjoys some adequate invariance property. Modelling vascular trees by elementary cylinders with elliptical cross-sections belongs to this group of methods [Bresler 1987, Fessler 1989]. Their main limitations are the high number of parameters that may be required by an accurate modelling of a complex object and the lack of robustness with respect to pathological cases. The influence of the prior information, incorporated in the choice of the basis functions, may indeed predominate over the measurements and mislead the solution, even if stochastic modelling techniques may bring some flexibility.

The second group of methods emphasizes the robustness of the representation by using non-parametric modelling. The object is simply described by the set of the sampled voxels; the difficulty is then computational because of the huge number of parameters to be estimated. This problem has no simple solution, even in the 2D case, since the projection matrix that relates the measurements to the unknown object does not exhibit the structural properties (Toeplitz, circulante, ...) that allow derivation of fast algorithms. Existing methods therefore use estimation schemes that are simple enough to allow an effective implementation [Haaker 1984, Kruger 1987]. The counterpart of this simplicity of the estimation method is a lack of robustness with respect to the measurement noise [Rougée 1988].

Thus it appeared interesting to develop a regularizing reconstruction method that has the flexibility of the non-parametric approaches but offers a good robustness to the noise and requires a reasonably small amount of computations. The basic idea is to take full advantage of the sparseness property by restricting to only those voxels belonging to the object the reconstruction stages that are computationaly intensive. This idea is implemented with a two step reconstruction scheme. In the first step, the region of support of the object is determined given the projections. In the second step, the density value of the voxels within the region of support is estimated. But before applying this classical detection / estimation scheme, the raw data must be calibrated and corrected.

2. Pre-processing of the measurements

Whatever the method used to reconstruct a 3D object, the location of the projection of a given voxel in any view must be known. This requires accurate knowledge of the projection process which includes not only the geometrical conic projection but also the geometrical distorsions introduced by the measurement process. Both points can be modelled and treated separately.

An II introduces a geometrical distorsion between the ideal projection image formed on a plane and the actual measurement. The reasons are the spherical shape of the input screen and the effect of the earth's magnetic field on the electrons' trajectory. The global distorsion can be modelled by a displacement field mapping the ideal image onto the actual one. This field can then be identified using the image of a known reference grid, and the correction of any image is done by resampling the image pixel by pixel using the interpolated inverse field.

The conic projection is easily modelled using homogenous coordinate systems. The coefficients of the transformation matrix applying the 3D space onto a given projection plane can be identified using the corresponding undistorted image of a reference phantom that presents a sufficient number of fiducial points.

3. 3D Reconstruction Algorithms

3.1 Detection

The purpose of this first stage is to determine, given the projections, which are the voxels that belong to the object of interest. These will be called "full", the others "empty". The detection method must avoid negative false alarms, i.e. declaring a given voxel empty when it is full. This could be done at the possible expense of a higher rate of positive false alarms, i.e. declaring a voxel full when it is empty. One expects that the estimation step will attribute later a null value to these positive false alarms. The computational savings provided by the detection is larger when the object of interest is sparser and the detection itself is easier when the projections are also sparse. This property can be obtained by subtraction techniques. Possible subtraction and detection schemes suited to the processing of the two classes of objects of concern: vascular trees and bones, are considered now.

Subtraction and detection for vascular reconstruction. The logarithmic subtraction between images taken before and after the injection of iodine, commonly used in Digital Subtracted Angiography (DSA), provides a good image of the vessels only when there is no move between the two pictures. This procedure can be used here for each view angle and the resulting subtracted images turn to be the inputs for the 3D reconstruction. During the detection step, a voxel is declared empty if there is at least one subtracted image in which this voxel does not superimpose with any full voxel. This can be implemented using statistical hypothesis testing. For each voxel and for each view, a decision is taken if the measured values around the location of the projected voxel correspond to background or not. This decision is made with a given error probability knowing the noise statistics [Trousset 1989].

Subtraction and detection for selective bone reconstruction. In this case, the emphasis is on the subtraction stage since a simple segmentation of the bones in the projections cannot provide the required attenuation value. Dual energy (DE) techniques must be used for this purpose.

The principle is to distinguish between two materials, e.g. bone and soft tissues, using the different dependence of their attenuation coefficient on the photon energy. At the energy levels used in medical imaging, the absorption of a given material depicts two main phenomena: the Compton scattering and the Photoelectric absorption; it can therefore be assumed to be a function of the energy in a 2D space. This 2D space can be entirely described by the absorption behaviour of two distinct reference materials, such as aluminum and lucite. The total absorption of an inhomogeneous medium along a given ray path can thus be expressed as the sum of the absorptions through "equivalent thicknesses" of the reference materials. Each equivalent thickness is approximated by a polynomial expansion of the absorption measurements made at two different energies. The coefficients of the expansion depend on the reference materials only; they can be estimated with a calibration procedure where different known thicknesses of the reference materials are superimposed and exposed at both energies [Lehman 1981, Brody 1981].

The response of an II suffers from spatial variations at low frequency which are function of the energy. Rather than correcting these nonuniformities in each measurement, we take them into account directly in the DE calibration and combination process. For this purpose, the coefficients of the two polynomial expansions are taken dependent on the location of the pixel. In practice, these coefficients are estimated on a regular mesh covering the image and interpolated for the other locations [Picard 1990]. Then, for the object of interest, two "equivalent thickness images" (Aluminum and Lucite) are generated from two exposures at low and high energy by applying for each pixel the corresponding polynomial expansions. A selective projection of bones only is obtained from the two reference material images by a simple linear combination. Example of such a decomposition is given Fig. **1**: Fig. **1-a** and **1-b** shows respectively the high and low energy exposures of a hand phantom, made with actual bones and lucite emulating soft tissues, and Fig.**1-c** shows the selective bone image resulting from the dual energy decomposition.

For 3D selective bone reconstruction, two exposures of the object must be taken at low and high energy respectively for each X-ray source position. They are combined according to the above technique and thus a set of selective projections of bones only is generated. The detection of the region of support of the bones can then be done using the method described before for the vascular trees. Alternatively, a segmentation can be done first in the 2D selective projections to locate the 2D regions of support, then a logical back-projection determines the region of support in the 3D space.

3.2 Estimation

The purpose of this second step is to estimate the value of each voxel belonging to the region of support. Two different approaches to solve this problem are presented. The first one is based on the well known Algebraic Reconstruction Technique (ART) [Gordon 1970], the second one is a multiscale implementation of an iterative procedure such as ART.

Three-dimensional ART. ART stands on the Kaczmarz method which iteratively computes the generalized inverse of a system of linear equations. Although the asymptotic solution is unstable when the problem is ill-posed, regularization is obtained by stopping the iterations according to a given rule and by introducing prior information (in the present case, non-negativity and knowledge of the region of support) on the solution [Saint-Félix 1988]. The support information is used by applying the estimation step to only the voxels belonging to the detected support, the value of the voxels outside the support being fixed to a background value after the detection step.

Multiscale estimation. When estimation is based on an iterative technique such as ART, extra computational savings can be achieved by a multiscale approach. The underlying idea is that the first few iterations estimate essentially the low frequencies of the volume, and thus may not need to be performed at full resolution. The 3D object can therefore be described at different levels of resolution 2^j, $j_{min} \leq j \leq j_{max}$, and the estimation is performed successively for the increasing values of j. Time saving in the global estimation task is expected because the complexity of ART is linear with respect to the number of processed voxels, and thus one iteration at resolution 2^{j-1} is 8 times faster than one iteration at resolution 2^j.

Let us detail this approach. An initial estimation phase is first performed at the lowest resolution j_{min}, by means of one or several iterations of ART (or any other iterative estimation method). Then, for any j, $j_{min} \leq j < j_{max}$, the estimation obtained at resolution j is interpolated to a 2^{j+1} grid and used as initial value for the estimation at resolution $j+1$, which is also achieved by one or more iterations of ART. This multiscale approach can be used either with a constraint of support or for the reconstruction of a whole parallelepipedic volume. In the case where a constraint of support is to be applied, the detection is performed at the highest available resolution j_{max}, in order not to lose any critical detail, such as for instance small vessel branches in the vascular case. The region of support is then reduced to resolution j_{min} before the beginning of the estimation step.

Additional time saving can be achieved by eliminating false alarm voxels from the region of support, i.e. voxels that do not belong to the object of interest but have been erroneously detected. In the multiscale implementation, this can be done by performing at each resolution j a segmentation of the estimated volume and by removing from the support all the voxels which do not belong to the segmented object. Because SNR is higher in the estimated volume than in the measured projections, false alarm voxels due to noise in 2D projections can be removed in this way. Furthermore, "empty" voxels located in concave parts of the object (for instance, the inside of the skull in selective bone reconstruction) can also be removed.

4. RESULTS

The reconstruction methods described in this paper have been tested on sequences of 2D conic X-ray projections acquired on two differents kinds of devices: an experimental bench [Trousset 1989] and a standard digital angiography system.

4.1 Bone imaging

The abovementioned hand phantom was used to test the selective reconstruction method. A set of 120 dual-energy projections pairs (50 and 100 kVp) was acquired on the experimental bench. A comparison was made between non selective reconstruction of the whole hand from the 50 kVp projections and selective reconstruction of the bone from the dual energy pairs. Both reconstructions were done at resolution 256 (physical voxel size: 0.7 mm) with the monoscale method. The selective method provided a speed-up factor of 5.3 for the estimation step (5.3 being the ratio of the number of voxels detected in each case, see Fig.**2-a** and **2-b**), while producing roughly the same quality for the reconstruction of bone (Fig.**2-c** and **2-d**).

4.2 Vascular imaging

The reconstruction of a cadaver heart was performed with the multiscale estimation method. A set of 31 projections of the heart with post-mortem injection of constrast agent in the coronaries arteries was acquired on an angiography system (Fig.**3**). As it was not possible to acquire also a mask image (i.e. an image without constrast agent) the whole myocardium had to be reconstructed, and not only the coronary arteries. Results are presented in Fig.**4**, in the form of surface display of the reconstructed arteries. Fig.**4-a** (resp. **4-b**) presents the results obtained with the monoscale approach after 1 (resp. 2) iterations(s) of ART at resolution 256 (physical voxel size: 0.6 mm). Fig.**4-c** was produced with the multiscale method, used with 1 iteration at resolution 128 followed by 1 iteration at resolution 256. Fig.**4-d** was obtained similarly, but with 2 iterations at 128 resolution. The images presented in Fig.**4-b**, **4-c** and **4-d** have approximatively the same quality and are superior to the one of Fig.**4-a**. The conclusion is then that the multiscale approach saves a processing time corresponding roughly to the time of 1 iteration of ART at the highest resolution. This is specially interesting in the case of sparse objects reconstruction, where the introduction of a strong region of support constraint permits the use of a very limited number of iterations [Rougée 1989].

Another conclusion to be drawn from the results obtained on this phantom is the very great importance of 3D vizualisation tools adapted to angiography. Cut planes of the reconstructed volume (Fig.**5**) bring a good feeling of the 3D anatomy of the myocardium but carry little 3D information about the coronary network itself. Traditional surface display methods cannot perform a correct segmentation and thus a correct rendering of very small vessels (Fig.6-a). A method which may be more specifically adapted to angiography is maximum voxel ray tracing, first introduced in MR angiography [Laub 1988], in which each ray going from the observer eye to the 3D volume is analysed and only the maximum value along this ray is written to the 2D display image (Fig.6-b).

5. Conclusion

The problem of 3D reconstruction from X-ray projections is an ill posed problem, which implies that the reconstruction methods must depend on the kind of object to be reconstructed. We have introduced a methodological framework for the reconstruction of sparse high-contrast objects, based on a detection/estimation scheme. Various implementations of this scheme have been described, and their efficiency illustrated on the reconstruction of geometric and anatomic phantoms. The next step will be to validate our approach on real clinical data.

Acknowledgments

The authors thank Dr. Holmes, Dr. Schwartz and Dr. Wondrow from Mayo Clinic for helping in acquiring the projection data, and J. Knoplioch for providing 3D visualization tools.

6. REFERENCES

Bresler Y. and Macovski A., "3-D reconstruction from projections with incomplete and noisy data by object estimation", IEEE Trans. ASSP-35, 1139-1153, 1987.

Brody W., Blutt G., Hall A. and Macovski A., "A method for selective tissue and bone visualisation using dual energy scanned projection radiography", Med.Phys. 8-3, 353-357, 1981.

Crawford C.R., Colsher J.G., Pelc N.J. and Lonn A.H., "High Speed Reprojection an its Applications", SPIE Vol. 914, 311-318, 1988.

Fessler J.A.and Macovski A., "3-D Reconstruction of Vessels with Stenoses and Aneurysms from Dual Bi-Plane Angiograms", SPIE Vol. 1092, 22-32, 1989.

Gordon R., Bender R. and Herman G.T., "Algebraic reconstruction technique for three-dimensional electron microscopy and X-ray photography", J. Theo. Biol. 29, 471-81, 1970.

Haaker P., Klotz E., Koppe R., Linde R.and Moller H., "A new digital tomosynthesis method with less artifacts for angiography", Med. Phys. 12, 431-436, 1985.

Kirillov A.A., "On a problem of I.M. Gel'fand", Soviet. Math. Dokl. 2, 268-269, 1961.

Kruger R., Reinecke D., Smith S. and Ning R., "Reconstruction of blood vessels from x-ray subtraction projections: limited angle geometry", Med. Phys. 14, 940-949, 1987.

Laub G.A. and Kaiser W.A., "MR Angiography with Gradient Motion Refocusing", Journal of Computer Assisted Tomography, 12(3), 377-382, 1988.

Lehman L., "Generalised image combinations in Dual kVp Digital Radiography", Med.Phys. 8-5, 659-667, 1981.

Nashed A.M., "Operator theoretic and computational approaches to ill-posed problems with applications to antenna theory", IEEE Trans. AP-29, 220-231, 1981.

Picard C., Trousset Y., Rougée A. and Saint-Félix D., “Quantitative 2D dual energy imaging: towards 3D selective reconstruction”, Proc SPIE Medical Imaging IV, Newport Beach, 4-9 Feb, 1990.

Rangayyan R., Dhawan A.P. and Gordon R., "Algorithms for limited-view computed tomography: an annotated bibliography and a challenge", App. Opt. 24, 4000-4012, 1985.

Rougée A., Hanson K.M. and Saint-Félix D., "Comparison of 3-D tomographic algorithms for vascular reconstruction", Proc. SPIE 914, 397-405, 1988.

Saint-Félix D., "Problèmes inverses en imagerie: difficultés, méthodologie et algorithmes", Proc. 2ème Atelier Scientifique TIPI, CNRS, XXII.1-22, 1988.

Smith B.D., "Image reconstruction from cone-beam projections: necessary and sufficient conditions and reconstruction methods", IEEE Trans. MI-4, 14-25, 1985.

Trousset Y., Rougée A., Picard C. and Saint-Félix D., "3-D Reconstruction of High Contrast Objects from Limited X-Ray projections", SPIE Vol. 1137, 1989.

Tuy H.K., "An inversion formula for cone-beam reconstruction" SIAM J. Appl. Math. 43, 546-552, 1983.

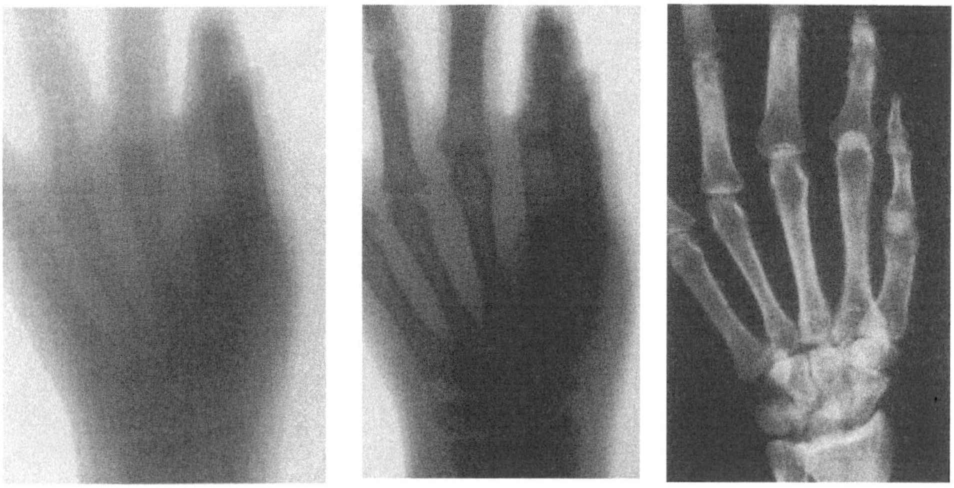

(a) (b) (c)

Fig.**1** - Dual energy substraction result for the hand phantom: raw low energy (**a**) and high energy (**b**) radiography and "bone only" computed projection (**c**)

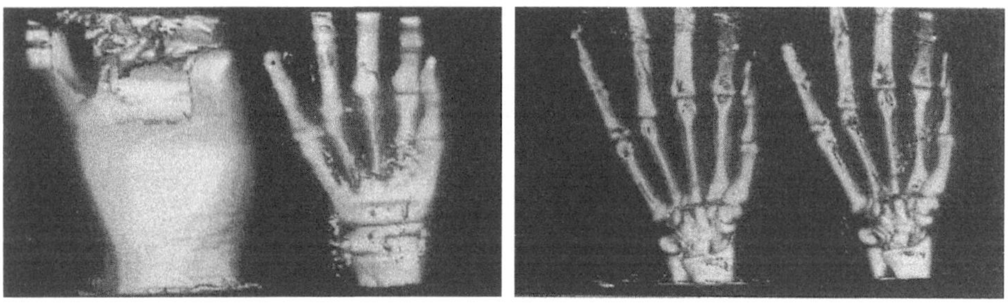

(a) (b) (c) (d)

Fig.**2** - Surface displays of the region of support: (**a**) non selective case, (**b**) selective case and of the reconstructed bone: (**c**) non selective case, (**d**) selective case

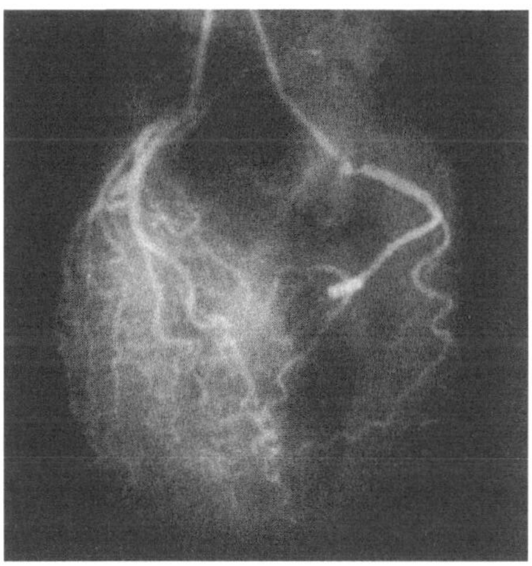

Fig.**3** - Original measured projection (resolution 512) of the cadaver heart

(a) (b)
(c) (d)

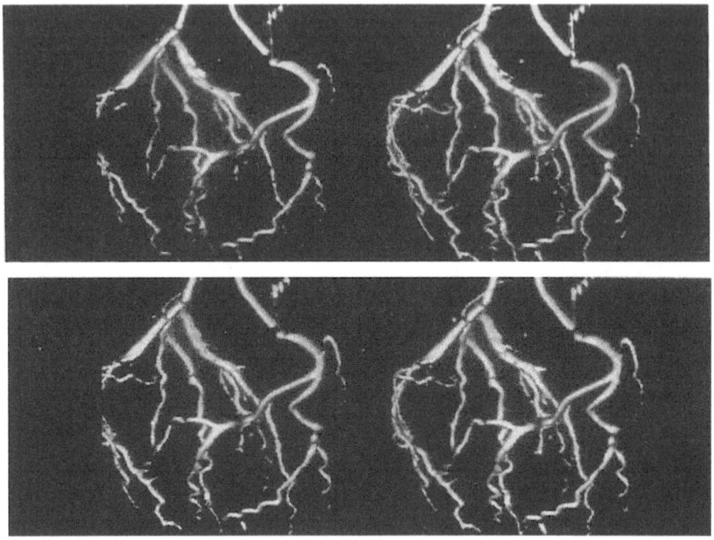

Fig.**4** - Surface display of the reconstructed heart: (**a**) 1 iteration of ART, resolution 256, (**b**) 2 iterations of ART, resolution 256, (**c**) Multiscale method, 1 iteration at resolution 128 followed by 1 iteration at rtesolution 256, (**d**) similar to (**c**) with 2 iterations at resolution 128.

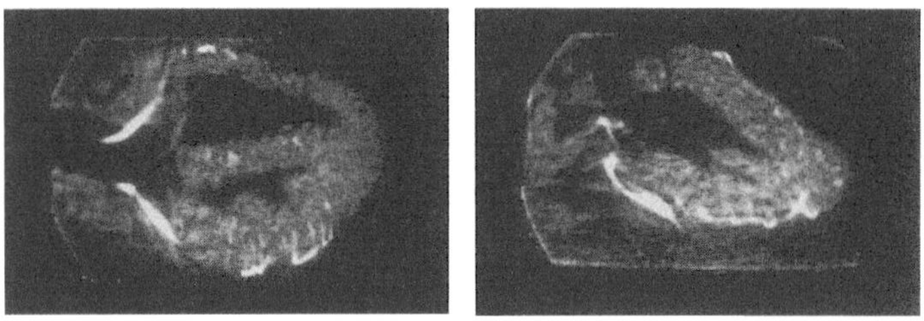

Fig.**5** - Cut planes in the reconstructed cadaver heart

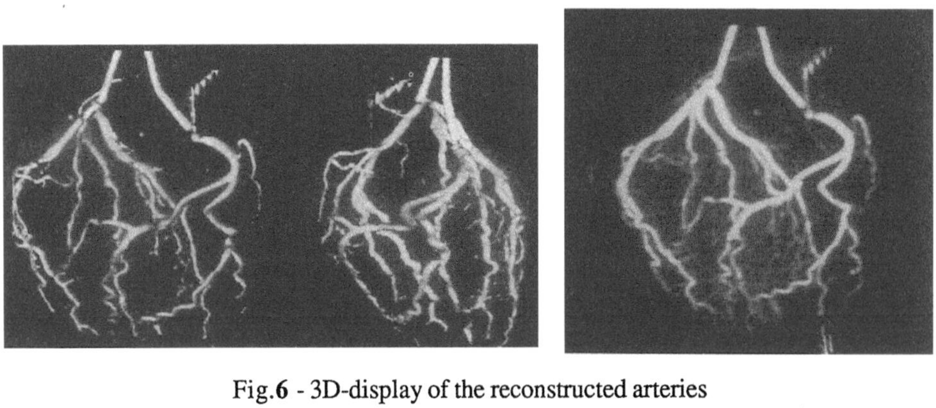

Fig.**6** - 3D-display of the reconstructed arteries
(**a**) Surface display (**b**) Maximum Voxel ray-tracing

Matching Free-form Primitives with 3D Medical Data to Represent Organs and Anatomical Structures

Jean Sequeira and Franck Pinson

IBM Paris Scientific Center
3 et 5 Place Vendôme, 75001 Paris, France

Abstract

Three-dimensional representations of organs and anatomical structures can be derived from sets of two-dimensional data (images) in different ways. Most approaches are based on the use of voxels (to provide digital representations of volumes) or triangles (to sample surfaces that interpolate sets of contours). The approach we have developed is different: it is basically interactive and thus it takes advantage of user's expertise to define specific and highly-structured geometrical models of organs. The first problem is to define free-form primitives as basic components for the creation of three-dimensional models. The second problem is to make these primitives match as well as possible with actual data: an interactive procedure enables the user to roughly position the primitive onto the data; then, an automatic process selects the relevant data and optimizes the shape of the primitive. In this paper, we describe this interactive approach and the problems it raises; then, we focus on the description of an original solution to optimize the shape of free-form primitives through actual data.

Keywords: 3D modeling, data selection, shape matching

1. Introduction

Nowadays, not only researchers in medicine but also physicians and surgeons can be efficiently helped by the use of new imaging techniques such as X-Ray Scanning or Magnetic Resonance. Images of organ and anatomical region cross-sections are computed as sets of values which represent the distribution of their

NATO ASI Series, Vol. F 60
3D Imaging in Medicine
Edited by K. H. Höhne et al.

physical and chemical properties. A set of such images (e.g. parallel cross-sections) can be visualized to study volumetric distribution but a better way would be to structure all this information into a three-dimensional model and display it on a graphics terminal. This geometrical model will then be a support for any further analysis, measurement and simulation of the organ in question.

In order to obtain three-dimensional representations of organs and anatomical regions, there are two possible approaches: the first one consists in directly computing volumetric information from all physical measurements; the second one uses sets of cross-section images computed independently.

The first approach gives a homogeneous model, the precision and validity of which can be given mathematically. However, the definition of functional organ structures (arteries, cavities, etc.) needs to go further in the modeling process. Much research work has been focused on this approach (Peyrin 1987) but, for purely technological reasons (state of the art devices mainly produce two-dimensional images), most recent developments in this area use parallel cross-section images as a support for three-dimensional reconstruction.

Two kinds of approaches for generating organ models from sets of cross-section images are classically used:

- *Voxel-based approaches:* pixels defining areas in parallel cross-sections are associated with volume elements (called voxels) which give a digital representation of volumes,
- *Triangulation approaches:* triangles are used to define a surface which interpolates the set of contours detected in two-dimensional images.

Another approach, based on interactivity, is described in this paper. But, before going into more detail, some general principles on geometrical modeling and display of anatomical structures must be given.

There are many different ways to generate geometrical representations of three-dimensional scenes. In the case of Computer Aided Design, for example, the user creates mechanical parts from a set of surface or volume primitives. In the case of medical imaging, the problem is quite different; geometrical models have to give an accurate representation of actual data which is not homogeneously distributed in space. It is therefore necessary to take this data into account during the processing.

Before processing a geometrical model, the initial volume of data has to be reduced to relevant data only. This is done by going through an "Image Processing" step: images are segmented into sets of contours (or regions associated with these contours). Then follows the modeling step which consists in setting a geometrical three-dimensional structure on the data that comes from the previous step (Image Processing).

Displaying procedures then enable physicians to watch the generated models from any part of space, as if they had the organ (or the anatomical region) they want to study in front of them. These procedures, which depend on the kind of model to be displayed, go through the three following steps (Foley 1982):

- a *geometrical transformation*: three-dimensional models have to be projected on a plane at right angles to the user's viewing direction and associated with the graphic screen,
- a *hidden part removal* process: those parts of the model which are not seen from the viewpoint position have to be removed from the list of visible surfaces,
- a *surface rendering*: the color of each pixel on the screen has to be computed from visible surface information and light source positions.

"Voxel based" and "triangulation" approaches process automatically three-dimensional models of anatomical structures. But this can be a limitation in the case of very complex situations or if we need to represent part of these structures. For these reasons, we have designed an interactive approach which takes external help into account to process the geometrical models.

2. An interactive approach based on the use of free-form primitives

This approach uses interactivity as a basis for the modeling process as in Computer Aided Design. Thus, it takes advantage of the medical scientist's expertise to bring solutions to specific modeling problems. For example, ambiguous situations, such as those met in triangulation approaches, can be easily solved by specialists (physicians, surgeons, etc.). In addition, the process can be oriented by the user so that the result is adapted to specific applications (diagnosis, surgery, etc.).

The user defines a three-dimensional model of organs and anatomical regions by adapting geometrical primitives to that part of the initial data which is relevant to the application. These primitives enable the representation of complex shapes and are characterized by their topological properties.

We first use Image Processing (or "Volume Processing" if we start with a homogeneous set of voxels) to extract relevant points from initial data (e.g. points on contours). We then interactively select those points which are pertinent to our application (e.g. points on arteries if we are studying the artery network). This final set of points will be the geometrical support for modeling and displaying the anatomical structures we want to analyze: from now on it will be the only reference set for all the processing.

This geometrical support is used as a visual support: moving it in three-dimensional space (and displaying it on a graphics terminal) helps the user to understand how it is organized (e.g. how the cavities should be connected) and to decide which kind of structure to set over this data. This is done by picking "free-form primitives" from a geometrical knowledge base and by indicating data to be structured with these primitives. This geometrical support is also used as a mathematical support to optimize the shape of these primitives so that they represent the pertinent points as well as possible. This step occurs automatically each time the user acts (interactively) on the modeling process.

Free-form primitives used in this approach are defined as sets of free-form patches (e.g. cubic B-Splines): relations between these patches define their topological properties (Sequeira 1987). Geometrical transformations on these primitives enable the user to give them the right shape: global transformations (e.g. translations, rotations, similarities) are used to locate these primitives rapidly and ergonomic deformations (curving, twisting, etc.) are used to reach their final shape, accurately and efficiently.

3. Problems to be solved in this approach

The main problem, in the case of automatic approaches, is to structure actual data into relevant geometrical models which take into account the topology of the anatomical regions (e.g. to connect contours located in different cross sections). In this approach, we don't have to solve such kinds of problems because users can provide solutions at any time by means of interactivity. Thus, the main difficulty is to define a powerful ergonomic user interface which makes it easy to identify two- and three-dimensional entities.

Three kinds of problems are specific to this approach and need specific solutions:

1. **The definition, representation and management of free-form primitives**: the set of three-dimensional primitives has to be structured in such a way that users can easily characterize any kind of topological situation (by picking the right primitive and deforming it until it roughly reaches the shape of raw data). In addition, these primitives, which are defined as sets of free-form surfaces, can be used for the representation of volumes and have to describe the functional structure of organs.
2. **The matching process**: once a free-form primitive has been selected and roughly positioned to visually identify a subset of points, we need to extract those data which have been selected and to optimize the shape of the primitive so that it represents the data as well as possible; this step has to be automatic and is called the *matching process*.
3. **The improvement of communication with users**: the system can ask for the help of the user at any time during the modeling process but if the number of interactions is too high this approach becomes inefficient. The introduction of *Artificial Intelligence* elements can reduce this number so that interaction with the user is only required when the system needs an expert's solution to solve a specific problem.

So far, we have provided basic solutions to all these problems. But we need to go further in the development of these solutions, especially for the first and third points mentioned above. We have not yet defined very complex primitives (basic ones are sufficient in many cases) or introduced Artificial Intelligence procedures in the communication interface. Nevertheless, a general robust solution to the matching process has been provided, as described below.

4. Matching free-form primitives with 3D medical data

We are now going to describe the *Matching Process* in more detail. Although an illustration in 3D space will be given at the end of this section, all the discussion will be developed on 2D entities for a better understanding of its basic principles (which apply in R^n space).

The problem is as follows:

- For a set of points and a free-form primitive (e.g. a curve or a surface) on which we can provide local deformations by means of control points (e.g. a B-Spline or a Beta-Spline), we want to select those points located close to the primitive and then to adapt the shape and the complexity of this primitive to the selected subset of points.

This is done in three steps:

1. **data selection**: the distribution of points in space is analyzed and those which are considered as correlated with the primitive are selected,
2. **shape normalization**: the definition of the primitive can be modified (by changing the distribution of its control points) so that its complexity is similar to that of the selected subset,
3. **shape optimization**: control points are moved to give the primitive its final shape (i.e. the appropriate shape to represent the subset of selected points).

4.1. Data selection

Let us consider a set of points and a closed curve in an image. The points located *close to* the curve are selected by analyzing the distribution of distances between all these points and the primitive, then keeping all the points whose distance to the primitive is less than a **threshold distance** D_{th} . The problem is now to determine the value of D_{th} .

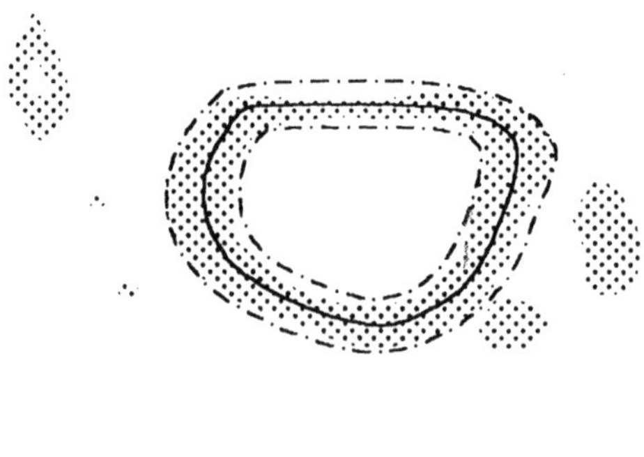

First, all the points which are *too far* from the primitive are removed (e.g. a box containing the primitive is found, another box which is twice as big is defined and just the points lying inside this second box are kept). Then, the distance of each point to the primitive (i.e. the smallest distance between this point and all the points of the primitive) is evaluated and, finally, the **Repartition Function** of these values (i.e. the integral of their histogram) is computed.

If the primitive is positioned correctly (i.e. not far from a subset of points) this function has a typical shape characterized by a breakpoint, as we can see in the following drawing (because most of the points are close to the primitive and the other points are quite uniformly distributed in space).

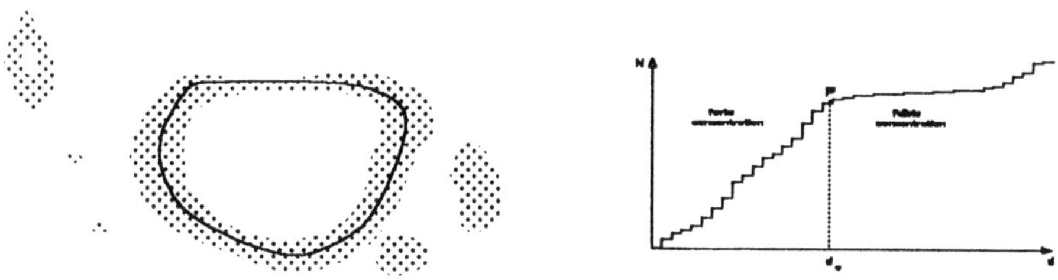

A segmentation of this Repartition Function gives this breakpoint which characterizes the threshold distance. This segmentation is performed by using a classical *Split and Merge* algorithm.

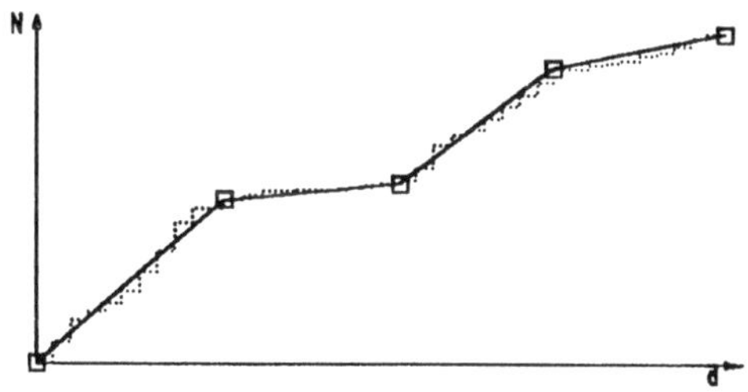

In this paragraph we have described an efficient method to determine which points are correlated to a free-form primitive when this primitive is not too far from the subset of points to be selected. Actually, the problem is a bit different because this hypothesis is not always verified:

- we want to reduce the interaction time to give the users a powerful communication tool (it implies that the users will not shape the primitive onto the data correctly),
- because of interactivity, errors can occur at any time (the system must then check the validity of the designation).

The segmentation of the Repartition Function and, more precisely, the respective positions of its two first segments provide an interesting tool to analyze the validity of designation (Pinson, 1989) and, possibly, to slightly modify the primitive (using a translation, scaling or rotation to position it better) before calling the main procedure (described above).

4.2. Shape normalization

As mentioned earlier, each data point is associated with its projection on the primitive. This set of relations gives a structure to the subset of selected points (e.g. in two-dimensional space, we can consider it as a contour on which we can move).

The normalization step consists in modifying the distribution of control points in order to get a uniform distribution of associated data points (i.e. the number of data points associated with the part of the primitive between two control points is a constant).

Free-form primitives are normalized by inserting and/or removing control points. In the case of uniform cubic B-Splines (many primitives are defined as sets of such surfaces), we can perform a global redistribution by using the associated cubic splines (as illustrated below). First the corresponding nodes are computed, then these nodes are redistributed and, finally, the new set of control points is computed.

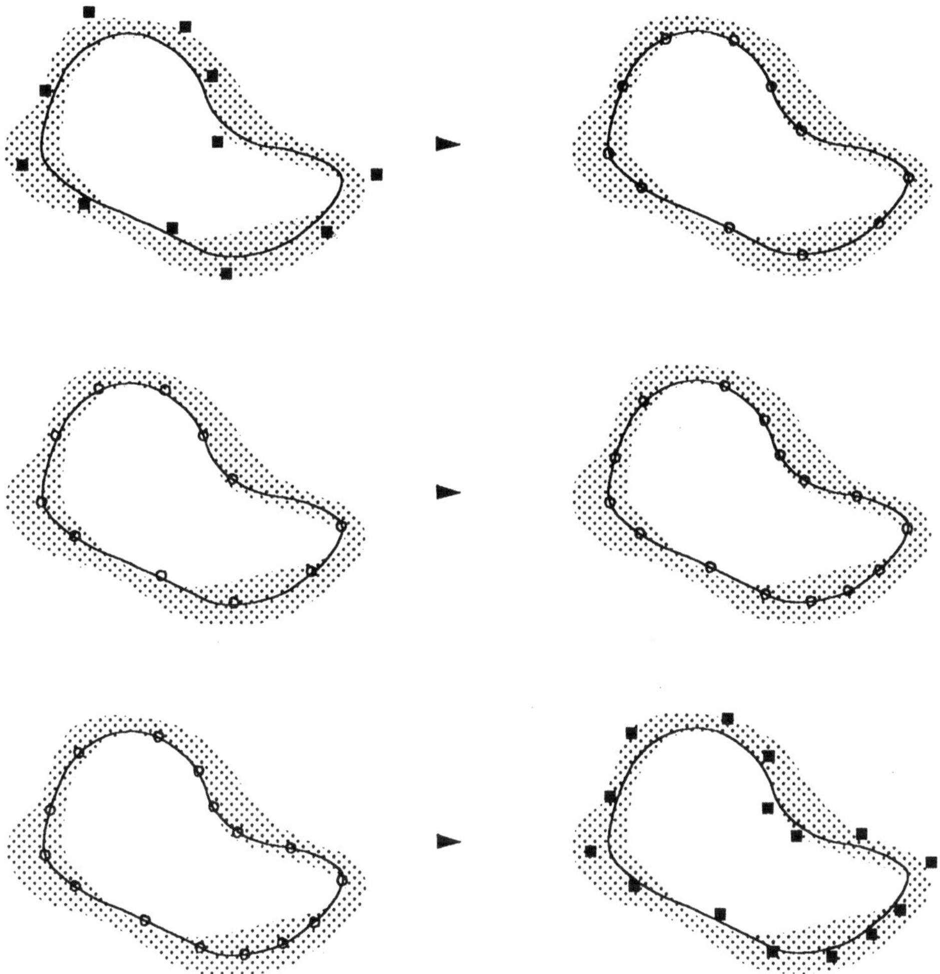

4.3. Shape optimization

The method we have developed to optimize the shape of primitives takes advantage of their local deformation property. This method is based on a progressive adaptation of the primitive to the set of data points, knowing that local deformations will be propagated slightly in their immediate neighborhood (and not at all, further away). An illustration of the process (in the two-dimensional case) is given below.

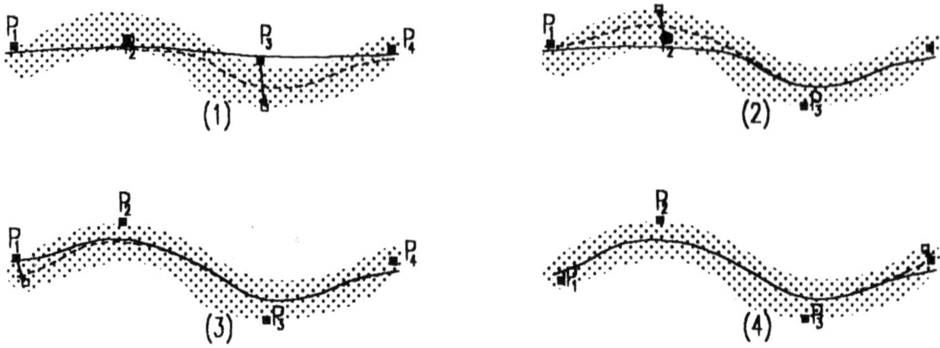

In order to evaluate the quality of the shape optimization, we have to define an error criterion E. Because of the primitive local deformation property we define E as the upper bound of all local criteria E_i :

$$E = \sup_{i=1,\dots,n}\{E_i\}$$

where E_i is the local criterion associated with the control point P_i. E_i is computed as follows (the sum being extended to all points in relation with the deformation support of P_i):

$$E_i = \frac{1}{N_i}\sum_{j=1}^{N_i} d_e^2(M_j, H_j)$$

where N_i is the number of data points taken into account and $d_e(M_j, H_j)$ is the distance between the data point M_j and its projection H_j on the primitive.

These elements are illustrated below (the curve is a uniform cubic B-Spline), the deformation support being drawn in a dotted line.

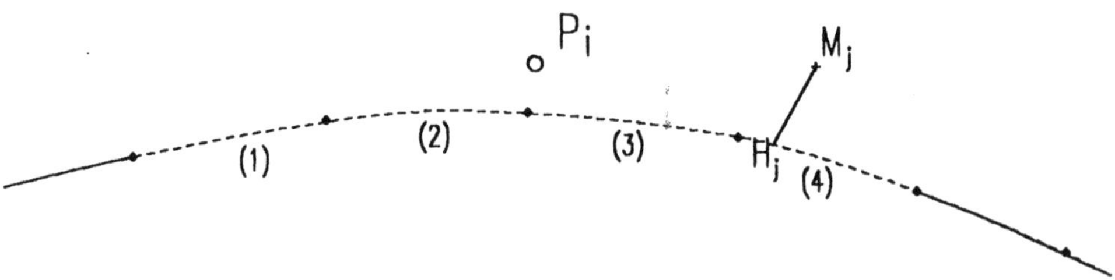

Moving the control point P_i modifies the value of all criteria whose deformation support intersects that of P_i. In fact, the effect of this operation on other local criteria is not very important (this problem has been widely studied in (Pinson, 1989)). Because of this property, we have developed a shape optimization process that consists in a sequence of local optimizations.

We consider the criterion E_i just as a function of the point P_i: then we look for the position of P_i that minimizes E_i. In fact, we look for a solution on the line going through P_i and which is at right angles to the primitive.

Thus, the process is as follows: the potential displacement of each control point (i.e. the displacement which would optimize its local criterion) is evaluated, control points are sorted by decreasing values of such displacements, and the first control point of this sorted list is moved. Then, these values and the sorted list are updated (only a few values have to be updated). This process continues until the first value in the sorted list (i.e. the highest displacement) is lower than a given error.

5. Some results

In this interactive approach, the matching process is used both in two- and three-dimensional spaces. First it enables the user to identify those contours (in images) which are relevant for the application. Then, it is used to obtain a high-level representation of the anatomical structures to be modeled.

Pictures of the next page are an illustration of what is done in three-dimensional space. The problem was to get a model of the renal pelvis and the ureter. In order to do that, the user picked the right primitive (equivalent to a *topological tube*) and positioned it among the data. The last of the pictures below show the primitive when it reaches its final shape.

6. Conclusion

The approach described in this paper provides highly-structured models of human organs. Because the user retains a full control over the process, he can orient it to obtain specialized models for specific applications. Moreover, since it doesn't need important computing means (e.g. only a few points define a model) it can be implemented on a workstation. However, some problems still have to be solved to make this approach really efficient, especially concerning the ergonomics of the interaction.

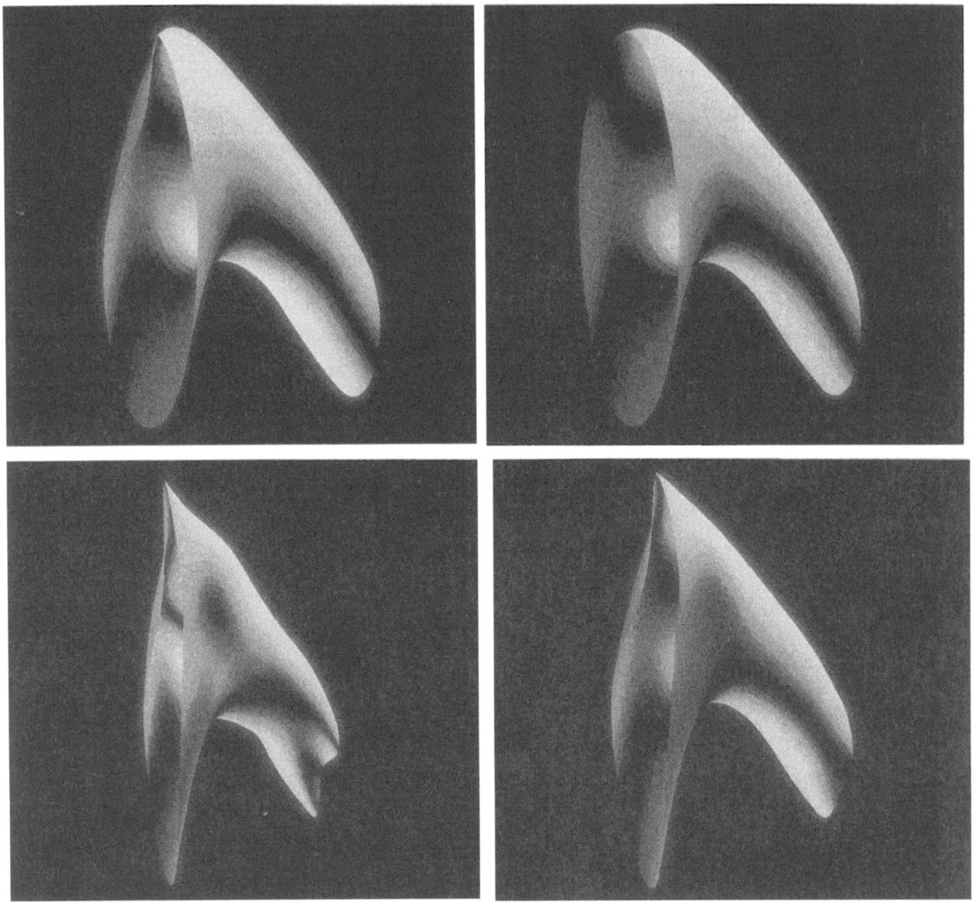

7. References

R.H. Bartels, J.C. Beaty, B.A. Barsky: *An Introduction to Splines for use in Computer Graphics and Geometric Modeling*, Morgan Kaufmann Publishers, 1987.

D. Bechmann: *Définition et Manipulation de Primitives Surfaciques de Forme Libre*, PhD (Université de Strasbourg), June 1989.

J.D. Boissonnat: *Shape Reconstruction from Planar Cross-Sections* Research Report N. 546 of INRIA, July 1986.

Ph. Cinquin: *Application des fonctions-spline au traitement d'images numériques* Thèse de Doctorat d'Etat (Université de Grenoble), September 1987.

P. Craven, G. Wahba: *Smoothing Noisy Data with Spline Functions*, Numerische Mathematik 31, 377-403, 1979.

J. Flifla, R. Collorec, A. Bouliou, J.L. Coatrieux: *Three-Dimensional Positioning from Multimodality Medical Imaging*, Tenth annual conference of IEEE/Engineering in Medicine and Biology Society, New Orleans, November 1988.

J.D. Foley, A. Van Dam: *Fundamentals of Interactive Computer Graphics* Addison-Wesley, 1982.

F. Peyrin, R. Goutte, M. Amiel: *Evaluation of a Truly 3D Reconstruction Algorithm Using 3D Deconvolution* Ninth annual conference of IEEE/Engineering in Medicine and Biology Society, Boston, November 1987.

F. Pinson: *Ajustement de Primitives d'Objets de Forme Libre sur un Ensemble de Données Réelles*, PhD (Université de Compiègne), June 1989.

C.H. Reinsch: *Smoothing by Spline Functions*, Numerische Mathematik 10, 177-183, 1967.

C.H. Reinsch: *Smoothing by Spline Functions II*, Numerische Mathematik 16, 451-454, 1971.

J. Sequeira: *An interactive Approach to Modelling Human Organs*, Ninth annual conference of IEEE/Engineering in Medicine and Biology Society, Boston, November 1987.

J. Sequeira: *Modélisation interactive d'objets de forme complexe à partir de données hétérogènes - Application à la représentation géométrique des organes du corps humain.*, Thèse de Doctorat d'Etat (Université de Besançon), November 1987.

J. Sequeira, P. Conort: *Geometrical Modelling and Display of Kidney Cavities*, Tenth annual conference of IEEE/Engineering in Medicine and Biology Society, New Orleans, November 1988.

J. Sequeira: *Modeling Human Organs with Free-form Surfaces*, Eleventh annual conference of IEEE/Engineering in Medicine and Biology Society, Seattle, November 1989.

Visualization

A Survey of 3D Display Techniques to Render Medical Data

Jean-Louis Coatrieux[1], *Christian Barillot*[2]

INSERM U-335,
[1]Laboratoire Traitement du Signal et de l'Image, Université Rennes I
Campus de Beaulieu, 35042 Rennes cedex, France
[2]Service de Neurochirurgie, Hôpital de Pontchaillou, 35033 Rennes cedex, France

Abstract

This paper gives a brief survey of 3D display techniques which were developped during the last fifteen years to render medical data. These methods are classified in two families *surface rendering* and *volume rendering*. Emphasis is done on the two most recent methodologies which become widely used to render medical volumetric data sets: *Ray-Tracing* and *Octree Encoding*.

Keywords: 3D Display, Volume Rendering, Ray-Tracing, Octree Encoding

1. Introduction

1.1 Aim of 3D Display in Medicine

Almost all physical processes have now been used to image human anatomy from X-ray computed tomography (CT), angiography, and positron emission tomography (PET) to magnetic resonance imaging (MRI). These medical imaging modalities have introduced significant changes in routine patient care, adding diagnostic evidence and expanding therapeutic efficiency. Nevertheless, the mechanisms of medical image understanding are complex. They rely particularly upon the knowledge of organ topologies and the relations they share. In other respects, the multiplicity of pathological processes may induce many structural and functional outlines. The morphology, functional data, and the different stigma of a lesional state are as many elements to integrate in the space. The image presentation modes, based on scan sections or radiographic views, do not completely meet the requirements of diagnostic interpretation, therapeutic decision, or surgical operation. They require the physician to do sequential examinations of images and mental 3D reconstruction.

NATO ASI Series, Vol. F 60
3D Imaging in Medicine
Edited by K. H. Höhne et al.

During the last fifteen years, these reasons have supported all the attempts developped toward 3D viewing, matching closely advances in medical imaging, computer technology, and computer graphics research. This area conforms with the constant concern for better viewing for better understanding and better understanding for better medicine.

From the recent advances on 3D display reviewed in this paper, the challenge of tomorrow is going to be focussed on the use of these methods. An important feature of the medical imaging devices is the heterogeneity which exists between them concerning the nature of the information as well as their managing capabilities (communication and archiving). This separateness restreints the capabilities to numerically and accurately use together these data. Beside Picture Archiving and Communication Systems (PACS), it turns out that 3D imaging must issue solutions to display these data with representation modes and geometrical references independant to the source systems. The second challenge to make use of these methods is to develop effective application layers above them. Dedicated applications have actually more chance to put in practice 3D display methods since they handle better the complementarity existing between the display tools, the interactive tools and the medical practice and knowledge.

1.2 Overview of 3D Display Methods

At the present time, two main different ways to display 3D medical data can be defined. The first one can be called *Surface Rendering* and the second one *Volume rendering*. Surface Rendering was the first used method in 3D medical imaging. Within this family two different kinds of representation can be defined : "*Polygon Oriented*" and "*Voxel Oriented*". These methods have become widely used from the end of the '70's and into the '80's because it became possible to run them on the more widely available faster mainframe computers. Although, for the last few years, the Volume Rendering methods have become more popular with the design of specialized machines in image processing with faster CPU and larger memory.

• *Surface Rendering Methods.* Within this family, the earliest method was the Polygon Oriented method, it appeared along the arrival of triangulation algorithms using planar contours [Keppel 1975], [Fuchs 1977]. These methods have been mostly applied to 3D representation of anatomical structures from parallel slices [Barillot 1988], example is shown on plate 3.

A classical approach consists firstly from the binary sequence of the source data to extract a set of contours representing the surface. The tiling of this surface is done by joining the different points belonging to the contours with triangles in order to optimize a criterion (volume, area of triangles, length of segments, ...) within a planar graph. This method gives a surface data base formed by polygons which can be displayed with standard computer graphics display algorithms. Boissonnat [Boissonnat 1988] has recently proposed another method which is based on the Delaunay triangulation. This new approach carries out a solution to the difficult problem of the 3D connectivity of a set of contours [Barillot 1988].

Using a polygonal approach to represent surfaces, some works have also been done to display 3D structures apart from parallel slices original data set such as display of vessel network from angiograms [Barillot 1985a].

Later, some surface rendering techniques with a voxel oriented approach were developped [Herman 1979] [Artzy 1981] [Barillot 1985b] [Heffernan 1985] [Luo 1986]. These new techniques have improved the accuracy and the reliability of the 3D representation. For these methods, either binary volume or grey scale volume can be used. With both volume the data must be interpolated from the original sequence in order to obtain an homogeneous matrix of voxel along the three main axis. Within the volume (binary or not), a surface tracking algorithm is executed to extract the components of the surface (voxels or face of voxels) in order to build up a surface database. The display of this surface is done by rotating the surface elements along the point of view and processing visibility and shadowing by computing the orientation of the surface at each surface element, with a 3D gradient operator [Barillot 1985b] [Höhne 1986].

• *Volume Rendering Methods.* Volume display techniques have appeared more recently and have the ability to display either surfaces with shading or part of the volume with its grey level values or both together [Meagher 1982] [Schlusselberg 1986] [Höhne 1987] [Trousset 1987] [Herman 1988] [Robb 1989]. The main feature of these techniques is to display data directly from the grey scale volume. The selection of the data which will appear on the screen is done during the projection of the voxels as a function of different attributes such as grey level or gradient values (for surface segmentation), spatial coordinate (for volume cutting), volume set operation (union, intersection, difference of volume) or transparency for instance. The two most representative methods "*Ray Tracing*" and "*Octree Encoding*" are further described in this paper.

2. Ray-Tracing for Volume Rendering of Medical Data

One of the most efficient ways to implement a volume rendering method is to use a Ray-Tracing algorithm. Unlike Octree Encoding, no preprocess needs to be done in order to set up a hierarchical structure on it.

Based on the methods developped in the early 80's [Roth 1982] [Kajiya 1983], ray-tracing applied to medical data involved several processes. First of all, the geometrical transformations align the data along the viewing direction. Then two more processes select the voxels to be displayed, *i)* the spatial segmentation process which performed a 3D clipping according to the actual volume of interest and *ii)* the active segmentation which consists to display surfaces from the raw volume according to a particular function which can be simply related to the voxel grey values met by a ray throughout this volume.

2.1 Geometrical Aspects

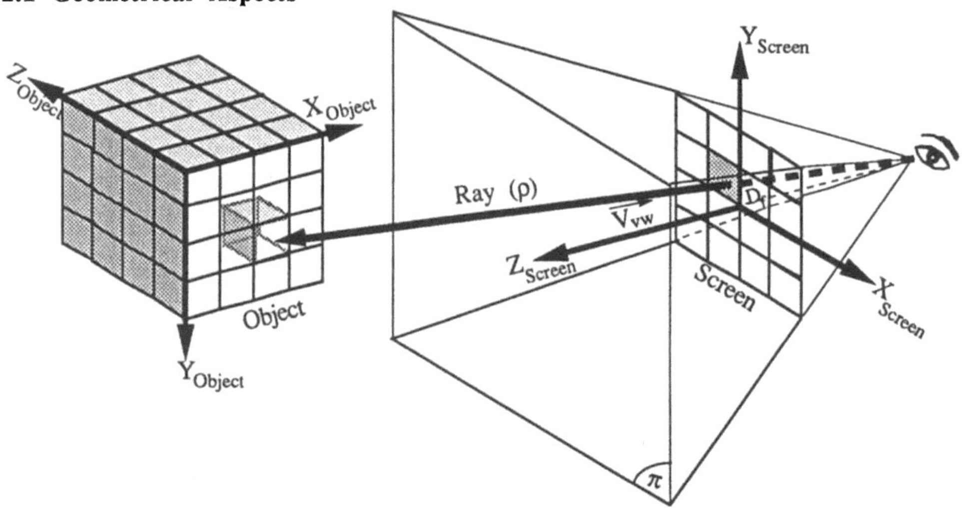

Figure 1:Camera model for ray-tracing with perspective view

• *Camera Model* The most common camera model deals with a source point (called "*Eye*"), a focal point (where the eye is looking) and a matrix of pixels (the *screen*). The object to display (called *scene*) is in front of the camera within a truncated volume called the viewing pyramid (Figure 1). Therefore, from these parameters, a viewing direction is defined by the vector coming from the eye and going to the focus point (V_{VW}).

The position of the screen in front of the camera is defined by the observation vector (V_{VW}) and the focal length D_f. When the focal length is infinite, the view is considered parallel because all rays are parallel to the viewing direction. The screen features (size, orientation, focal length) define a solid angle (π) representing the viewing pyramid.

This general model is applicable not only for a ray-tracing display but also for polygon oriented display. A requirement of a ray-tracing model is to define the geometry of the rays cast through the scene. Therefore, to connect the source point of the scene, for each pixel of the screen, a ray (ρ) is defined as a straight line coming from the source point and passing through the pixel. To compute the picture, the pixel value is assigned according to an intensity measured by the ray throughout the scene. For instance, for shaded surface display, the ray's intensities are measured and stored according to the current lighting model (intensity and light source orientation, reflections, textures, surfaces orientations, ...) where the rays intersects the scene.

As we can see on Figure 1, the most common coordinate system for the camera has the screen in the (x-y) plane, the unclipped scene in the +z space (positive depth), the clipped scene and the source point (i.e. the eye) in the -z space. This system is commonly called screen coordinate system or screen space. It may be convenient to have the viewing vector congruent with the z axis and increment of depth along the positive part of the z axis.

• *Ray definition.* A basic parametric equation of a ray (ρ) coming from the screen is:

$$\rho(t) = \left\{ \begin{array}{l} x_t = x_0 + D_x \cdot t \\ y_t = y_0 + D_y \cdot t \\ z_t = z_0 + D_z \cdot t \end{array} \right\} \qquad (1)$$

So, for each ray, the value (x_0, y_0, z_0) is given by the position of the pixel on the screen and the direction vector (D_x, D_y, D_z) by the vector between the pixel and the eye position. In order to speed up the display process the parallel view is often chosen, hence a ray passing through the pixel $p(x_0,y_0)$ becomes:

$$\rho(x_0, y_0, t) = \left[x_t y_t z_t\right] = \left[x_0 y_0\, 0\right] + \left[0\,0\,1\right] . t = \left[x_0 y_0\, t\right] \qquad (2)$$

• *Geometrical Transformations.* Usually, two kinds of cartesian coordinate systems are used ; the *scene coordinate system* or *scene space* and the *screen coordinate system* or *screen space* (Figure 2).

With regards to the amount of information to be treated, it is more efficient to transform the screen's position (hence the ray's equations) according to the *scene coordinate system* (where the volume is defined), thus a "*screen to scene*" transformation is used.

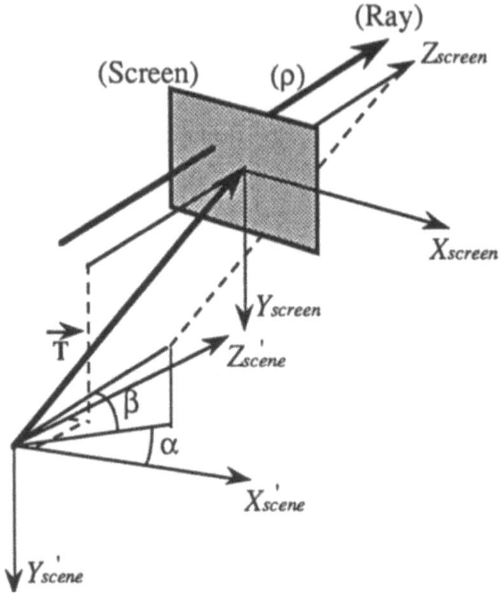

Figure 2: Coordinate systems used for the geometrical transformations

Therefore, after transformation, the equation of a ray becomes:

$$\rho(x_0, y_0, t) = \left[x'_t y'_t z'_t\right] = \left[x_0 y_0\ t\right] \bullet [R] + T \qquad (3)$$

where

$$R = R\left(\alpha/_y, \beta/_x\right) = \begin{bmatrix} r_{11} r_{12} r_{13} \\ r_{21} r_{22} r_{23} \\ r_{31} r_{32} r_{33} \end{bmatrix}$$ is the rotation matrix according to figure 2.

$T = \left[T_x T_y T_z\right]$ is the translation vector according to figure 2.

In order to simplify this equation, a point on the ray can be evaluated as a function of the point before it, on this ray ; the chain coding of the ray can then be expressed at the instant t_i as :

$$\left(\rho(x_0, y_0, t_i) - \rho(x_0, y_0, t_{i-1})\right) = \left\{ \begin{array}{l} x'_{t_i} - x'_{t_{i-1}} = (t_i - t_{i-1}) \bullet r_{31} \\ y'_{t_i} - y'_{t_{i-1}} = (t_i - t_{i-1}) \bullet r_{32} \\ z'_{t_i} - z'_{t_{i-1}} = (t_i - t_{i-1}) \bullet r_{33} \end{array} \right\} \qquad (4)$$

So, now we can have the general equation of the ray at the instant t_i within the scene coordinate system :

$$\rho(x_0, y_0, t) = \begin{cases} \rho(x_0, y_0, t_0) = \left[x'_0 y'_0 z'_0\right] = \left[x_0 y_0 0\right] \bullet [R] + T \\ \rho(x_0, y_0, t_i) = \rho(x_0, y_0, t_{i-1}) + \left[r_{31} r_{32} r_{33}\right] \bullet \Delta t \end{cases} \quad (5)$$

where ($\Delta t = t_i - t_{i-1}$) is the *ray sample.*

Look-up tables can be used to implement these equations [Robb 1988].

2.2 Spatial Segmentation

This stage consists to define the volume of interest (VOI) wherein the rays have to be computed. A very common way to perform this process is to define a bounding box which surrounds the VOI and by projection on the screen selects the set of pixels about which rays will be cast. The size of this box is usually set to the dimensions of the volumetric data set.

Another way to define the volume of interest is to set up a polygonal structure to bound this volume [Robb 1988]. A particular advantage of this method is its ability to define a more complicated shape of the VOI. Hence, during the ray casting process, this polygonal structure is projected first on the screen in order to set up the 3D clipping region selecting the screen's pixels about which rays are cast and to measure the actual length of each ray. Another advantage of this approach is its ability to define cutting planes within the volumetric data set. Indeed, when a polygon or part of it clips the volume, a pixel related to this image region can easily be assigned by the grey level value of the first voxel met by a ray.

This feature figures out a convenient tool to perform a "*digital dissection*" of the data set. This yields to the definition of multiple VOI's from the same volume since a few amount of information (i.e. the polygonal structure) can describe one volume of interest. Hence, these different VOI's can be displayed alone or together according to their own display parameters.

2.3 Active Segmentation and Rendering.

This stage consists to display particular surfaces from the volume of interest. To perform this, one function is assigned to each pixel of the screen, this function is often related to the voxel grey level values met by a ray throughout the volume, it could also be related to particular label values assigned to the volume by a pre-segmentation process.

One of the most spectacular properties of this process is its ability to display a new object (i.e. a new surface) as easily and rapidly as to change the viewpoint. Several approches can be used, they can be classified as reflection oriented display methods (like a photograph) or transmission oriented display methods (like a radiograph).

• *Reflection oriented display.* These methods differed from each others by their actual segmentation process (binary or fuzzy) and their rendering (depth, deph gradient or gradient shading). The most common segmentation process is binary and consists to classify the object of interest by using a binary function of the voxel grey level values (e.g. a plain threshold) which selects the *in-* and the *out-object*. Thus when a ray goes throughout the volume, the active segmentation function returns a *true* value when this ray meets the first voxel with a grey level value above the threshold. Three main qualities of rendering can then be applied :

• Depth shading : A pixel value is only related to the depth of the first voxel along the ray belonging to the VOI according to the current threshold.

• Depth gradient shading : This rendering is just an enhancement of the previous method. A pseudo gradient is computed upon the screen's depth values to estimate the surface orientation which is used by the well-known lighting models (e.g. Gouraud or Phong's models). This appproach was first introduced by Frieder [Frieder 1985].

• Gradient shading : This rendering technique consists to estimate the surface orientation on each voxel detected on the surface by computing the gradient vector related to these voxel neighborhoods [Barillot 1985b] [Luo 1988]. From this normal vector estimation the same well-known lighting models can be used.

• Fuzzy Segmentation : One of the major problems to display a surface from a 3D volume is to decide whether or not a given point (e.g. voxel) belongs to this surface. This binary decision often introduces artifacts, a common error is for instance the presence of holes within the display of a bony surface when the structure is rather composed by a low density bony structure (cartilage or thin bone surrounded by air for instance).

Get rid off this unsteady decision means improve the segmentation process. A solution would use an image processing approach to enhance these weak structures in order to improve the binary decision segmentation. Nevertheless, that kind of technique is probably more useful for a surface rendering method upon which you know exactly the structure according to the user choice without pre-processing or pre-segmentation.

A new approach specifically applied to a volume rendering method [Drebin 1988] [Levoy 1988] consists to assume that the surface to display cannot be simplified as a binary density function. An alternative is to use a so called *fuzzy function* where for a range of grey level values the surface is not present, for another range the surface is partially present and for a third range the surface is fully present. This merging of information is called *volumetric compositing* and is an approximation of the visibility calculations required to render a semi-transparent gel [Blinn 1982] [Duff 1985] (see plates 1a,b).

• Classification : The implementation of this method consists firstly to classified the voxels according to a range of voxel grey level values called *fuzzy range*. This classification assigns to a *fuzzy voxel* (i.e. a voxel belonging to this fuzzy range) a parameter α function of its amount of *true-structure* and *none-structure*. Assuming that at the top of the fuzzy range, a voxel belongs to the true-structure (classification 1) and at the bottom of the same range, it belongs to the none-structure (classification 0) ; a voxel belonging to the fuzzy range will be classified according to its rank within these values (e.g. linear repartition between classification 0 and classification 1). The classification α of the voxel v according to its value(e.g. density) d can be expressed as:

$$\alpha_d^v = \left\{ \begin{array}{cc} 0 & \text{if } P(v \in S) = 0 \\ \alpha_{d_0} \cdot \dfrac{d^v - d_0}{d_1 - d_0} + \alpha_{d_1} \cdot \dfrac{d_1 - d^v}{d_1 - d_0} & \text{if } 0 < P(v \in S) < 1 \\ 1 & \text{if } P(v \in S) = 1 \end{array} \right\} \quad (6)$$

where

d_1 and d_0 are resp. the top and the bottom values of the fuzzy range.

$P(v \in S)$ is the probability for the voxel v to belong to the structure S.

This classification can be improved by adding surface information. Assuming that two voxels with the same structure information (e.g. grey level value) should not have the same influence for the display whether there are close to the border or not, the classification can then be weighted by a surface information (e.g. gradient intensity). The gradient classification coefficient $\alpha^v{}_\nabla$ can be computed by replacing in (6) the grey level information d by the gradient intensity ∇. The combination of both coefficients ($\alpha^v = \alpha^v{}_d \cdot \alpha^v{}_\nabla$) can be used as the classification value assigned to the voxel v.

• Rendering : The merging of the voxel luminescences and opacities is performed using a transparency model. For two surfaces (one semi-transparent and one opaque) the resulting luminescence can be expresssed as :

$$I_{tot} = I_{tsp} \bullet \alpha + (1-\alpha) \bullet I_{opq}$$

where I_{tsp} and I_{opq} are the luminescences of both surfaces and α the opacity coefficient of the transparent surface.

Assuming that a classification coefficient behaves like an opacity value, the *volumetric compositing* is just an extension of this transparency model. This compositing can be done from back to front as well as from front to back. Therefore for a pixel on (x_0,y_0) the resulting compositing C is:

$$C(x_0,y_0) = \sum_{z=1}^{Z} \left[I_z \bullet \alpha_z \bullet \left(\prod_{\zeta=1}^{z-1} (1-\alpha_\zeta) \right) \right] \qquad (7)$$

where v_z are the voxels involved in the compositing along the ray $\rho(x_0,y_0)$, α_z their classification (i.e. opacity) values and I_z their luminescences.

This expresssion can be simplified by looking at the luminescence compositing C before and after a particular voxel v_z :

$$C_z = C_{z+1} \bullet (1-\alpha_z) + I_z \bullet \alpha_z$$

• *Transmission oriented display.* These methods do not involved any surface indentification process. The rays go totally throughout the VOI and the pixel values are computed by using integration functions. Two different display methods are usually proposed: the *brightest pixel display* where the pixel value is the brightest grey level value along the ray (method often used to display MRA data sets) and the *weighted summation display* where the pixel value is an average of the voxel grey level values along the ray.

3. Octree Representation

Many algorithms for octrees have been proposed. They are issued from and applied to a large number of engineering fields. Their interest comes from the ability of hierarchical structure to represent and manipulate uniform data regions as single entities. Easy and fast access to the data is so-provided together with a significant reduction of memory space [Reddy 1978] [Hunter 1978] [Jackins 1980] [Maegher 1980]. Volumetric operations as required in medical imaging are quite simple to design and the full use of recursive structure with only arithmetic computations leads to efficient implementation on special purpose hardware.

3.1 Basics

The basic concept consists to model a cubical volume of $2^n x 2^n x 2^n$ voxels through a treelike structure. The octree representation of a data base is obtained by a recursive subdivision process where the entire volume is decomposed into octants, these octants being divided themselves into suboctants

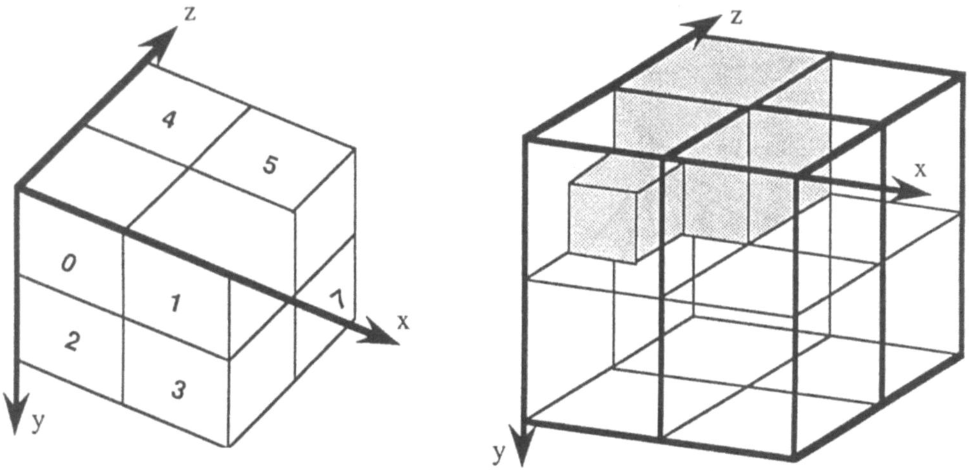

Figure 3:Labels and binary object

until some a priori defined properties are verified or until the resolution limit is reached. The octants, represented by nodes within the tree, are assigned labels from 0 to 7 (Figure 3). The root node depicts the full space array. The resulting octal tree is specified by intermediate nodes and leaf nodes containing respectively heterogeneous and homogeneous cubes according to the property value. A binary object is shown in Figure 3 where the node types are defined by Empty (outside the object), Full (inside the object) and Partial (node to be further reduced). The decomposition process is determined by the properties of the data within a cube. For multilevel database, such as CT Scanner, the criterion can make use of density values. In surface representation based on polygons, it will be related to the presence of planes, edges or vertices within the region [Ayala 1985].

The information to be stored in the nodes has also to be defined. It fully depends on the application purposes : density values, surface normal vectors, links to the neighbours etc. This choice requires to find a good trade off between the amount of data stored (to avoid multiple processes on the same data) and the octree size. If we decide to store as less information as possible, the tree size will be minimum but more processing will be required (computation of gradients and neighbourhood search...) and computation times will increase. Otherwise, huge octree size leads to the non-optimal utilization of the computer memory resources (paging faults...).

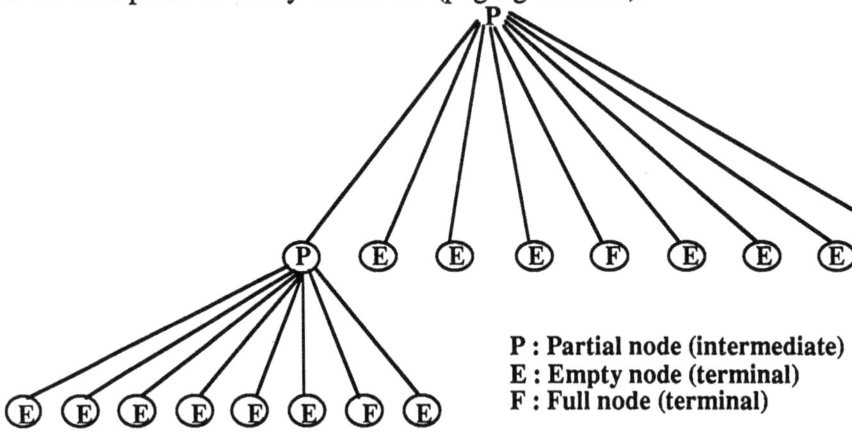

Figure 4:Octree encoding

3.2 Encoding and Conversion

Once the tree and the data associated to the nodes have been defined, a computer representation must be chosen : two main approaches exist using a tree or a list representation. A pointer-based approach is the most natural ; it is based on a recursive definition of the tree. Each node is composed by a p-fields structure, eight of them contain pointers to the child-structures. The other fields are used to encode information concerning the father and the properties associated to the node. Two methods improving the encoding have been proposed : the first one consists to reduce the number of node-structures by coding several nodes within the same data structure (8 nodes [Jannin 1987], 64 nodes [Meagher 1987]. It is then possible to not represent Empty nodes (data outside of the chosen density window). The tree size becomes smaller but the node sizes become variable according to the amount of Empty children in a node. The other way is based on reducing the pointer size by using a specific memory allocation (relative addressing).

Regardless of these improvements, pointers are very memory consuming, that explain why solutions without pointers have been carried out. Coded trees

[Cohen 1985], [Sandor 1985] are made by a list of all tree nodes arranged according to a given tree traversal mode. Linear octrees [Gargantini 1982] take only into account Full nodes. Each node is represented by a code called "location key" allowing its localization within the tree. This second category of methods induces significant reduction of the tree size by keeping only the necessary information but an access to a given node requires travelling all through the previous nodes in the list. Furthermore, this representation barely allows tracking objects throughout the tree. A classical decomposition mode in eight octants has been chosen due to algorithmic and implementation simplicity [Jannin 1988]. Because the ability to display the "inside" of an object was considered so important with regards to the application purposes, it was decided to store density values in terminal nodes. Surface normal vectors have also been associated to these nodes because they are necessary to provide a realistic and accurate rendering of surfaces. Furthermore the computation of gradient vectors are very time consuming (search of neighbours, calculation of the gradient itself) so it is performed only once during the octree construction process for each node belonging to the object surface. A pointer strategy was also carried out, allowing an optimal use of the octree recursive and hierarchical properties : easy travelling through the nodes, fast and easy (random) access to the nodes, well-suited to an implementation on a dedicated hardware (parallel and multiprocessor architecture). The chosen structure called *"One pointer by intermediate node"* stores information on terminal children with these nodes. Therefore, the bottom level of the tree is destroyed.

The proposed node structure is composed of nine fields ; a one byte field contains information characterizing the node itself (belonging to the surface object, type of node...), and eight fields containing pointers to the eight child-nodes, if they are partial. Either all information concerning the child nodes or part of it can be stored at the location of a child pointer, according to the child properties.

A node may have one of the following types :

-*"Partial"* : it defines a non homogeneous octant belonging (at least partially) to the object. A pointer to its node structure is stored in the ith field of the parent node (for a child node "i").

-*"Empty"* : the node represents a cube which is outside the object according to some segmentation criteria. A null value (filling characters) is stored in the ith field of the parent node.

- *"Full"* : these node types definie the voxels belonging to the surface, internal visible or non visible voxels, internal visible or non visible cubes. According to these types, the information attached to the nodes are density values and/or components of the surface normal vectors.

Other improvements have been achieved : the normal vector encoded on two bytes allows the storage of the density on every nodes. For each node, maximum and minimum values of the current region can be stored, hence speeding up the octree segmentation process [Amans 1986] [Antoine 1988].

The rules for building octrees depend on the formats of their input data such as object models (coming from constructive solid geometry or boundary representation) or silhouette images (reconstruction of an object from its projections) and so on [Chen 1988]. We are dealing in medical imaging with parallel serial sections.Construction algorithms have to be suited to the limited computer memory space : a slice by slice approach allows a progressive construction of the octree. A quadtree construction of each plane section and a progressive insertion of them within the octree can be used [Yau 1983]. Our octree conversion algorithm provides a direct construction : sets of four consecutive slices by using arrays where structure's addresses, which represent the nodes, are buffered until the eight brothers at a given level have been obtained. Among these four sections, the middle ones contain the voxels to be actually encoded and the lateral ones are just used to compute the local gradient involved by the segmentation process and the surface normal vector computation [Luo 1988]. Because an octree represents a cubic volume ($2^n*2^n*2^n$) and the number of slices is generally lower than the number of pixels within a plane section, an automatic filling procedure has to be carried out to generate the Empty nodes corresponding to the missing slices.

3.3 Display and Manipulation

The overall implemented scheme of the octree-based encoding makes use of a linear description to reduce the storage requirements. Conversely a pointer based representation is handled when rendering and analysis are performed. The display operations involve geometrical transformations to place the observer with regard to the object , hidden surface removal and surface shading computation. One solution consists to translate and rotate the octree in order to make the viewing direction (joining the observer's point of view and the point aimed in the scene) perpendicular to one of the new octree generic planes. The octree must be recomputed, i.e. the positions of all nodes even those which will not be visible [Meagher 1982]. Another way is to move the viewer around the scene and to project the object onto the screen. Finding the front Full octants that are visible can be carried out through a specific sequence to visit the octree nodes. It is well known that on the average, the front to back sequence is more efficient because once a part of the image plane has been painted, any other nodes (subtrees) projected onto it can be ignored. The image plane can be represented by a quadtree so that the projection is obtained by examining the relative positions of a projected octant and the quadrants in the image plane [Meagher 1987]. The bounding box (Figure 5) defines the smallest rectangle parallel to the axes of the screen which encloses the projection of an octant. The length of the sides of the bounding box are half the sizes of the bounding box of its parent node. The geometrical coordinates (dx, lx, dy, ly) of the bounding box are defined in an overlay space. This overlay is formed by four adjacent quadrants (fi) of equal size that cover up the projection of the considered octant. The state of the overlay is then read. If the four quadrants have been already painted, the procedure stops. If none or partial painting has been performed, the subdivision continues by looking for the child-overlay composed of four nodes selected between the 16 nodes of the parent overlay. These nodes are determined through the computation of the location of the left-higher corner of a child bounding box. The whole algorithm includes the special processing of the terminal nodes of the octree.

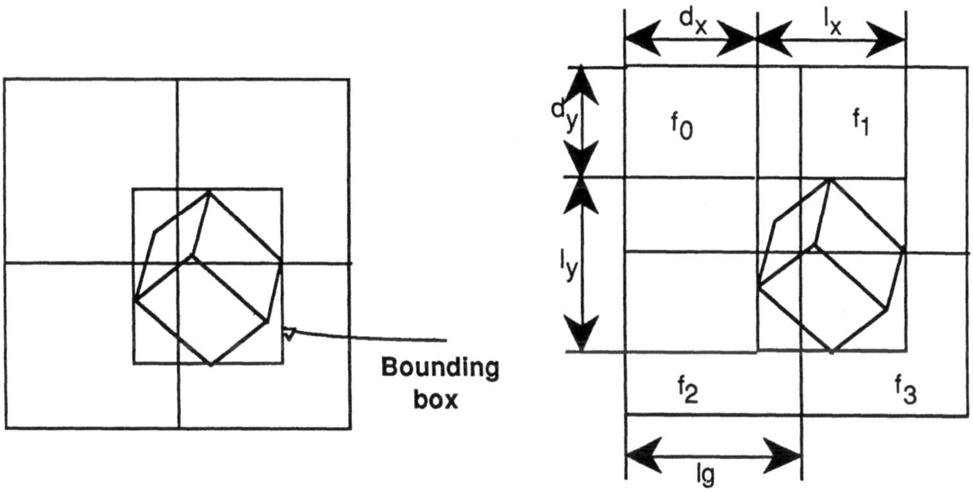

Figure 5:Overlay technique.

Efficient capabilities to extract and analyze the information of clinical relevance can be easily implemented. Set operations (union, intersection, difference) between two objects are performed by a parallel tree traversal on both octrees by making simultaneous comparisons between nodes.

Cutting plane requires the comparison of all vertices of the cubes corresponding to the octree nodes with the cutting plane. The nodes cut by the plane are decomposed and the process follows to their children. If a terminal Full node is cut by the plane, its virtual coordinates are used to continue the process. All these operations are executed node by node (region by region) and not voxel by voxel. The octree construction induces the loss of the neighbourhood notion in this representation. To find it again and to develop operations using the examination of the neighbourhood (segmentation, search for connected components...), some simple algorithms researching the neighbourhood of a node within the tree along a direction have been developped. They are based on the research of the first common ancestor of both nodes and a going down on the tree with the way obtained by the mirror of the way used to climb up the tree. Volume measurements can also be obtained by computing the number of nodes or the size of the represented cube, surface ones by the number of sides of cubes and their size. The density, mean value and centre of gravity measurements can be computed by using the density stored in the tree.

4. Discussion

Most research efforts have been directed in the last two decades towards designing efficient methodologies for 3D display. Rather than attempting to survey all the techniques, we have focussed our own works on volume rendering. However, some pointers to the recent litterature have been provided which emphasize the variety of approaches [Fuchs 1989] [Chen 1989] [Pizer 1989] [Coatrieux 1989]. This juxtaposition clearly evidences that there is not an unique way to deal with 3D medical imaging. Even if volume rendering can be considered as a more general display modality than surface rendering, the main issue is to define the most appropriate method according to the data acquisition modes and the medical relevance. For example, surface representation may remain of interest for shape conformation required in atlas matching. The fewer number of manipulated elements, the capability to introduce mechanical and mathematical properties during modelling yield

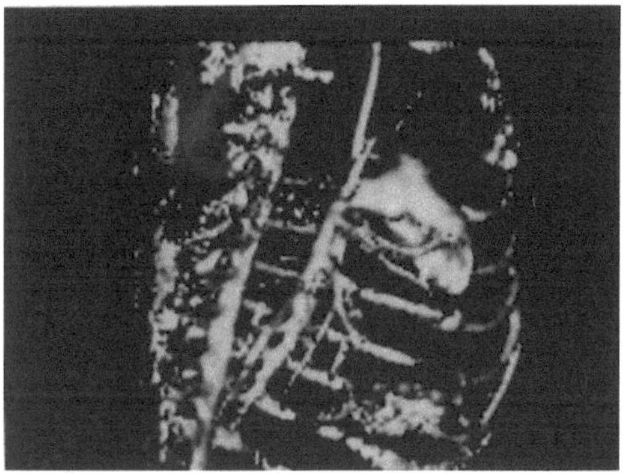

Plate 1a

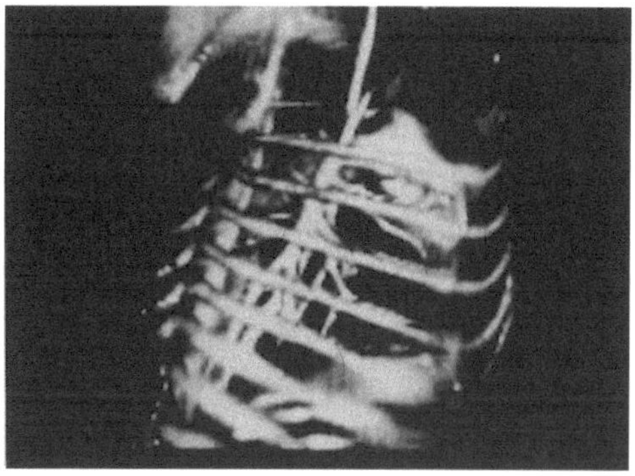

Plate 1b

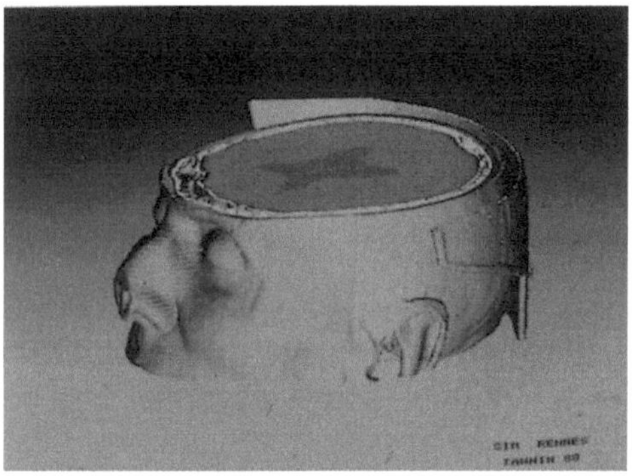

Plate 2

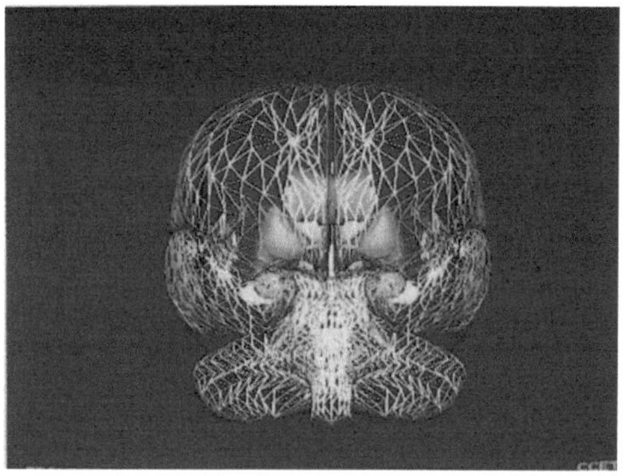

Plate 3

Plates 1a,b: Volume Rendering of a dog chest with a *Ray-Tracing* algorithm. Emphasis of the volumetric compositing approach to display the same structure (binary segmentation on (a) and fuzzy segmentation on (b)). The data base was provided by the BRU dept of Mayo Clinic.

Plate 2: Volume Rendering of a human head with an *Octree* algorithm. Data base acquired from CT with a 512x512x200 resolution.

Plate 3: Surface rendering of a brain model by using polygons. 3D modeling of intra-cerebral structures from planar contours provided by a stereotactic atlas, the 3D data base was reconstructed by using a triangulation algorithm [Barillot 1988].

attractive solutions for fitting procedures and faster production of target objects [Sequeira 1987]. Even if a manual segmentation has to be achieved (it is also a recurrent problem in volume rendering), the full interactivity that can be expected by using the current technology is an important advantage. So, open problems in surface rendering must be addressed among which the ability to define the optimal distribution of points on surfaces and the correct reconstruction of complex objects from restructured point sets (see [Boissonnat 1988] for more details).

Volume rendering techniques must also be improved to enhance the effectiveness of the display by a proper selection of fuzzy classification values (or appropriate choice of "colors" and "opacities" in [Levoy 1988] [Drebin 1988]. Most often, software packages have been proposed which severely increase the time lag required to produce an image and to modify the objects. Parallel processing hardware [Goldwasser 1987] [Kaufman1988] and VLSI design [O'Donnell 1988] have been considered . The octree encoding is particularly suited for parallel computing. Two different architectures have been recently reported [Toumoulin 1989] and a hypercube array based on transputers [Jannin 1990].

Another field of research can be explored through the merging of techniques. It covers more or less complex situations. Octree encoding can be used for example to determine a precise bounding box of the 3D object and so providing faster travelling along the ray path. Combined displays of geometric objects with slices derived data bases as well as superimposition of computed data with morphological ones are required in a number of applications such as surgery simulations or radiotherapy planning.

These comments point out the dual efforts to be carried out in 3D medical imaging. On one side, new methodological areas have to be addressed through the optimization of the most computer intensive steps while enhancing the efficiency of rendering. In addition, the clinical assessment of 3D representation will be widely accepted through the avaibility of user friendly applications.

5. Conclusion

This short survey on 3D visualization techniques reveals that quite encouraging results have been provided in the last two decades. Emphasis has been given to volume rendering techniques, i.e. ray tracing and octree encoding, however surface representations are also of relevance. The future of 3D medical imaging depends on our ability to handle and to improve the overall stages involved in the patient management. They include the image acquisition, the processing ressources, the combination of data and methods as well as the development of pre- and post-operative planning tools.

6. Acknowledgements

The authors are indebted to Luo Li Min, Jannin Pierre, Gibaud Bernard, Le Noach Richard and Toumoulin Christine for their valuable contributions to this research activity. They are grateful to Professor Scarabin J.M. for his early vision of the medical interest of three dimensional imaging techniques. Special recognition is also given to Dr. Richard A. Robb for its support regarding to the development of the ray-tracing algorithm.

7. References

Amans J.L., Antoine M., Darier P.: Processing and display of three dimensional arrays of numerical data using octree encoding. Proc. 2nd Intern. Image Symp, Nice, pp. 823-830, (April 1986)

Antoine M. : Les techniques algorithmiques du codage octree appliquées à l'analyse des volumes tridimensionnels de données numériques. PHD Thesis, Université de Grenoble, (1988)

Artzy E., Frieder G., Herman G.T. : The Theory, Design, Implementation and Evaluation of a Three-Dimensional Surface Detection Algorithm. Comp. Graph. & Im. Proc., Vol.15, pp.1-24, (1981)

Ayala D., Brunet P., Juan R., Navazo I. : Object representation by means of nonminimal division quadtrees and octrees. A.C.M. Trans on Graphics, 4, 1, pp. 41-59, (1985)

Barillot C., Gibaud B., Scarabin J.M., Coatrieux J.L.: Three-Dimensional Reconstruction of Cerebral Blood Vessels. IEEE Computer Graphics & Applications, Vol.5(12), pp.13-19, (1985 a)

Barillot C., Gibaud B., Luo L.M., Scarabin J.M. : 3-D Representation of Anatomic Structures from CT Examinations. Proc. SPIE Biostereometrics'85, Vol.302, pp.307-314, Cannes (1985 b)

Barillot C., Gibaud B. et al. : Computer Graphics in Medicine: A Survey. CRC Critical Reviews in Biom. Eng., Vol.15(4), pp.269-307, (1988)

Blinn J.F. : Light Reflection Function for Simulation of Clouds and Dusty Surfaces. Computer Graphics, Vol.16(3), pp.21-29, (1982)

Boissonnat, J.D. : Shape reconstruction from planar cross-sections. Computer Graphics & Image Proc. 44, 1-29, (1988)

Chen, H.H., Huang. T.S. : A survey of construction and manipulation of octrees. Comput. Vis Graph Image Proc, 43, 409-431, (1988)

Chen, L.S., Sontag, M.A. : Representation, display and manipulation of 3D digital scenes and their medical applications, Comput Vis Graph Image Proc, 48, 190-216, (1989)

Coatrieux, J.L., Jannin, P., Toumoulin, C., Lenoach, R. : Representation tridimensionnelle en imagerie médicale : bases théoriques, Journées Francophones d'Informatique Médicale, Nîmes, 19-21, (mars 1990), to appear.

Cohen, Y., Landy, M.S., Pavel, M. : Hierarchical coding of binary images. IEEE Trans Pat Anal Mach Intel, 7, 3, 284-298, (1985)

Drebin R.A., Carpenter L., Hanrahan P. : Volume Rendering. Computer Graphics, Vol. 22(4), pp.65-74,.(1988)

Duff T. : Compositing 3-D, Rendered Images. Computer Graphics, Vol.19(3), 41-44, (1985)
Frieder G., Gordon D., Reynolds R. : Back-To-Front Display of Voxel-based Objects. IEEE Comp. Graph. & Appl., Vol.18(3), pp.52-60, (1985)
Fuchs H., Kedem Z.M., Uselton S.P. : Optimal Surface Reconstruction from Planar Contours. Comm of the ACM, Vol.20, pp.693-702, (1977)
Fuchs, H., Levoy, M., Pizer, S.M. : Interactive visualization of 3D medical data. Computer 46-51 (1989)
Gargantini, I. : Linear octrees for fast processing of three dimensional objects, Comput Graph Image Proc, 20, 356-374, (1982)
Goldwasser, S.M., Reynolds, R.A. : Real-time display and manipulation of 3D medical objects : the voxel processor architecture, Comput Vis Graph Image Proc, 39, 1-27, (1987)
Heffernan P.B.., Robb R.A. : A New Method for Shaded Surface Display of Biological and Medical Images. IEEE Transactions on Medical Imaging, Vol.MI-4, pp.26-38,(1985)
Herman G.T., Liu H.K. 3-D Display of Human Organs from Computed Tomograms. Comp. Graph. & Im. Proc., Vol.9(1), (1979)
Herman G.T. : Three-dimensional imaging on a CT or MR scanner. J. Computer Assist. Tomogr. 12, 450-458 (1988)
Höhne K.H., Bernstein R. : Shading 3D Images from CT Using Grey-Level Gradients. IEEE Trans. on Medical Imaging, Vol.5(1), pp.45-47,(1986)
Höhne K.H., Riemer M., Tiede U. : Viewing Operation for 3-D Tomographic Grey Level Data. Proc. CAR'87, pp.599-609, Springer-Verlag Ed., (1987)
Hunter, G.M. : Efficient computation and data structures for graphics, PHD Thesis, Dept Elect Comput Science, Princeton University, Princeton, NJ, (1978)
Jackins, C., Tanimoto, S.L. : Oct-trees and their use in presenting three dimensional objects, Comput Graph Image Proc, 14, 249-270, (1980)
Jannin, P., Gibaud, B., Barillot, C., Scarabin, J.M., Bouliou, A. : Octree encoding with volume and surface information in medical imaging. In: Proc. 9th IEEE EMBS Conference, Boston 1987
Jannin, P. : Modélisation, visualisation et manipulation de données volumiques par arbre octal en imagerie médicale, PHD Thesis, Université de Rennes I, juillet 1988
Jannin, P., Prévost, G., Le Noach, R., Coatrieux, J.L. : Conception d'architectures spécialisées pour des stations de travail interactives à base d'octree en imagerie médicale, Actes Forum des Jeunes Chercheurs, Paris, (mai 1990)
Kajiya J.T. : Ray Tracing. Tutorial Notes, ACM SIGGRAPH'83, (1983)
Kaufman, A., Bakalash, R. : Memory and processing architecture for 3D voxel based imagery, IEEE Computer Graphics, 10-23, (1988)
Kehtarnavaz, N., Simar, L.R.,De Figuereido, R.J. : A syntactic/semantic technique for surface reconstruction from cross sectional contours. Comp. Graph. & Im. 42, 399-409 (1988)
Keppel E. : Approximating Complex Surfaces by Triangulation of Contours Lines. IBM J. Res. Develop. 19, pp.2-11, (1975)
Levoy M. : Display of Surfaces From Volume Data. IEEE Comp. Graph. & Appl., pp.29-37,(1988)
Luo, L.M. : Reconstruction tridimensionnelle en imagerie médicale à partir de coupes parallèles. Application en imagerie scanner X et RMN, PHD Thesis, Université de Rennes I, 1986
Luo, L.M., Coatrieux, J.L. : Surface normal for 3D object display in cuberille environment, Proc 10 th IEEE EMBS Conference, New Orleans, 420-421, nov. 1988
Meagher, D. : Octree encoding : a new technique for the representation manipulation and display of arbitrary 3D objects, IPL-TR-80-111, Rensselaer Polytechnic Institute, Troy, NY, Oct 1980
Meagher D. : Geometric Modeling Using Octree Encoding. Comp. Graph. & Im. Proc., Vol.19(2), 129-147, (1982)
Meagher, D. : The manipulation analysis and display of 3D medical objects using octree encoding techniques, Special Issue on Computer Graphics, J.M. Scarabin, J.L. Coatrieux Eds, Innov Technol Biol Med, 8, 1, 23-36, (1987)

O'Donnell, M. : Applications of VLSI circuits to medical imaging, Proceedings IEEE, 76, 1106-1114, (1988)
Pizer, S.M., Fuchs, H., Levoy, M., Rosenman, J.G., Davis, R.E., Renner, J.B.: 3D display with minimal predefinition. Proc Computer Assisted Radiology 723-736 (1989)
Reddy, D.R., Rubin, S. : Representation of three-dimensional objects, CMU-CS-78-113, Comput Sci Dept, Carnegie-Mellon University, Pittsburgh, April 1978
Robb R.A., Barillot C. : Interactive 3-D Image Display and Analysis. Proc. of SPIE 18th Conf. on Hybrid Image and Signal Processing, pp.173-202, Orlando 1988
Robb R.A., Barillot C. : Interactive Display and Analysis of 3-D Medical Images. IEEE Trans. on Medical Imaging, Vol.8(3), pp.217-226, (1989)
Roth S.D. : Ray Casting for Solid Modeling. Computer Graphics and Image Processing, Vol. 18(2), pp.109-144,.(1982)
Sandor, J. : Octree data structures and perspectives imagery, Computer and Graphics, 9, 4, 393-405, (1985)
Schlusselberg D.S., Smith W.K., Woodward D.J. : Three-Dimensional Display of Medical Image Volumes. Proc. of 7th NCGA, Vol.III, pp.114-123, Anaheim 1986
Sequeira, J. : Modélisation interactive d'objets de forme complexe à partir de données hétérogènes. Application à la représentation géométrique des organes du corps humain, PHD Thesis, Université de Besançon, 1987
Toumoulin, C., Prevost, G., Knoplioch, J. : Interactive three dimensional image workstation, Proc 11th IEEE EMBS Conference, Seattle, 354-535, Nov 1989
Trousset Y., Schmitt F. : Active Ray-Tracing for 3D Medical Imaging. In: Proc. Eurographics'87, pp.139-149, Amsterdam, 1987
Yau, M.M., Srihari, S.N. : A hierarchical data structure for multidimensional digital images, Comm A.C.M., 26, 7, 504-515, (1983)

Rendering Tomographic Volume Data: Adequacy of Methods for Different Modalities and Organs

Karl Heinz Höhne, Michael Bomans, Andreas Pommert, Martin Riemer, Ulf Tiede, Gunnar Wiebecke

Institute of Mathematics and Computer Science in Medicine (IMDM)
University Hospital Eppendorf, University of Hamburg
Martinistraße 52, 2000 Hamburg 20, F.R.G.

Abstract

A large variety of different algorithms for rendering radiological volume data has been described in the past. It has turned out that not every method is equally well suited for all kinds of modalities and organs. We therefore compare algorithms for different applications. The algorithms include Z-buffer gradient shading, gray level gradient shading, transparent gray level gradient shading, and two mutations of the marching cubes algorithm. The object classes are bone (CT), head (MRI), heart (MRI), vessels (MRI), ultrasound, and isodoses in radiotherapy. Properties of the algorithms, such as the achieved realism, the flexibility, and the computational cost are discussed. As a general result we found that the best visualization is achieved when different algorithms appropriate to the objects can be applied within one image.

Keywords: 3D-imaging, volume rendering, gray level gradient shading, marching cubes algorithm, generalized voxel model, computer tomography, magnetic resonance imaging, ultrasound, radiotherapy planning

1. Introduction

Three-dimensional (3D) visualization techniques are increasingly considered useful for the interpretation of medical volume data, such as those obtained by X-ray computed tomography (CT) or magnetic resonance imaging (MRI). The algorithms described for 3D image presentation are based on a large variety of principles: opaque or transparent rendering,

NATO ASI Series, Vol. F 60
3D Imaging in Medicine
Edited by K. H. Höhne et al.

binary or fuzzy surface segmentation, voxel or polygon data structures, surface normal calculation from depth buffer, gray level gradients or polygon inclinations, and use of different illumination models [Goldwasser et al. 1985, Barillot et al. 1985, Höhne + Bernstein 1986, Höhne et al. 1987a, Höhne et al. 1987b, Lorensen + Cline 1987, Levoy 1988, Drebin et al. 1988, Baker 1989, Levoy 1989, Höhne et al. 1990, Kaufman et al. 1990]. It is thus not surprising that the visualization of the same data using different techniques might result in different images, which might even differ in their diagnostic interpretation. For clinical applications, it is of course important to be sure of the fidelity of a 3D image, or at least to know about its limitations. Even though this is a very important issue, only a few systematic investigations have been published to date [Tiede et al. 1987, Magnusson et al. 1988, Pommert et al. 1989, Pommert et al. 1990a]. Most recently this issue has been addressed in more detail by the authors using simulation studies [Pommert et al. 1990b, Tiede et al. 1990].

Unfortunately a quantitative measure for rendering quality can only be given if the object contours are known precisely. In most cases especially with MRI or soft tissue with CT, results using binary segmentation are questionable. It is therefore the objective of this paper to discuss qualitatively the adequacy of different rendering methods for different organ classes and modalities. In doing this we are aware of the fact that the quality of rendering does not only depend on the method, but also on the data acquisition. We have taken the best available data, because it does not make much sense to talk about shading if the data is strongly undersampled or deteriorated by patient movement. Aspects of data acquisition for 3D-imaging are discussed by [Fishman et al. 1989].

2. Applied Principles of Volume Visualization

In tomographic medical imaging, typical raw data comes as a spatial sequence of images containing 256^2 or 512^2 pixels each, with a resolution of 4096 different gray values. In our implementation, we compress the data to slices of 256^2 pixels with a dynamic range of 256 intensity values. In order to achieve cubic volume elements, we perform a linear interpolation of the gray values between these slices. The isotropic cube of gray levels thus obtained is the basic data structure for the described algorithms. When further attributes are added to the gray values we term this structure the *generalized voxel model* [Höhne et al. 1988].

The volume is investigated with a ray casting algorithm, using the program system VOXEL-MAN, developed at our institute [Höhne et al. 1987b]. The tested algorithms are the same as those described in [Pommert et al. 1990b]. The images presented here are rendered according to one of the following basic procedures:

- *binary segmentation*: for every ray, a single surface voxel is determined by a criterium such as an intensity threshold.
- *fuzzy segmentation*: an opacity is assigned to each voxel by some criterium such as membership to an intensity range. Several voxels hit by a ray may thus contribute to the image.

In both cases, surface normal vectors are calculated with the algorithms described below. For illumination we use the Phong model which calculates the brightness S of a surface element at (x, y, z) as

$$S(x,y,z) = K_a + \left(K_d\left(\vec{L}\cdot\vec{N}\right) + K_s(\vec{R}\cdot\vec{V})^n\right)\left(1 - K\frac{z}{z_{max}}\right)$$

where K_a, K_d and K_s are the fractions of ambient, diffuse and specular reflection, K is a depth factor, n is an exponent for modeling highlights, $\vec{N}$ is the surface normal vector at location x, y, z. $\vec{L}$, $\vec{V}$ and $\vec{R}$ are normalized vectors in directions of light source, observer and reflected ray, respectively, and z_{max} is the maximum depth of the 3D scene.

All images presented in this paper were calculated with white light from a single source, situated in front of the object. We used $K_a = 0.0$, $K_d = 0.8$, $K_s = 0.2$, $K = 0.7$, and $n = 5$.

3. Shading Methods for Binary Segmentation

3.1 Z-Buffer Gradient Shading (ZG)

This method, proposed by [Chen et al. 1985], calculates the surface normal vectors from the depth image of an object, stored in the so-called z-buffer. For every pixel, depth information is considered in a 3×3

environment. We used an improved version of this algorithm [Tiede et al. 1987, Tiede 1988a]:

Given the z-buffer $Z(x,y)$ the components of the surface normal are approximated from the gradient vector $\nabla z = \left(\frac{\partial z}{\partial x}, \frac{\partial z}{\partial y}, 1\right)$. The partial derivative $\frac{\partial z}{\partial x}$ is approximated as the weighted sum of the backward and forward differences of the neighbouring pixels in the z-buffer:

$$\frac{\partial z}{\partial x} \approx W(\mid \delta_{b_x} \mid)\, Sign(\delta_{b_x}) + W(\mid \delta_{f_x} \mid)\, Sign(\delta_{f_x})$$

$$W(\delta) = \begin{cases} \frac{\pi}{4}\left(\frac{\delta-a}{b-a}\right) & \text{, if } a \leq \delta \leq m \\ 0 & \text{, otherwise} \end{cases}$$

$$\begin{aligned} \delta_{b_x} &= Z(x,y) - Z(x-1,y) \\ \delta_{f_x} &= Z(x+1,y) - Z(x,y) \end{aligned}$$

The computation of $\frac{\partial z}{\partial y}$, δ_{b_y} and δ_{f_y} follows the same principle. The thus obtained spherical vector is converted to cartesian coordinates. We used $a = 0$, $m = 8$ and $b = 12$.

3.2 Gray Level Gradient Shading (GG)

Much better results can be obtained if the whole gray scale information present in the data is considered. As a consequence of the tomographic acquisition process, the gray levels in the 3D-neighbourhood of a surface voxel represent the relative proportions of different materials (usually two, e.g. air/fat , or soft tissue/bone) inside these voxels. As described by [Barillot et al. 1985] and independently by [Höhne + Bernstein 1986] and [Tiede et al. 1988b], the resulting gray level gradients can be used to calculate the surface inclinations.

Our algorithm computes the three components of a gradient G for a surface voxel at (i,j,k) directly from the gray levels g of its 6 neighbours along the main axes as

$$\begin{aligned} G_x &= g(i+1,j,k) - g(i-1,j,k) \\ G_y &= g(i,j+1,k) - g(i,j-1,k) \\ G_z &= g(i,j,k+1) - g(i,j,k-1). \end{aligned}$$

Normalisation of G yields the surface normal. The gray level gradient may also be calculated from all 26 voxels in a $3 \times 3 \times 3$ neighbourhood, weighted according to their distance from the surface voxel [Tiede et al. 1988b]. The algorithm is then identical to the Zucker-Hummel operator used for edge detection [Ballard + Brown 1982].

Due to the high dynamic range of the gray level gradients, this method leads to a more realistic impression of the objects than the z-buffer shading method, even when using only 6 neighbours (see 5.2).

3.3 Adaptive Gray Level Gradient Shading (AGG)

For thin objects even a 6 voxel neighbourhood may be too large. Therefore we tried an adaptive selection of the neighbourhood [Pommert et al. 1989]. An algorithm which adapts to the thickness of the object using only 3–6 adjacent voxels, computes G_x as follows (G_y, G_z similarly): if the gray value of the surface voxel at (i, j, k) is greater (smaller) than the gray value of its neighbours at $(i-1, j, k)$ and $(i+1, j, k)$, it takes the gradient between the surface voxel itself and the neighbour with the lower (higher) gray value, otherwise it takes the gradient between both neighbours.

As has been shown in [Pommert et al. 1989], this algorithm significantly improves the accuracy of the computed surface normals. This principle can be generalized by choosing the neighbourhood from all 26 neighbours.

3.4 Marching Cubes Algorithm (MC)

The algorithm presented by [Lorensen + Cline 1987] creates a polygonal structure of an object surface from the input volume. Basically, an octuple of $2 \times 2 \times 2$ neighbouring voxels is considered. Depending on whether these voxels are within or outside an object (gray levels above or below a threshold), a structure of up to 4 triangles is placed within this cube. The result is a surface representation with subvoxel resolution. Surface normals are calculated at the triangle vertices via linear interpolation of the gray level gradients at the cube edges.

In the original algorithm the surface position is defined by polygons while gray level gradients are used for the definition of surface inclination. We implemented an even simpler variation, called *marching cubes segmentation* (MCS), which uses the marching cubes algorithm for creating a polygon surface that is rendered traditionally.

A purely polygonal surface description has the advantage that all methods developed in computer graphics may easily be applied. For small objects, zooming smoothes jagged edges and also eliminates staircase artefacts. If the Phong interpolation scheme is applied iteratively, an even larger neighbourhood of adjacent polygons is used for computing the surface normals [Tiede et al. 1990] (MCS-P).

4. Shading Methods for Fuzzy Segmentation

In contrast to all other methods previously mentioned, the algorithms described by [Levoy 1988, 1989] and [Drebin et al. 1988] are based on fuzzy segmentation. These algorithms have the following in common:

- certain opacities are assigned to certain intensity ranges (e.g. for bone)
- voxel opacities are weighted with the associated gray level gradients, thus enhancing surface voxels
- transparent views are produced, shading all visible voxels with the gray level gradient method.

This class of algorithms is usually called *volume rendering* although it is not the only way to visualize the content of a volume. It is claimed that the smooth and continuous weighting functions introduce less artefacts than any binary operations such as thresholding, and that especially the visualisation of small objects like thin bones can thus be improved. We implemented an algorithm which we call *transparent gray level gradient shading*, that determines the intensity in the image plane $I(x, y)$ using the following recursive equations:

$$I(x, y) = F(x, y, 0, 1)$$

$$F(x, y, z, l) = \begin{cases} l \cdot \text{ background value}, & \text{if } l \leq t \vee z \geq z_{max} \\ l \cdot O(x, y, z) \cdot S(x, y, z) + \\ F(x, y, z+1,\ l \cdot (1 - O(x, y, z))) & \text{otherwise} \end{cases}$$

t	minimum remaining fraction of incoming light
z_{max}	maximum depth of 3D scene
l	fraction of incoming light
$O(x,y,z)$	$W(g(x,y,z)) \cdot G(x,y,z)$ opacity
$W(g(x,y,z))$	weight assigned to the gray level g (between 0 and 1)
$G(x,y,z)$	$\sqrt{G_x^2 + G_y^2 + G_z^2}/\sqrt{3 \cdot 255^2}$
	magnitude of gray level gradient for 6 neighbours
$S(x,y,z)$	surface shading component (see section 2)

Similar to a recent proposal by [Levoy 1989], our algorithm traverses the volume in a front to back manner and allows a stopping criterium. Thus computing time can be greatly reduced. Still, computational costs are much higher than for binary segmentation.

5. Applications

For our study we have chosen a set of image classes which is certainly not complete but is representative for today's 3D imaging applications. These are: *bone (CT), head (MRI), heart (MRI), vessels (MRI), ultrasound,* and *isodoses in radiotherapy.* The images were produced with the standard capabilities of VOXEL-MAN.

5.1 Criteria of Assessment

An ideal volume imaging mode should be able to simulate all modes of exploration that an anatomist has, giving one in addition radiological eyes. Furthermore, there are other non-visible properties like metabolism, temperature, and isodoses which would certainly be helpful if they could be visualized together with the anatomy.

Translated into the language of computer graphics, the rendering algorithms should have the following capabilities:

First of all **surfaces** of distinguishable objects should be visually displayed. Only infrequently does one assess just a single object, so the display of **multiple objects** has to be provided. If objects obscure one another one must be able to remove interfering objects by **cutting**. If one wants to see what is inside the objects, **cutting while exhibiting**

the original gray values is necessary.

Another way to look inside a structure is **transparency**. Here one distinguishes essentially two modes: **surface transparency** that only takes into account a surface opacity or **volume transparency** that assigns opacities to the whole volume. In the following critical evaluation, we discuss which of these methods is best suited for different classes of imaging and which possible combinations allow the best exploration of the anatomical objects.

5.2 Bone from CT

Fig. 1 shows the effect of the different rendering methods in visualizing a fracture of the tibia (CT, 107 Slices, 1 mm slice distance). ZG-shading (fig. 1, top left) gives a rather rough image but nevertheless exhibits all relevant details of the fractures. Very small details such as sutures cannot be recognized. Interestingly, the texture caused by discretisation is not considered harmful by the clinicians, because it delivers an additional 3D-cue. The AGG-shading (fig. 1, top center) delivers a clearer image. Despite this fact, clinicians are sometimes irritated by the absence of the typical texture of bones. It is a decisive advantage of the voxel model that the surrounding tissue can be visualized by cuts (figs. 1, 2). This can also be done using TGG-shading as shown in fig. 1 top right, where the outer tissue has been made transparent. Our experience, however, shows that cutting gives a better 3D-understanding. The blurring introduced by TGG-shading removes jagged edges. The main problem of TGG-shading is the assignment of opacities to gray values, which must be done in a heuristic way and thus allows to produce an infinite number of different images of the same object.

The marching cubes algorithm MC (fig. 1, bottom left) has similar properties as AGG but lacks the capability to show gray level cuts. As a surface oriented method it allows easy zooming the polygons and thus enables the reduction of jagged edges. The MCS-P algorithm delivers nearly identical results but it is easier to handle, because it does not rely on gray level gradients (fig. 1, bottom center). Since it delivers a purely polygonal data structure it enables manipulations such as surface smoothing by iteration of the Phong interpolation (fig. 1, bottom right).

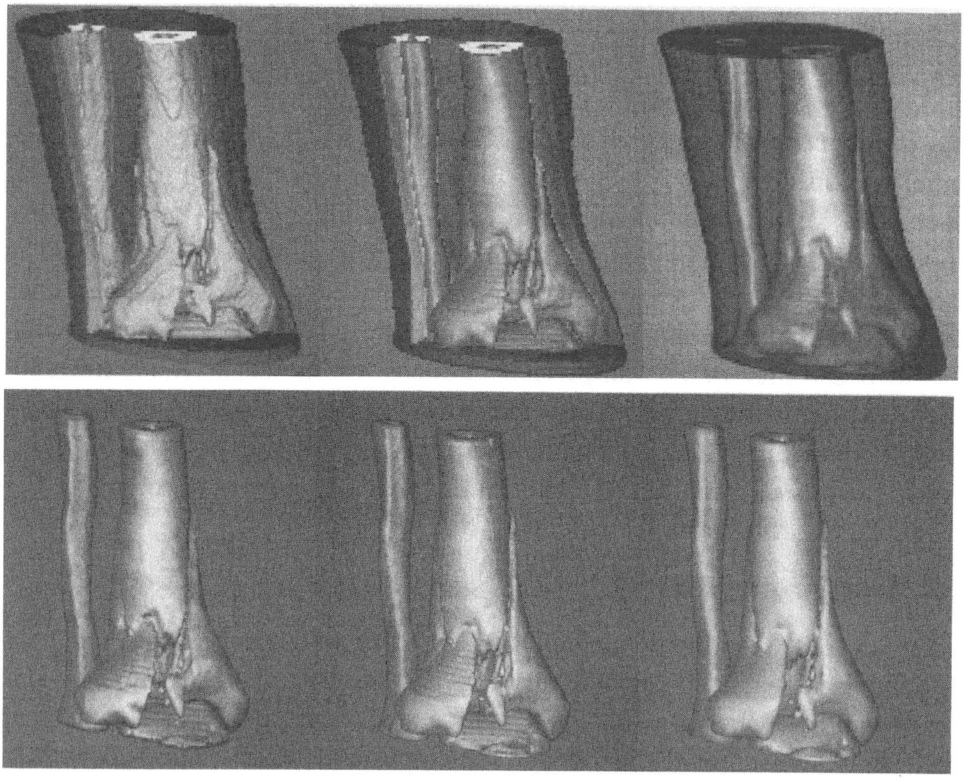

Fig. 1 Multiple fracture of a tibia (CT). Upper row: ZG-shading, AGG-shading with 26 neighbours, TGG-shading with 6 neighbours. Lower row: original marching cubes algorithm (MC), marching cubes segmentation with Phong interpolation (MCS), MCS with 5 iterations (MCS-P5).

5.3 Head from MRI

Fig. 3 shows the effect of different rendering methods to the head of a patient with a brain tumor (MRI, flash, 128 slices). The contours were determined using the Marr-Hildreth operator with additional morphological filtering [Bomans et al. 1990]. ZG shading (fig. 3, top left) shows modest but sufficient rendering of the skin while eye, tumor, and brain do not reflect reality very well. AGG-shading (fig. 3, top right) improves the situation considerably. At sharp edges, however, artefacts do occur. These can be reduced by adding a small amount of transparency using TGG-shading (fig. 3, bottom left). A slightly improved impression is achieved by the MC algorithm and MCS algorithms that deliver practically identical images (fig. 3, bottom right). The improvement is due to the possibility of zooming, that reduces artifacts. It must however be stated again that there is no easy possibility of producing composite images containing both cuts and volume transparency. At least for diagnostic purposes, cuts still are the main source of information, while the 3D-surfaces serve more for orientation.

5.4 Heart from MRI

As an example for fuzzy surface segmentation, we demonstrate the case of a human heart. It turned out that an automatic segmentation was not feasible. Segmentation by hand was no alternative for 26 slices at each of 12 heart phases. In order to uncover the heart, the thorax was removed by a mask defined interactively. Fig. 4 (top) shows the result of GG-rendering of the heart after binary segmentation. Only the gross structures are visible and the whole picture looks rather unnatural. However, if the surfaces are considered as a coordinate frame for the location of cuts, such display has its justification. A more 'natural' appearance is achieved with TGG rendering, although the images are blurred to some extent. Yet an exact location of the objects is no longer possible (fig. 4, bottom).

5.5 Vessels from MRI

Special problems arise with the rendering of small objects such as vessels from MRI, because the determination of surface normals is no longer possible when the neighbourhood becomes small [Wiebecke et al. 1989]. The simplest and most effective way of visualizing the vessels is still the maximum projection (fig. 5, top left). Its drawback is that it does not give any depth cue in static imaging. Our proposal is therefore to implant the maximum image into the surrounding morphological structures (fig. 6). The lack of a sufficiently large neighbourhood can be overcome with

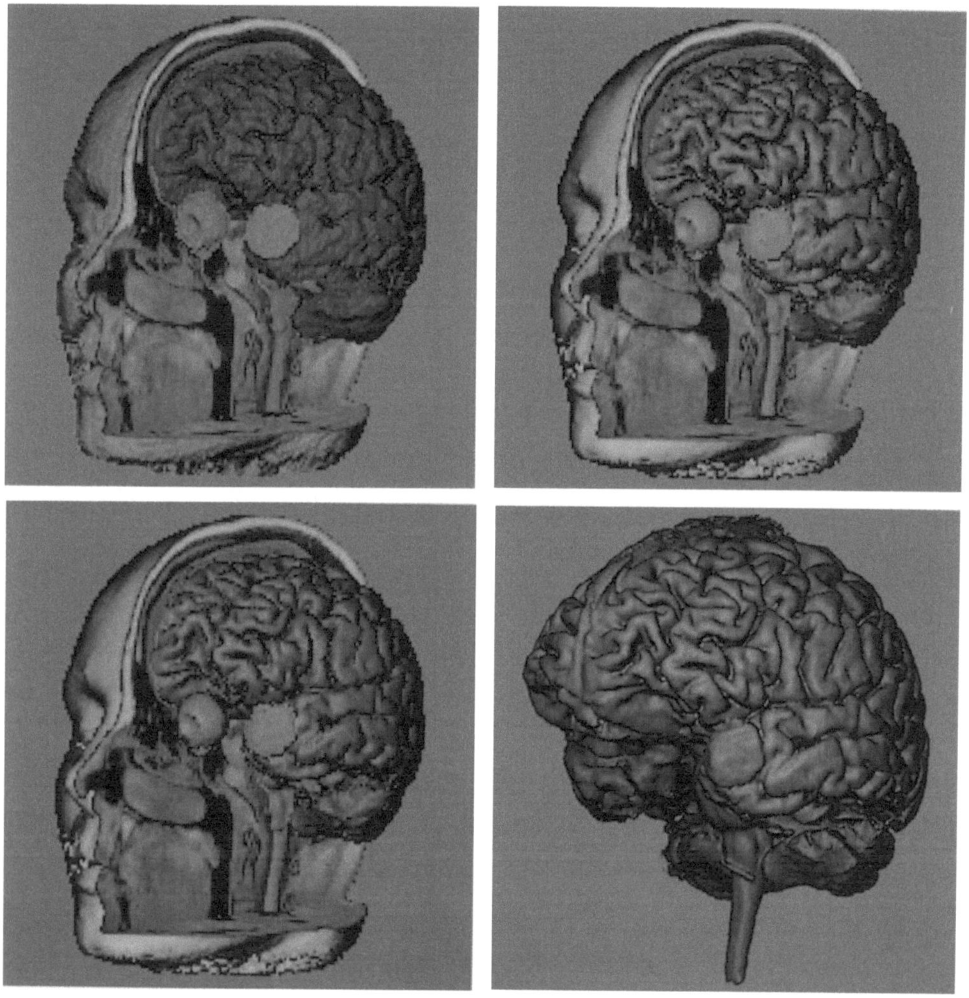

Fig. 3 Human head (MRI) with a tumor (green). Top left: ZG-shading. Top right: AGG-shading with 26 neighbours. Bottom left: TGG-shading. Bottom right: marching cubes segmentation with Phong interpolation (MCS).

TGG shading, by whose use errors in surface normal determination may be smoothed out to a certain extent (fig. 5, top right). As an experiment we applied TGG shading only around the maximum (fig. 5, bottom left). In the shown example this produced a slight improvement. It has been proposed to apply morphological filters to the vessels before rendering [Hu et al. 1989]. We have applied a grayscale dilation to our data set (fig. 5, bottom right). In fact the large vessels get more plasticity, but the smaller ones are deteriorated. In addition the lumen is changed.

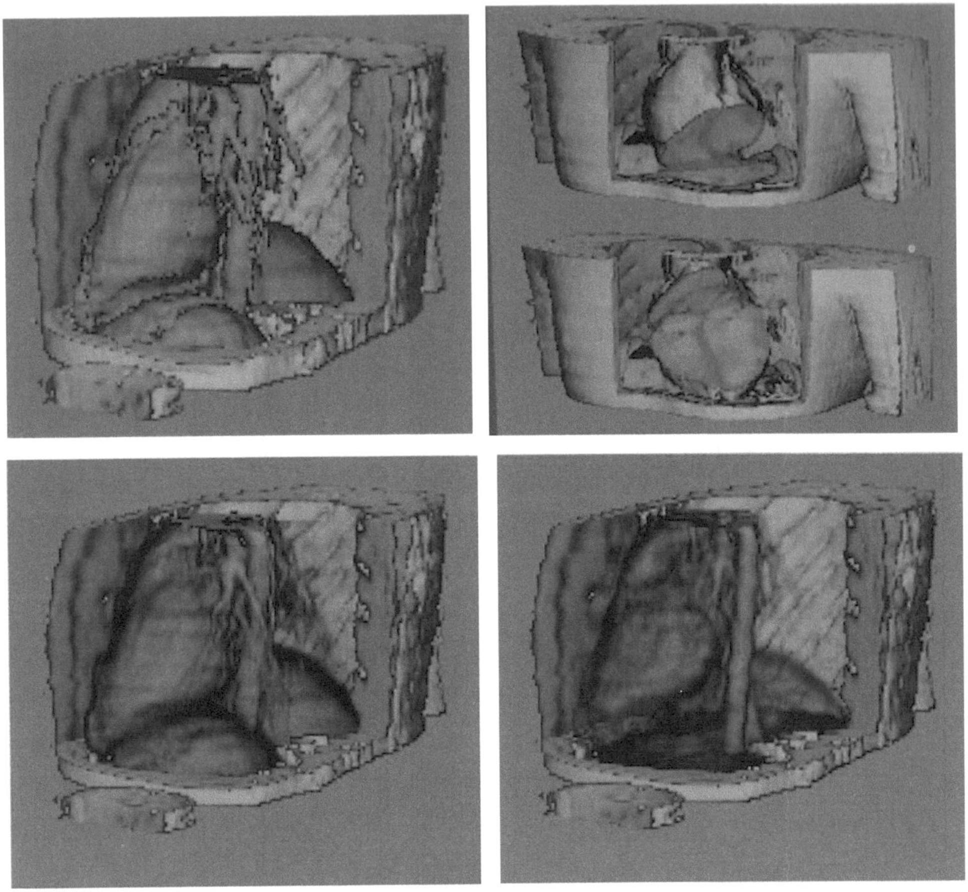

Fig. 4 View of a beating heart (MRI). Top: AGG-shading with 26 neighbours. Bottom: TGG-shading with two different opacity assignments.

5.6 Ultrasound

Increasingly ultrasound data are available as sequential parallel cross-sections, especially in transrectal and transesophagal ultrasound. We did some experiments with a data set of a prostate containing 24 slices with a distance of 2.5 mm. Since a non-trained observer can find the contours of the prostate only with difficulty, the 3D-images require that the reviewer be experienced. Our preliminary results do not allow a final assessment of the quality of the different rendering methods. Fig. 7 shows examples, which at least our urologists consider a decisive improvement in comparison to 2D-imaging. Needless to say, the 3D-perception is further improved by real time rotation.

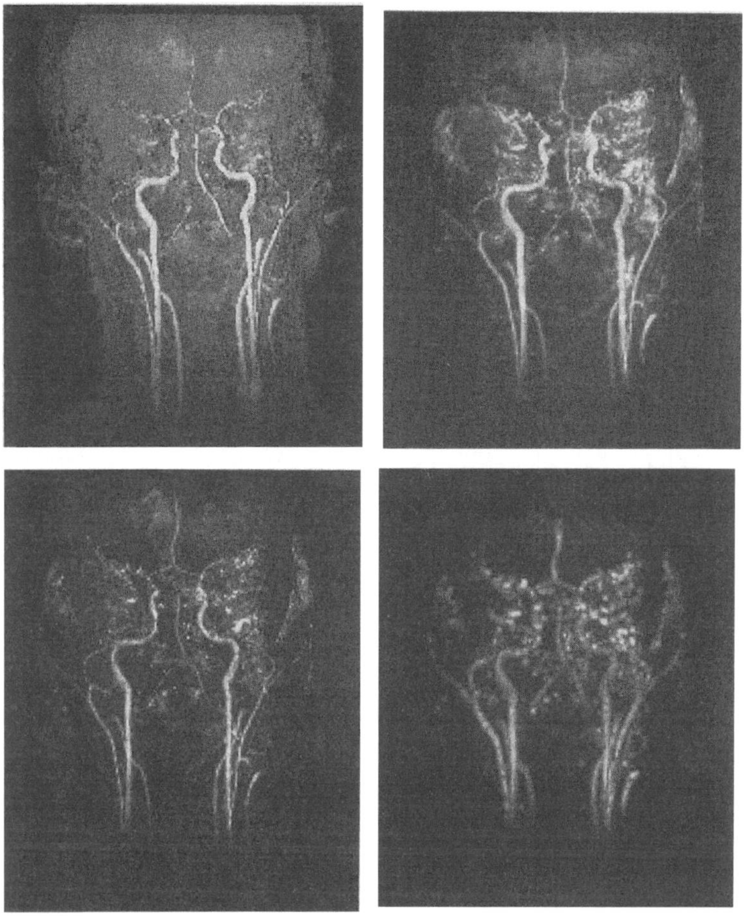

Fig. 5 Blood vessels of a head (MRI). Top left: maximum projection. Top right: TGG-shading. Bottom left: TGG-shading around maximum. Bottom right: TGG-shading after morphological filtering.

5.7 Isodoses from radiotherapy planning

Conventionally, 3D-radiotherapy planning is performed on the basis of surface oriented rendering [Schlegel 1990]. Both isodoses and objects are visualized as surfaces. As a drawback the doses can be visualized only for objects that are segmented.

Voxel-based rendering techniques allow the visualization of isodoses on cut planes that show the real morphology. The cut in fig. 8 (bottom)

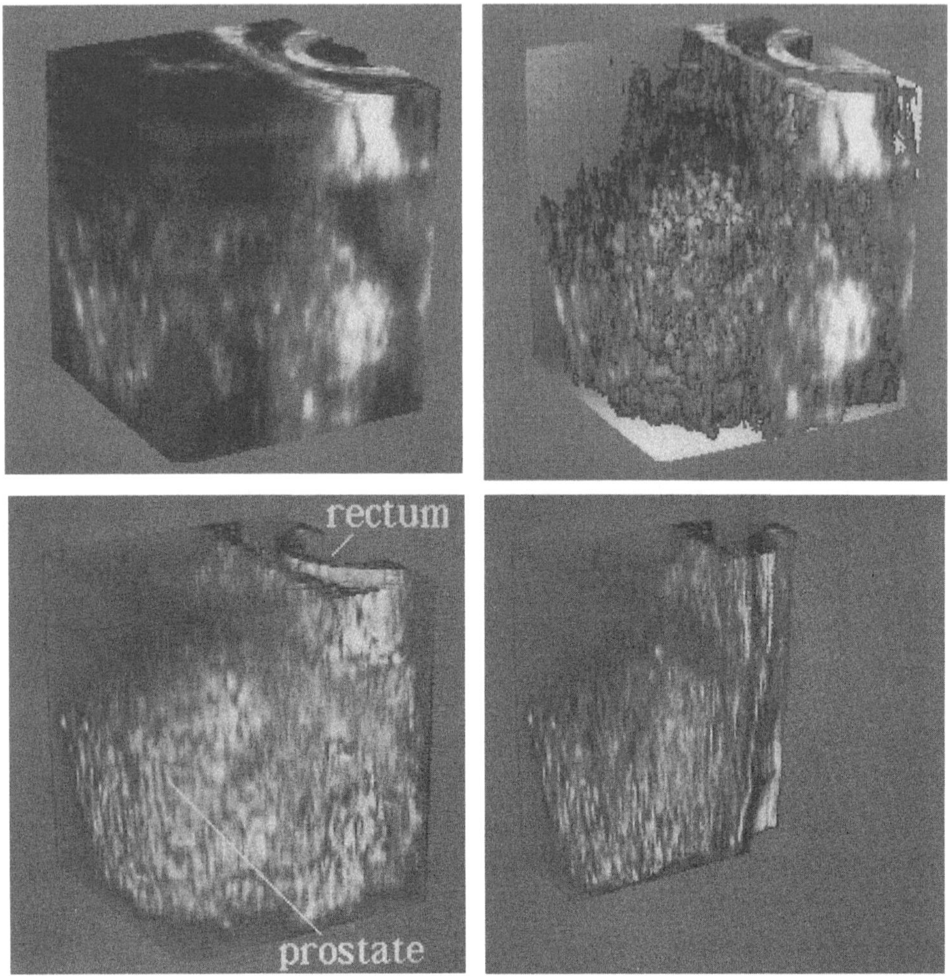

Fig. 7 Prostate from transrectal ultrasound sonography (TRUS). Top left: data cube with original gray values at its borders. Top right: AGG-shading with 26 neighbours. Bottom left: TGG-shading. Bottom right: TGG-shading with a cut.

Fig. 2 Multiple fracture of a tibia (CT). Composite image with AGG-shading.

Fig. 6 Blood vessels of a head (MRI). Maximum projection implanted into surfaces and cuts.

Fig. 8 Radiotherapy planning for a pelvis. Top: 60 % isodose contour as a surface. Bottom: cut through the volume enclosed by the 60 % isodose exhibiting the dose distribution as color and the morphological structures as brightness.

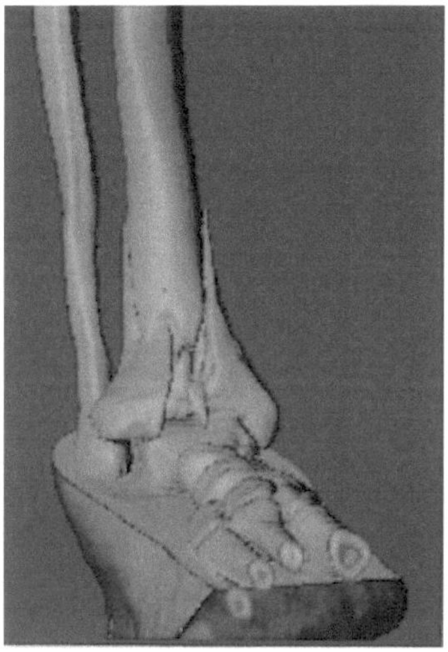

Fig. 2

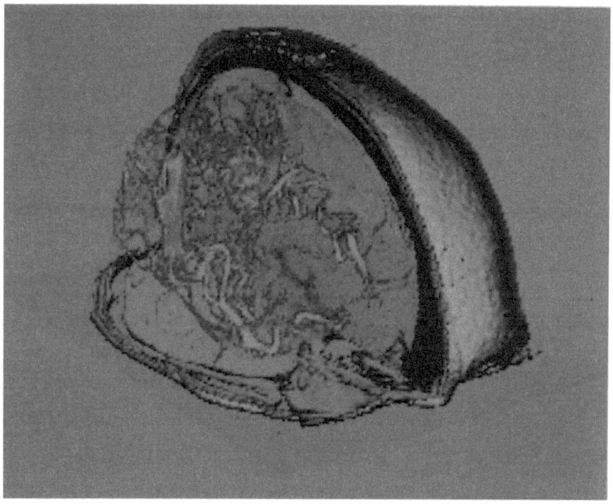

Fig. 6

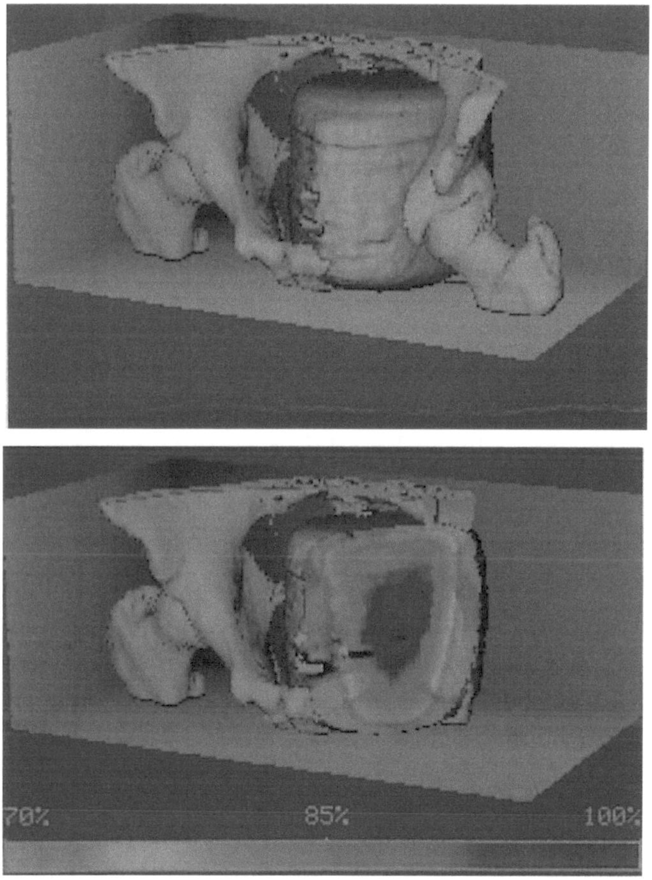

Fig. 8

exhibits the dose distribution as color and the morphological structures as brightness. Morphological structures such as bones may still be shown as surfaces. Selected isodoses may also be visualized as surfaces. Fig. 8 (top) shows the 60 % isodose shaded with AGG-shading. It must be pointed out that the images of fig. 8 were produced without the need of outlining any contour by the radiotherapist.

6. Conclusions

As a general finding it can be stated that at present there appears to be no single rendering method that is optimal for all tomographic objects. Generally, as far as pure visualization is concerned the voxel-based methods exhibit much more flexibility than surface oriented techniques. For modeling, however, surface oriented techniques are indispensible. For linking the voxel to the polygon world, the marching cubes idea provides an excellent tool.

Different modalities and organs obviously require different tools of visualization. For bone in CT we found gray level gradient shading optimal. The marching cubes algorithm is equally good but does not (or at least does not easily) allow the production of composite images, containing gray level cuts or volume transparency.

The situation is somewhat different for MRI, where the head (especially the brain surface) was used as an example. Here the marching cubes algorithm delivers the best results. Gray level gradient shows artefacts at edges, that can be reduced by adding some transparency using transparent gray level gradient shading.

If there are segmentation problems, semi-transparent visualization is the only way out as demonstrated by the example of a beating heart. There is currently no general solution for the visualization of small objects such as vessels from MRI. Transparent gray level gradient shading and/or additional processing with morphological filters do not decisively improve the quality of simple maximum projections.

Concerning ultrasound, we do not yet know which rendering methods are best. Although the preliminary images are not very well perceivable for the layman, they are considered useful by the clinicians, especially if rotation is provided as a depth cue. A severe drawback in TGG-shading is that the assignment of opacities to gray values is very unstandardised at present.

One direction for further research should therefore be the establishment of reproducible opacity assignments. For broad practical application, the design of adequate user interfaces certainly remains a generally unsolved problem. Since we saw that there are advantages both in the voxel-based and in the polygon methods, the combination of these methods in hybrid systems is certainly a worthwhile direction of future research.

Acknowledgement

The research work was supported in part by the Werner Otto Foundation, Hamburg. We are grateful to Prof. G. Witte (Dept. of Radiology) and Dr. J. Wening (Dept. of Traumatology) who provided the CT data, and to Dr. L. Chumbley, B. Pflesser, R. Schubert, and J. Wulff for helpful discussions and practical assistance. The MRI data sets from which the perspective views were generated, were kindly provided by Siemens, Erlangen. The ultrasound data were produced in cooperation with Dr. A. Schneider (Dept. of Urology). We thank Dr. E. Tabbert (Bezirkskrankenhaus Schwerin) for the cooperation in the radiotherapy application.

References

H. H. Baker: Building Surfaces of Evolution: The Weaving Wall. *Computer Vision 3* (1989), 51–71.

D. H. Ballard, C. M. Brown: *Computer Vision.* Prentice Hall, Englewood Cliffs, 1982, 82–83.

C. Barillot, B. Gilbaud, L. M. Luo, J. M. Scarabin: 3-D Representation of Anatomic Structures from CT Examinations. In *Biostereometrics '85, Proc. SPIE 602.* 1985, 307–314.

M. Bomans, K. H. Höhne, U. Tiede, M. Riemer: 3D-Segmentation of MR-Images of the Head for 3D-Display. *IEEE Trans. Med. Imaging* (1990), in press.

L. S. Chen, G. T. Herman, R. A. Reynolds, J. K. Udupa: Surface Shading in the Cuberille Environment. *IEEE Comput. Graphics Appl. 5*, 12 (1985), 33–43.

R. A. Drebin, L. Carpenter, P. Hanrahan: Volume Rendering. *Computer Graphics 22* (1988), 65–74.

E. K. Fishman, D. Magid, D. R. Ney, J. E. Kuhlman: Three-dimensional imaging: orthopedic applications. In J. K. Udupa, G. T. Herman (Eds.): *Computer Aspects of 3D Imaging in Medicine: A Tutorial,* Lewis Publ., Chelsea, MI, 1989.

S. M. Goldwasser, R. A. Reynolds, T. Bapty, D. Baraff, J. Summers, D. A. Talton, E. Walsh: Physicians Workstation With Real Time Performance. *IEEE Comput. Graphics Appl. 5* (1985), 44–57.

K. H. Höhne, R. Bernstein: Shading 3D-Images from CT Using Gray Level Gradients. *IEEE Trans. Med. Imaging MI-5,* 1 (1986), 45–47.

K. H. Höhne, R. L. DeLaPaz, R. Bernstein, R. C. Taylor: Combined Surface Display and Reformatting for the 3D-Analysis of Tomographic Data. *Investigative Radiology 22* (1987a), 658-664.

K. H. Höhne, M. Riemer, U. Tiede: Viewing Operations for 3D-Tomographic Gray Level Data. In H. U. Lemke et al. (Eds.): *Computer Assisted Radiology, Proc. CAR '87.* Springer, Berlin, 1987b, 599–609.

K. H. Höhne, M. Bomans, U. Tiede, M. Riemer: Display of Multiple 3D-Objects Using the Generalized Voxel-Model. In R. H. Schneider, S. J. Dwyer (Eds.): *Medical Imaging II, Image Data Management and Display, Proc. SPIE 914, Part B.* Newport Beach, 1988, 850–854.

K. H. Höhne, M. Bomans, A. Pommert, M. Riemer, C. Schiers, U. Tiede, G. Wiebecke: 3D-Visualization of Tomographic Volume Data Using the Generalized Voxel-Model. *Visual Computer 6,* 1 (1990), 28–36.

X. Hu, K. K. Tan, D. N. Levin, S. G. Galhorta, C. A. Pelizzari, G. T. Y. Chen, R. N. Beck, C. Chen, M. D. Cooper: Volumetric Rendering of Multimodality, Multivariable Medical Imaging Data. In C. Upson (Ed.): *Proc. Chapel Hill Workshop on Volume Visualization.* Dept. of Computer Science, University of North Carolina, Chapel Hill, NC, 1989, 45–49.

A. Kaufman, R. Yagel, D. Cohen: Intermixing Surface and Volume Rendering. *This volume.*

M. Levoy: Display of Surfaces from Volume Data. *IEEE Comput. Graphics Appl. 8,* 3 (1988), 29–37.

M. Levoy: Design for a Real-Time High-Quality Volume Rendering Workstation. In C. Upson (Ed.): *Proc. Chapel Hill Workshop on Volume Visualization.* Dept. of Computer Science, University of North Carolina, Chapel Hill, NC, 1989, 85–92.

W. E. Lorensen, H. E. Cline: Marching Cubes: A High Resolution 3D Surface Construction Algorithm. *Computer Graphics 21,* 4 (1987), 163–169.

M. Magnusson, R. Lenz, P. E. Danielsson: Evaluation of Methods for Shaded Surface Display of CT-Volumes. In *Proc. 9th Int. Conf. on Pattern Recognition (ICPR).* IEEE Publishing, 1988, 1287–1294.

A. Pommert, U. Tiede, G. Wiebecke, K. H. Höhne: Image Quality in Voxel-Based Surface Shading. In H. U. Lemke et al. (Eds.): *Computer Assisted Radiology, Proc. CAR '89.* Springer, Berlin, 1989, 737–741.

A. Pommert, M. Bomans, U. Tiede, K. H. Höhne: Simulation Studies for Quality Assurance of 3D-Images from Computed Tomograms. In A. Todd-Pokropek, M. A. Viergever (Eds.): *The Formation, Handling and Evaluation of Medical Images.* NATO ASI Series F: Computer and Systems Sciences, Springer, Berlin, 1990a, in press.

A. Pommert, U. Tiede, G. Wiebecke, K. H. Höhne: Surface Shading in Tomographic Volume Visualization: A Comparative Study. In *Proc. First Conf. on Visualization in Biomedical Computing.* IEEE Publishing, Atlanta, GA, 1990b, in press.

W. Schlegel: Computer Assisted Radiation Therapy Planning. *This volume.*

U. Tiede, K. H. Höhne, M. Riemer: Comparison of Surface Rendering Techniques for 3D-Tomographic Objects. In H. U. Lemke et al. (Eds.): *Computer Assisted Radiology, Proc. CAR '87.* Springer, Berlin, 1987, 610–614.

U. Tiede: *Entwurf, Implementation und Vergleich verschiedener Repräsentationsformen für die 3D-Darstellung von medizinischen Objekten.* Diplomarbeit, Fachbereich Informatik, Universität Hamburg, 1988a.

U. Tiede, M. Riemer, M. Bomans, K. H. Höhne: Display Techniques for 3D-Tomographic Volume Data. In *Proc. NCGA '88, Vol. III.* Anaheim, 1988b, 188–197.

U. Tiede, K. H. Höhne, M. Bomans, A. Pommert, M. Riemer, G. Wiebecke: Investigation of Medical 3D-Rendering Algorithms. *IEEE Comput. Graphics Appl.*, 2 (1990), in press.

G. Wiebecke, M. Bomans, U. Tiede, K. H. Höhne: 3D-Visualisierung von schwer segmentierbaren tomographischen Volumendaten. In H. Burkhardt et al. (Eds.): *Mustererkennung 1989, Proc. 11. DAGM-Symposium.* Informatik-Fachberichte 219, Springer, Berlin, 1989, 433–439.

Intermixing Surface and Volume Rendering

Arie Kaufman, Roni Yagel, and Daniel Cohen

Department of Computer Science
SUNY at Stony Brook
Stony Brook, NY 11794-4400

Abstract

Two complementary methodologies for volume visualization, surface rendering and volume rendering, can be employed constructively together. In many situations, as in 3D medical imaging, the voxel-based sampled image need to be visualized together with synthetic surface-based objects such as surgical cuts, prosthesis, scalpels, and radiation beams. Four approaches for intermixing geometric models with sampled 3D medical data are presented. Details are provided on a hybrid approach that employs a Z-merging algorithm.

Keywords: Volume rendering, surface rendering, ray casting, medical imaging, Z-merging

1. Introduction

Volume visualization is the process of generating 2D images from volumetric data and it involves mechanisms to reveal and explore the inner or complex structures of volumetric data. The primary source of volume data is empirical, generated either by a sampling device (e.g., CT, MRI, confocal microscope) or by a computer simulation that generates discrete sampling points of a simulated reality. Such data sets appear in different topological arrangements and quality levels, requiring enhancement by 2D or 3D image processing techniques and reconstruction into a (regular) volume data set. The volume data set is commonly stored in a cubic frame buffer which is a 3D array of voxels within the 3D discrete digital voxel space.

An alternative source of volume data is 3D continuous geometric models commonly represented as a display list of geometrically-defined objects. A geometric model is *3D scan-converted (voxelized)* by 3D scan-conversion algorithms that generate a digital discrete representation (i.e.,

NATO ASI Series, Vol. F 60
3D Imaging in Medicine
Edited by K. H. Höhne et al.

a set of voxels) that "best" represents the synthetic model within the discrete voxel space [Kaufman and Shimony 1986], [Kaufman 1987], [Kaufman 1988]. These 3D scan-conversion algorithms bridge between volume visualization techniques and the traditional modeling techniques used in computer graphics and computer aided design, which depend primarily on parametric surface patches and linear and quadratic objects.

In order to visualize a volume data set, the volume primitives are transformed (projected) into 2D discrete pixel space and stored as a raster image in a frame buffer. This process, which is termed *volume rendering*, involves both the viewing and the shading stages. The alternative to volume rendering is a process that employs traditional surface-based techniques in which objects are defined in terms of surfaces (polygon mesh, B-splines, NURBS). The geometric primitives are rendered to the screen by a process called *surface rendering* or *geometry rendering*. In order to exploit the software experience in geometric rendering and the powerful hardware of the geometry engines for the benefit of volume visualization, volume data must be first converted into geometric primitives in a process called *surface reconstruction* employing techniques such as *(iso)-contouring*, [Fuchs, Kedem, and Uselton 1977] *(iso)-surfacing* [Cline et al. 1988], and *surface extraction* [Frieder et al. 1985].

It seems that surface rendering and volume rendering are two rival methodologies each suited for the rendition of different types of data. Surface rendering is capable of real-time rendering of dynamic objects, a capability which in not currently possible using volume rendering techniques. Volume rendering, on the other hand, is capable of displaying volume entities which might not consist of surfaces at all as well as showing the information inside those volume entities. Volume rendering is a direct display technique of volume primitives, while surface rendering first converts the data into an intermediate surface representation before the rendering.

In some situations, as in surgical planning and radiation therapy planning, the voxel-based image data captured through an imaging device need to be visualized, manipulated, and analyzed along with objects such as osteotomy surfaces, surgical cuts, prosthetic devices, grafts, scalpels, injection needles, isodose surfaces, and radiation beams, which may not be available in digital form. For example, when a physician would like to simulate a surgical cut using an "electronic scalpel", he or she specifies the cut directly on the screen images utilizing a 2D or 3D input device [Kaufman and Yagel 1989]. to specify B-spline or NURBS surfaces. The 3D scan-conversion algorithm then determines which voxels are actually being cut, which ones are outside or inside the cut, and which ones are to be removed or lifted.

Mechanisms that can blend geometric objects with discrete space medical objects, are clearly of much value. The next section presents four different approaches for intermixing geometric models with 3D sampled data, followed by a comparative analysis of these approaches. The first two approaches represent extremes views supporting only one object representation scheme. The other two approaches allow the synthetic, surface-based model to be intermixed with the sampled data set forming a *hybrid data model.*

2. Intermixing Approaches

The surface-based approach favors a geometric object representation and thus only supports rendering of analytically-defined surfaces. Therefore, the volumetric data must be first transformed into surface-based description by applying a surface reconstruction algorithm [Fuchs, Kedem, and Uselton 1977], [Sunguroff and Greenberg 1978]. The opposite approach is the voxel-based approach in which the voxel is the only primitive used for object representation. Geometric objects are converted from their continuous mathematical definition into a volumetric representation by employing scan conversion algorithms [Kaufman and Shimony 1986], [Kaufman 1987], [Kaufman 1988]. This volumetric representation is merged directly with the sampled voxel-based data. Figs. 3 and 4 depict voxel-based images of a head intermixed with scan-converted geometric objects (e.g., scalpel, radiation beam).

Compromising between these two extremes of surface-only and volume-only approaches, are the point-based approach and the hybrid approach. The point-based approach, extends the surface-based approach by adding a point primitive to the set of geometric objects [Johnson and Mosher 1989]. This primitive consists of a 3D location, a normal value used by a geometry engine for illumination calculation, and some additional data values (e.g., color, opacity, density). When synthetic objects are scan converted into voxel representation the normal value can easily be calculated for each voxel. However, sampled volumetric data do not contain information about the surface inclination. Thus, a pre-processing stage of normal restoration is needed in order to convert each non-empty voxel to a point. This is usually done by examining a close neighborhood of the voxels in order to extract a surface and calculate its normal [Cline et al. 1988], [Herman and Udupa 1981], [Hoehne and Bernstein 1986].

Since common volumetric data sets consist of millions of voxels and as contemporary graphic hardware can not cope with such a magnitude of objects, a reduction in the amount of points is desired. This reduction is achieved by feeding the graphic engine only with those points (voxels) which are visible, that is, those voxels comprising the outer layer of the objects. One way to extract these voxels from the volume is by thresholding; voxels belonging to an object are assigned the value '1' while the others are assigned a '0'. The '1' voxels bordering with at least one '0' voxel are those possibly visible [Cline et al. 1988]. A more efficient and accurate method is based on starting from a seed point on the surface and then tracking its neighbors until the entire surface is extracted [Frieder et al. 1985].

It should be noted that the point-based approach makes use of the point primitive only for display purpose, while all the other volumetric manipulations are performed on the volume itself. Some operations require only minor alterations of the point list (e.g., scalpeling), while others require more extensive changes (e.g., cut planes), or even a total rebuilding of the point-list (e.g., change in the translucency parameters). The point primitive can be rendered either by using a geometry engine to render it as a point in 3D space [Cline et al. 1988] or by applying a *splatting technique* that computes the contribution of the projected point to a neighborhood of pixels [Westover 1989].

The fourth approach to the intermixing problem is the hybrid approach which supports the hybrid data model by rendering each part of it separately and then combining the surface rendered and the volume rendered images into a final 2D pixel image. Two flavors of this approach are described, the Z-merging and the ray-merging.

The *Z-merging* algorithm employs two separate (and possibly parallel) rendering processes - one executes (any) volume rendering algorithm on the volumetric data set while the other performs (any) traditional surface-based rendering on the geometric model. Each rendering process produces two buffers consisting of the shaded image along with the corresponding Z-buffer. The two image buffers are combined according to their depth values as described in Fig. 5. It should be noted that after the image is rendered for the first time, the system waits for a change that requires re-rendering and then renders only the necessary model. For example, when a surgeon simulates the positioning of a prosthesis, he can interactively move the (geometric) prosthesis over a static (volumetric) organ since only the geometric model is repeatedly re-rendered (see Figs. 1 and 2). The algorithm, as described in Fig. 5, is more appropriate for current technology where volume rendering is time consuming while surface rendering can be performed in real time. The algorithm is easily transformed to give equal weights for both rendering processes or to incorporate a more sophisticated compositing mechanisms to provide translucency (as in Fig. 2) and reduce aliasing [Duff 1985].

Another flavor of the hybrid approach is the ray-merging method which simultaneously traces rays through both the volumetric data set and the synthetic model producing two vectors of samples. These two vectors are combined to produce the final image value [Goodsell, Mian, and Olson 1989], [Levoy 1990].

---

volume_z and *geometry_z* are 2D arrays of depth values.
volume_image and *geometry_image* are 2D arrays of color values.
Screen is a 2D array of color values representing the screen frame buffer.

```
forever do
    Clear_Buffers(volume_z, volume_image);
    Render_Volume(volume_z, volume_image);
    while (no change in viewing direction or in volumetric data)
        Clear_Buffers(geometry_z, geometry_image);
        Render_Geometry(geometry_z, geometry_image);
        for each (i, j) in the screen do
            if volume_z[i][j] < geometry_z[i][j]
                Screen[i][j] := volume_image[i][j];
            else
                Screen[i][j] := geometry_image[i][j];
        end for each;
        wait for any change;
    end while;
end forever;
```

Fig. 5: *The Z-merging algorithm.*

---

3. Discussion

Although the surface-based approach coupled with a contemporary powerful geometry engine, provides an interactive visualization environment, it has some severe disadvantages. The source of the deficiencies is the surface reconstruction process that loses most the volumetric data keeping only a trace of the outer layer of the objects. Thus, it is impossible to perform many investigative procedures. For example, in this approach cut planes serve only as clipping planes without showing the values of the volume data on the cut plane itself. Moreover, this approach requires a pre-processing stage of surface reconstruction that must be repeated for every change in the volume. Another key flaw in this method is its underlying assumption that the data contains surfaces. Researchers dealing with applications where there is no notion of surfaces such as visualization of electron density maps, smoke flows, or radiation doses must compromise and display isovalue surfaces.

The main disadvantage of the voxel-based approach is that it requires a huge amount of storage (1 giga bits for a typical resolution) and extensive processing power to achieve real-time interaction. Furthermore, when drawing an object into a section of the volume memory the current values of the voxels occupying that section must be stored, so that in a subsequent translation of that object the content of the memory could be restored. In such a case the object is then re-scan converted in its new location, creating a new scene that must be re-projected and re-shaded. Current hardware can not support such operations in real time even for moderate resolutions.

Another difficulty with this approach arises from the difference in sampling frequencies of the scanned data and the geometric data. Most scan conversion algorithms produce binary objects. On the other hand, scanning devices have a limited sampling rate yielding objects that exhibit a partial volume effect. Shading methods relying on the partial volume effect for normal estimation [Hoehne and Bernstein 1986] fail to properly estimate surface inclination of binary objects. Our solution is to use different methods for normal calculation; the geometric surfaces are scan-converted with their computed normal values while a grey-level method is used to restore normal values in the volume (see Fig. 3 and 4).

The point-based method relies on a pre-processing stage of surface extraction and normal estimation, which must be repeated (to some extent) after every change to the volume data or to the translucency parameters. When only surface voxels are rendered, the final product tends to suffer from some surface-oriented maladies, for example, holes appear when the viewing point is too close to the objects.

Comparing the two variations of the hybrid approach we observe that ray-merging dictates the usage of ray casting as the hidden surface (or hidden voxel) removal algorithm. The Z-merging method, on the other hand, is a post-rendering process which is not influenced by the actual rendering algorithm. Considering the efficiency of both methods, Z-merging renders all data in the model while ray-merging does not render those parts occluded by objects in the other model. Since powerful geometry engines are available this drawback is noticeable only when the geometric model is very complex or when the volume must be frequently re-rendered. Nevertheless, the Z-merging method is more efficient since it renders a model only if such an operation is required while ray-merging always re-renders both the geometric and volumetric data sets after every change.

The Z-merging (and the restricted version of ray-merging [Goodsell, Mian, and Olson 1989]) can not handle some translucency combinations. Although it is possible to render translucent layers within the same data set, or between the two data sets as a whole, it is impossible to visualize a geometric object *inside*, and obscured by, a semi-transparent volume or visa versa.

The main appeal of the Z-merging approach is its adequacy for hardware realization. The merging mechanism is extremely simple, performing a local pixel-wise operation and thus it can be easily parallelized by interleaving the buffer memories and duplicating the merging mechanism. An efficient variant of the Z-merging algorithm simplifies the hardware implementation even further. Fig. 6 describes the Z-buffer based algorithm that initializes the Z-buffer used by the geometry engine with the volume Z-value map before surface rendering starts. The merging operation is achieved as a by product of the surface rendering Z-buffer based hidden surface removal algorithm.

We can safely assume that in a couple of years *volume engines* with memory and computing power will prevail with capabilities to generate, store, manipulate and display voxel-based images in real-time (e.g., [Kaufman et al. 1990]). With such a volume engine the voxel-based approach has the potential of yielding superb performance even for traditional surface based graphics. It should be observed that time consumption for voxel-based scan conversion of geometric objects is similar to the time consumed by the algorithms traditionally used in geometry engines, since both complexities are proportional to the area of the object. Moreover, in voxel-based 3D graphics, the scene is created (voxelized) once for multiple

```
forever do
        Clear_Buffers(volume_z, volume_image);
        Render_Volume(volume_z, volume_image);
        while (no change is in viewing parameters or volume data)
                copy volume_z to Z-buffer;
                copy volume_image to Screen;
                Render_Geometry(Z-buffer, Screen);
                wait for any change;
        end while;
end forever;
```

Fig. 6: *Z-buffer based intermixing.*

viewpoints and lighting conditions, while the geometry engine must repeatedly scan convert the scene after every slight change. This means that while geometry engines are sensitive to scene complexity, volume engines are sensitive only to the (constant) image resolution. This implies that if the geometric model at hand is complex (e.g., many polygons, anti-aliased, texture mapped) the voxel-based approach is preferable.

Based on the above discussion the hybrid model has the potential to provide a basis for future systems that exploit the advantages of both rendering methodologies. When performing animation, the system distinguishes between static and dynamic objects. The total throughput is improved by adopting the following scheme: the static objects are voxelized once, intermixed with the sampled data and rendered by the volume rendering engine. The dynamic objects remain in the display-list and are rendered by the geometry engine into the Z-buffer which is combined (using the Z-merging algorithm) with the volume projection. For example, the need to display a detailed artificial heart (composed of thousands of polygons) which is statically placed inside the patient chest degrades the performance of the geometry engine. Instead, it can be scan converted once into the volume and intermixed with the sampled chest. Moving a scalpel (dynamic object) over the volume data is a time consuming process in volume representation. Instead it is rendered by the geometry engine and intermixed with the static volume data through Z-merging.

References

Cline, H. E., Lorensen, W. E., Ludke, S., Crawford, C. R., and Teeter, B. C., "Two Algorithms for Three-Dimensional Reconstruction of Tomograms", *Medical Physics*, **15**, 3, 320-327, (May/June 1988).

Duff, T., "Compositing 3-D Rendered Images", *Computer Graphics*, **19**, 3, 41-44, (July 1985).

Frieder, G., Herman, G. T., Meyer, C., and Udupa, J., "Large Software Problems for Small Computers: An Example from Medical Imaging", *IEEE Software*, **2**, 5, 36-47, (September 1985).

Fuchs, H., Kedem, Z. M., and Uselton, S. P., "Optimal Surface Reconstruction from Planar Contours", *Communications of the ACM*, **20**, 10, 693-702, (October 1977).

Goodsell, D. S., Mian, S., and Olson, A. J., "Rendering of Volumetric Data in Molecular Systems", *Journal of Molecular Graphics*, **7**, 41-47, (March 1989).

Herman, G. T. and Udupa, J. K., "Display of Three Dimensional Discrete Surfaces", *Proceedings of the SPIE*, **283**, 90-97, (1981).

Hoehne, K. H. and Bernstein, R., "Shading 3D-Images from CT Using Grey-Level Gradients", *IEEE Transactions on Medical Imaging*, **MI-5**, 1, 45-57, (March 1986).

Johnson, E. R. and Mosher, C. E., "Integration of Volume Rendering and Geometric Graphics", *Proceedings of the Chapel Hill Workshop on Volume Visualization*, Chapel Hill, NC, 1-8, May 1989.

Kaufman, A. and Shimony, E., "3D Scan-Conversion Algorithms for Voxel-Based Graphics", *Proceedings ACM Workshop on Interactive 3D Graphics*, Chapel Hill, NC, 45-75, October 1986.

Kaufman, A., "Efficient Algorithms for 3D Scan-Conversion of Parametric Curves, Surfaces, and Volumes", *Computer Graphics*, **21**, 4, 171-179, (July 1987).

Kaufman, A., "Efficient Algorithms for 3D Scan-Converting Polygons", *Computers & Graphics*, **12**, 2, 213-219, (1988).

Kaufman, A. and Yagel, R., "Tools for Three-Dimensional User Interfaces", in *Work with Computers: Organizational, Management, Stress and Health Aspects*, M. J. Smith and G. Salvendy, (eds.), Elsevier Science Publishers B.V., Amsterdam, The Netherlands, , 468-475, September 1989.

Kaufman, A., Bakalash, R., Cohen, D., and Yagel, R., "Architectures for Volume Rendering - A Survey", *IEEE Engineering in Medicine and Biology*, April 1990.

Levoy, M., "A Hybrid Ray Tracer for Rendering Polygon and Volume Data", *IEEE Computer Graphics & Applications*, 1990, (to appear) .

Sunguroff, A. and Greenberg, D., "Computer Generated Images for Medical Application", *Computer Graphics*, **12**, 3, 196-202, (August 1978).

Westover, L., "Interactive Volume Rendering", *Proceedings of the Chapel Hill Workshop on Volume Visualization*, Chapel Hill, NC, 9-16, May 1989.

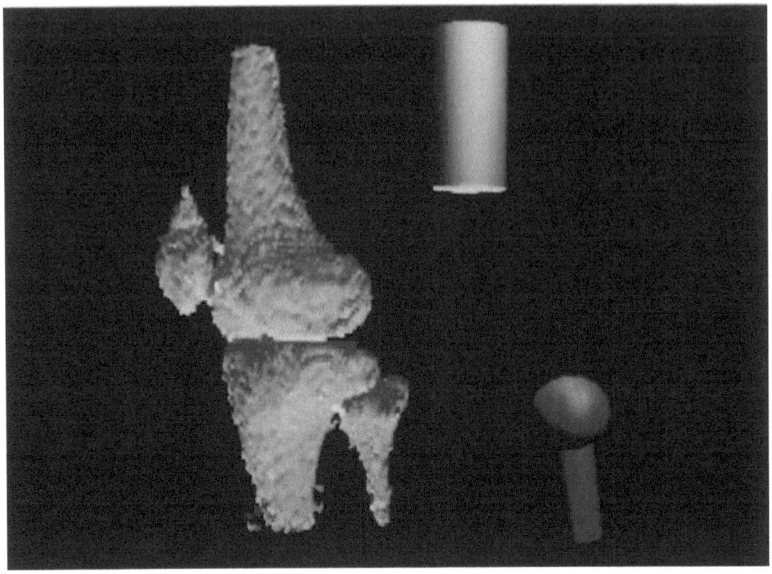

Fig. 1: A knee extracted from an MRI volumetric data along with a geometric model of a simulated prosthesis, rendered on a Silicon Graphics IRIS 4D/260.

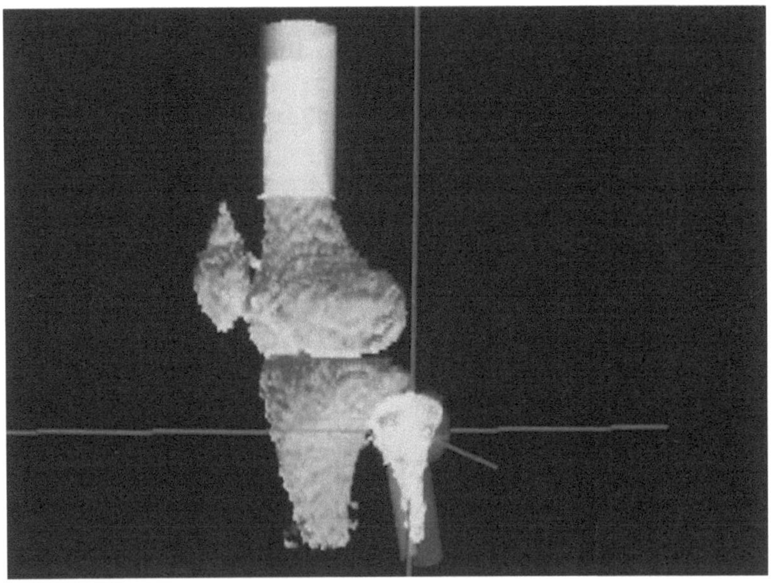

Fig. 2: The geometric model of the prosthesis is interactively positioned and intermixed with the knee using Z-merging enhanced with translucency.

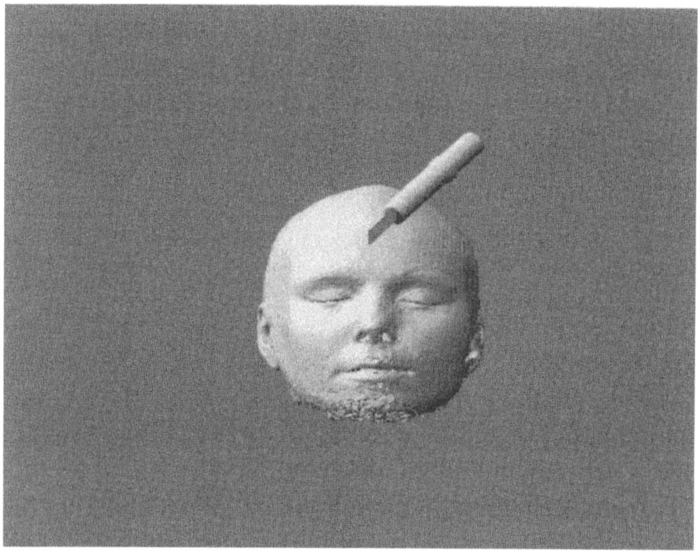

Fig. 3: Surgery planning: Voxel-based image of a patient head reconstructed from MRI scans, intermixed with a voxelized scalpel and rendered on Sun/4 workstation.

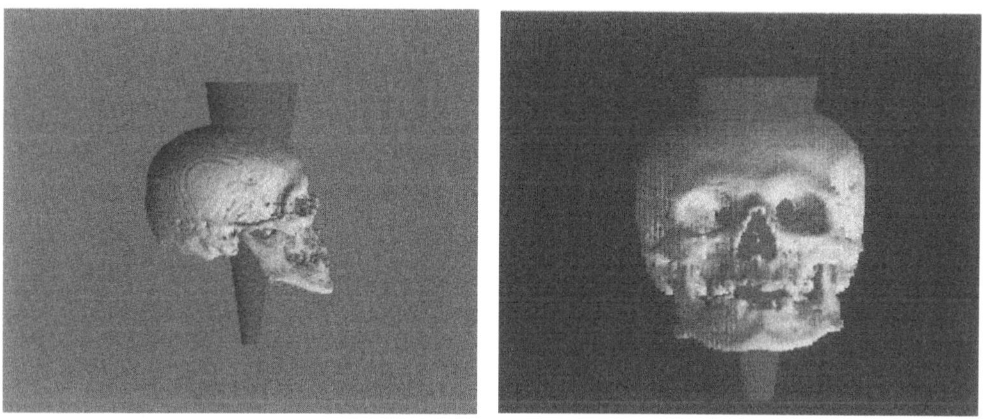

Fig. 4: Radiation therapy planning: Radiation beam is modeled as a cone that is voxelized and intermixed with a CT data set.

Combined 3D-Display of Cerebral Vasculature and Neuroanatomic Structures in MRI

Hans-Heino Ehricke, Gerhard Laub

Siemens Medical Systems, SMA1, Henkestr. 127,D-8520 Erlangen

Abstract

Neuroradiologists and neurosurgeons in many clinical situations take a vital interest in images revealing not only intracranial vasculature, but also other neuroanatomic structures. With MRI, high-quality volume datasets may be acquired containing the necessary anatomic information. 3D-postprocessing methods are required in order to generate integrated displays of vessels and brain structures. With respect to this issue in this paper a 3D-visualization approach is proposed which has been designed according to the principles of robustness, user-friendliness and clinical practicality. It includes measurement techniques for the acquisition of flow-compensated angiographic datasets and T1-weighted tissue volumes as well as image processing methods operating on the ray-tracing principle.

Keywords: Neurosurgery, planning tool, MR angiography, 3D-visualization, brain anatomy.

1. Introduction

Recently, flow-compensated 3D-gradient echo MRI pulse sequences have been developed and optimized allowing the acquisition of high-quality angiographic image data [Laub 1988, Ruggieri 1989]. Thus, for the first time, angiographic volume datasets with high vessel contrast (as compared to CT) have become available from which three-dimensional vessel coordinates can be derived.

Postprocessing of the data volumes is done by maximum intensity projection (MIP). The resulting projection images provide a 2.5D-view of the vascular tree, but fail to give sufficient information about its relationship to soft-tissue

NATO ASI Series, Vol. F 60
3D Imaging in Medicine
Edited by K. H. Höhne et al.

structures. In a number of clinical situations, knowledge of vessel and adjacent soft-tissue anatomy is of great importance, especially in the fields of neuroradiology and neurosurgery.

Up to now there has been only little research on achieving an integration of brain and vascular anatomy in 3D-imaging. On one hand, it concentrated on the combination of stereoscopic DSA-projections with CT- or MR-volume data for stereotactic neurosurgery planning [Henri 1989, Peters 1989, Peters 1986]. On the other hand, algorithms which operate on true volume datasets have been under investigation trying to match MR angiographic data with neuroanatomic soft-tissue information [Cline 1989, Ehricke 1990, Hu 1989].

In this paper we present an approach for generating combination displays revealing brain anatomy and intracranial vasculature in a three-dimensional manner. It involves not only postprocessing methods, but also MR image acquisition techniques which have proven suitable in this context. It has been designed with the aim of providing a robust, reliable and clinically practical tool for neurosurgery planning.

2. Data Acquisition

In order to acquire vascular datasets with high contrast even for peripheral cranial vessels we maximize inflow effects by a 2D sequential-slice method. Here we scan the volume to be imaged with a flow-compensated 2D-FLASH pulse sequence with a slice thickness of 2.0 mm to 3.0 mm and a slice shift of approx. 1.0 mm (fig. 1). Thus even slow flow, such as in venous anatomy, is visualized.

Brain tissue information is derived from T1-weighted images which are measured with a 3D-FLASH sequence. Measurement parameters include a flip angle of 40 degrees, echo times of 6 to 8 msec and a repetition time of approx. 40 msec.

The two datasets are acquired during a single session with equal in-plane resolution of up to 0.7 mm. The partition thicknesses vary: they are in the range of 0.7 to 1.5 mm for the T1-weighted dataset and 2.0 mm for the angiographic volume.

Measurement times depend largely on the thickness of the volume to be acquired. As an example the acquisition of a cube containing brainstem, cerebellum, pituitary gland and optic nerves (approx. 8 cm * 20 cm * 20 cm) takes 15 minutes including 5 minutes for 32 2D-slices (vessels) and 10 minutes for 64 3D-partitions (brain). For patients who are not able to keep still during data acquisition some fixation device may be used, such as those described in [Schad 1987] or [Tofts 1990].

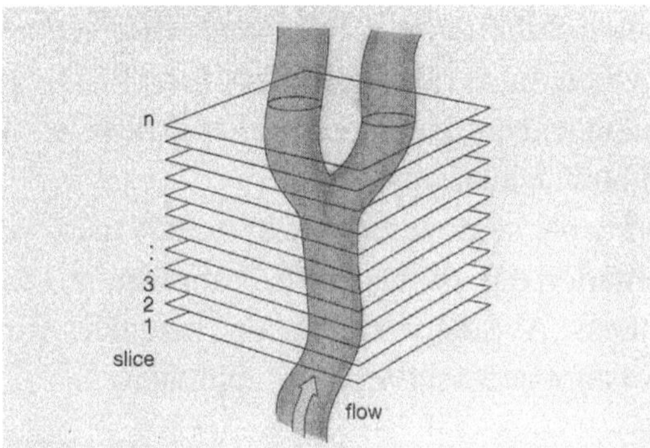

Fig. 1: Acquisition of angiographic data by 2D sequential-slice scanning.

3. Preprocessing and Segmentation

As a rule the measured datasets have a better in-plane than slice-to-slice resolution. In a first preprocessing step the data are interpolated in order to achieve an isotropic resolution, which typically is in the range between 0.7 and 1.0 mm. During the interpolation process the two data cubes are resliced to generate an identical volume partitioning.

3D-visualization of internal anatomic structures from MR-datasets requires the isolation of tissues to be displayed. Manual segmentation by tracking organ contour lines via mouse and cursor is an exhausting and time-consuming task. Therefore segmentation algorithms to automate this procedure are desirable, but their verification is one of the most serious problems in image processing [Marr 1980, Hildreth 1983]. Although a lot of research work has been done on this issue [Bomans 1987, Ehricke 1990a, Ehricke 1990b, Menhardt 1987, Pizer 1989] little success has been achieved. In our institution a variety of segmentation operators for MR-images has been tested and developed including region-based (region growing, region merging, polynomial classification, texture

analysis) as well as edge-based operators (local gradients, dynamic programming). A great deal of experience is necessary to select appropriate operators for a certain segmentation task defined in terms of anatomy, data quality, acquisition parameters etc. In order to provide a segmentation tool for an inexperienced user a workstation-based segmentation system has been developed incorporating some of the above mentioned operators into an environment with editing functions (drawing lines, rubbing). The system allows a semi-automatic segmentation of MR-datasets and can be used for a variety of segmentation problems. Furthermore a fully automatic segmentation system is under investigation which in a rule-based manner (PROLOG) makes use of domain-independant and expert knowledge in the field of image acquisition, image processing and human anatomy.
With the currently available semi-automatic system a user with a little training may achieve a segmentation e.g. of the whole brain from a 128 partitions dataset in approx. 30 minutes. A segmentation of vascular structures from the angiographic datasets is not necessary for our approach.

4. 3D-Visualization

4.1 Single Displays of Vessels or Brain

As basis for the generation of 3D-displays we make use of the ray-tracing principle [Höhne 1987, Levin 1989a]. As the original T1-weighted datasets have been segmented during the preprocessing phase the first peak in the intensity profile along each ray defines the brain surface. Brain surface voxels are depicted into the observer image plane with integral shading summing up a user-defined number of brain voxels along the ray. This technique which is rather robust in the presence of slight segmentation errors yields good results, especially for the visualization of the brain surface (fig. 2). The physical principle is simple: where the surface has an inclination nearly parallel to the ray a great number of dark greymatter or background (at object margins) voxels are used for the calculation of the integral value resulting in dark greyvalues for sulci and object margin pixels and light values for center pixels (gyri).

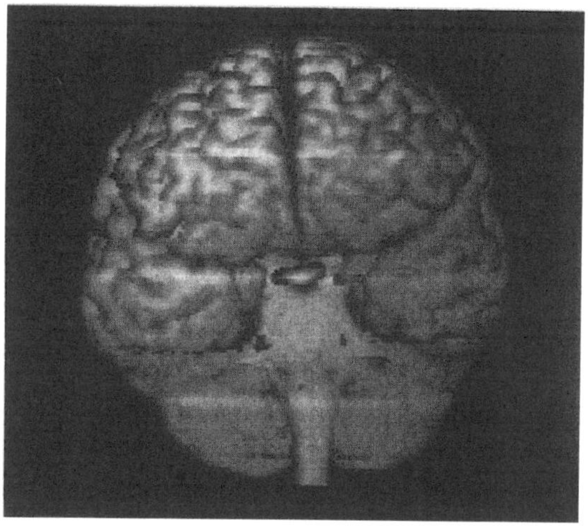

Fig. 2: Brain surface display generated with integral shading.

For the visualization of cerebral vascular anatomy maximum intensity projection has proven to be most suitable (fig. 3). Here along each ray only the maximum greyvalue is depicted onto the observer image plane (fig. 4). We do not use any shading method, because for cerebral vessels which have a thickness between one and a few pixels no real surface can be determined. Even transparent shading [Levoy 1988] does not yield better results for this application, but requires a segmentation of vessels to be visualized. Especially for smaller vessels this is not a trivial task and makes a great deal of user-interaction necessary.

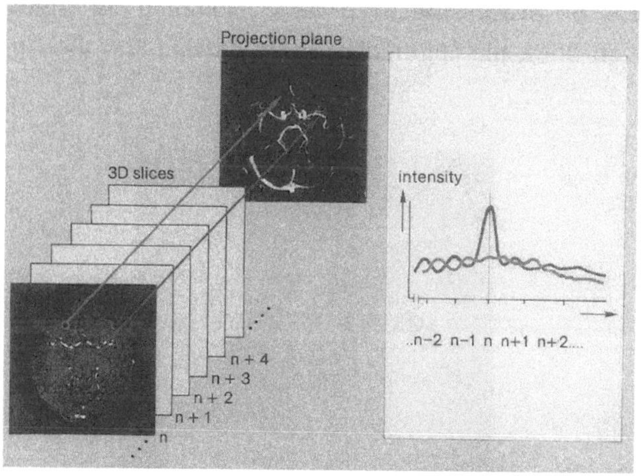

Fig. 3: Principle of maximum intensity projection.

MIP-projection images do not give a real 3D-perception. This may be provided by a cine-mode display showing the object in a rotating manner on a monitor. Another drawback of MIP is that the vessel contrast is decreased, because for background pixels the maximum noise value along the ray is used. This is true especially for datasets acquired with 3D-techniques. In images measured with the 2D-sequential slice method background noise is rather low and thus there is not a very serious decrease in contrast by MIP.

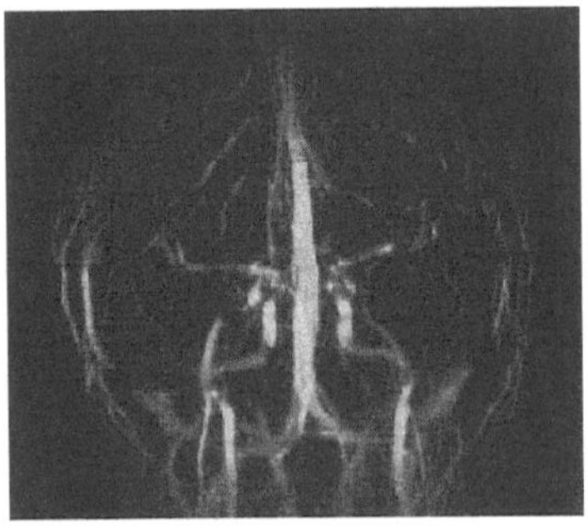

Fig. 4: Projection image generated with MIP.

4.2 Integrated Display

It is obvious that by a mere addition of corresponding MIP- and surface-images a realistic 3D-perception of vessels and brain anatomy cannot be achieved (figure 5). So the two display methods described above have to be combined in an earlier phase of processing. We have designed an algorithm operating simultaneously on both, the segmented brain dataset and the flow-compensated vessel volume.

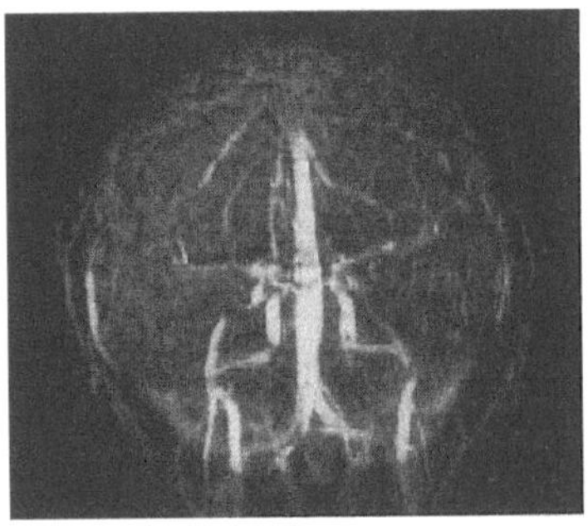

Fig. 5: Mere superposition of brain surface and MIP images.

Figure 6 illustrates the principle of the MIP-integration method. Along each ray through the volume in a first step the localization of the brain surface is computed in the segmented dataset from the first peak in the grey-value profile. The brain surface value for the ray is calculated by integrating several grey-values of brain tissue in viewing direction. The vascular contribution is derived from the vessel dataset as the maximum grey-value along the corresponding ray in the range between the observer position and the surface localization.
The two data values for brain surface and vessel contribution are projected onto the observer image plane by combining them with different weighting factors which take into account the presence or absence of brain or vessels in a pixel of the observer image plane. As can be seen in figure 7 this approach leads to a quite realistic 3D-perception of the brain surface and external vessels.

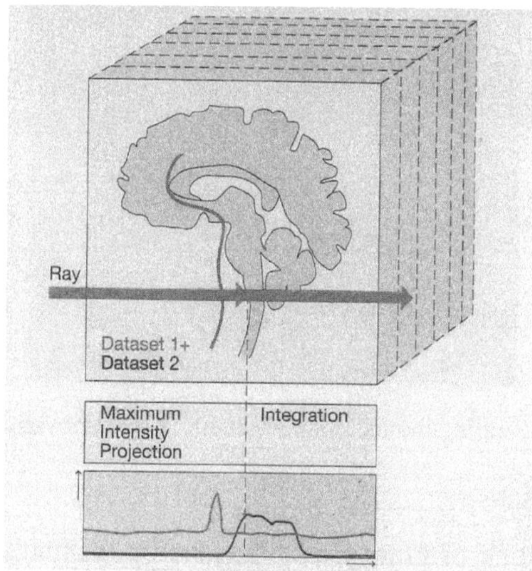

Fig. 6: Principle of MIP-integration.

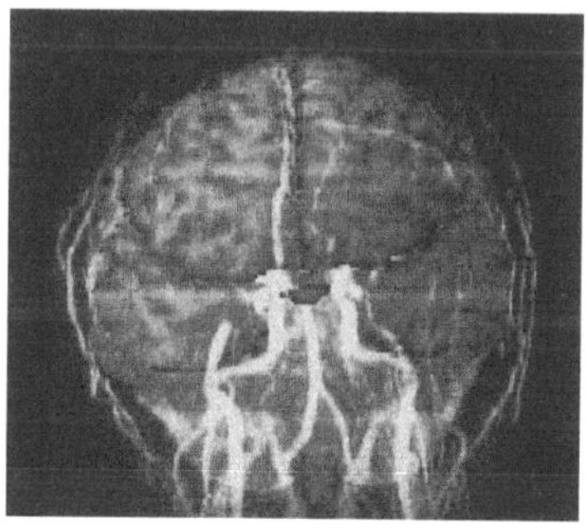

Fig. 7: Integrated 3D-display of brain and vasculature generated with MIP-integration.

4.3 Visualization of Intracerebral Vasculature

Visualization of intracerebral vessels can be achieved by three different methods. First, the transparency of the brain surface may be increased by widening the range for calculation of the vessel value along a ray starting with the observer position and ending a user-defined number of voxels in the brain tissue. For visualization of near-surface vessels which may be encountered during the process of neurosurgical exposure of an intracerebral lesion this may be quite useful.

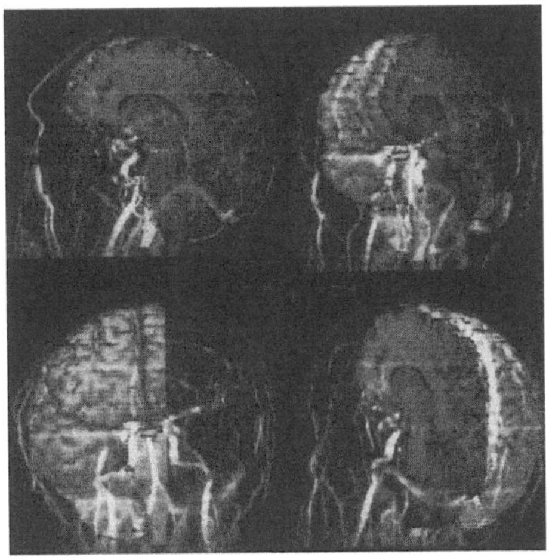

Fig. 8: Revealing intracerebral vasculature by removing brain tissue.

Second, a whole block of brain tissue may be interactively removed via mouse and cursor revealing hidden vessel structures, such as in fig. 8. Finally certain brain structures which a neuroradiologist or a neurosurgeon takes an interest in (e.g. a tumour in its anatomic environment) may be isolated during the segmentation phase and three-dimensionally displayed with adjacent vasculature. This may be helpful to reveal the infiltration and relationship of a tumour with surrounding tissue and vasculature. As an example we have concentrated on the inspection of brainstem, cerebellum, pituitary gland and optic nerves. Fig. 9 and 10 show 3D-views which have been generated by MIP-integration on basis of a high-resolution dataset with 0.7 mm in-plane and 1.1 mm resp. 2.0 mm interplane resolution.

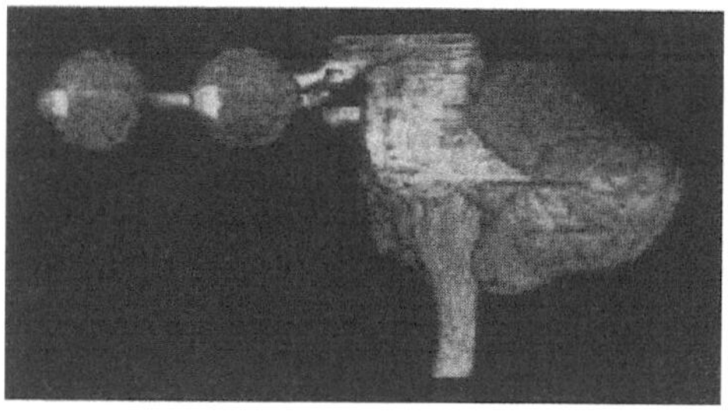

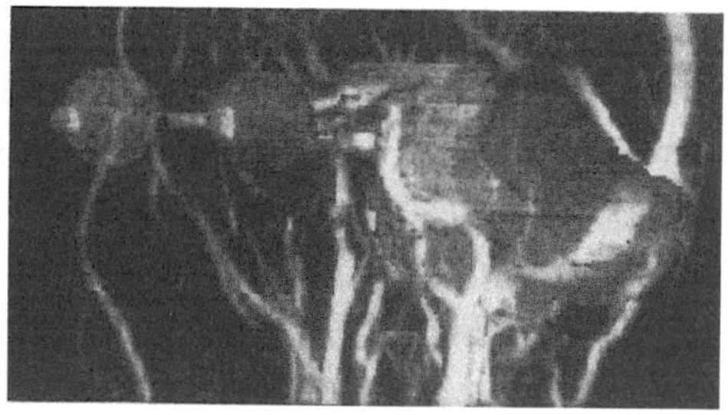

Fig. 9: 3D-display of brainstem, cerebellum, pituitary gland and optic nerves.

Fig.10: Integrated display revealing vascular and neuroanatomic anatomy.

5. Discussion

We have described an approach including image acquisition as well as image processing methods to reveal anatomic relationships between brain structures and cranial vasculature. We regard it as the most important message of this paper that we could demonstrate the usefulness of 3D image processing techniques for medical applications. 3D image processing in MRI provides not only nice images, as maintained by most critics, but can also give valuable assistance in various clinical situations. MIP-integration is a simple, robust, computationally inexpensive and reliable method which has been designed especially for the improvement of neurosurgical planning.

Simplicity on sides of image processing has become possible by the use of optimized image acquisition techniques. This demonstrates that by knowledge on the data acquisition process and its optimization for postprocessing goals new perspectives may be opened up for 3D image processing. The development of image processing techniques without taking into account modern acquisition methods may lead to rather sopisticated methods which are not clinically practical or relevant.

The most serious handicap of the MIP-integration method is the necessity to segment brain tissue from the original dataset. Although segmentation algorithms have been developed for an automation of the process, a considerable amount of user-interaction is still required. We are sure that by the introduction of user-friendly image processing workstations and the optimization of segmentation algorithms with methods of artificial intelligence, interaction time can be reduced to an amount which is clinically tolerable.

Our results suggest that this method, in combination with improved acquisition and processing capabilities, should provide a practical and clinically relevant technique to enhance the pre-operative value of MR imaging of the brain and cerebral vasculature. It is up to neuroradiologists and neurosurgeons now to make available their professional experience for the specification and development of an image processing tool that fits in with clinical routine work.

References:

Bomans, M.: 3D-Segmentation von Kernspin-Tomogrammen. Proc. 9th DAGM Symposium, pp 231-235, Springer, 1987.

Cline, H.E. et al.: Vascular Morphology by Three-Dimensional Magnetic Resoonance Imaging. Mag. Res. Im., vol. 7, pp 45-54, Pergamon Press 1989.

Ehricke, H.-H., Laub, G.: Gewebecharakterisierung in der dreidimensionalen Kernspintomographie mit Methoden der Texturanalyse. Proc. 6th Radiol. Sympos. Graz, Schneider et al. (eds), Ueberreuther, Berlin 1990.

Ehricke, H.-H.: Problems and Approaches for Tissue Segmentation in 3D-MR Imaging. Proc. SPIE 90 Medical Imaging IV, 1990, in press.

Ehricke, H.-H., Laub, G.: 3D-Visualization of Intracranial Vessels and Brain Anatomy in MRI. Proc. SPIE 90 Medical Imaging IV, 1990, in pess.

Henri, C. et al.: Three-Dimensional Interactive Display of Medical Images for Stereotactic Neurosurgery Planning. Proc SPIE 89 Medical Imaging III, vol. 1092, pp 67-74, 1989.

Herman, G.T., Liu, H.K.: Three-Dimensional Display of Human Organs from Computed Tomograms. Comp. Graph. a. Im. Proc., 9, pp 1-21, Academic Press 1979.

Hildreth, E.: The Detection of Intensity Changes by Computer and Biological Vision Systems. Comp. Vision,Graph. a. Im. Proc., 22, pp 1-27, Academic Press 1983.

Höhne, K.H.: 3D-Bildverarbeitung und Computer-Graphik in der Medizin. Informatik Spektrum, no 10, pp 192-204, 1987.

Hu,X. et al.: Volumetric Rendering of Multimodality Medical Imaging Data. Proc. Chapel Hill Workshop on Volume Visualization, Upson, C. (ed.), pp 45-49, 1989.

Laub, G., Kaiser, W.: MR Angiography with Gradient Motion Refocussing. J. Comp. Ass. Tom., vol. 12, pp 377-382, 1988.

Levin, D.N. et al: Surface of the Brain: Three-dimensional MR Images Created with Volume Rendering. Radiology, vol. 171, no 1, pp 277-280, April 1989.

Levin, D.N. et al.: The Brain: Integrated Three-Dimensional Display of MR and PET Images. Radiology, 172, pp 783-789, 1989.

Levoy, M.: Display of Surfaces from Volume Data. IEEE Comp. Graph. and Appl., vol. 8, no 3, pp 29-32, May 1988.

Marr, D., Hildreth, E.: Theory of Edge Detection. Proc. R. Soc. Lond., B207, pp 187-217, 1980.

Menhardt, W., Schmidt, K.-H.: Automated Interpretation of Transaxial MR-Images. Proc. CAR 87, H.U. Lemke et al. (eds), pp 386-390, Springer, 1987.
Pelizzari, C.A. et al.: Accurate Three-Dimensional Registration of CT, PET and/or MR Images of the Brain. J. Comp. Ass. Tom., 13(1), pp 20-26, Raven Press 1989.
Peters, T.M. et al.: Integrated Stereotactic Imaging with CT, MR and Digital Subtraction Angiographie. Radiology, 161, pp 821-826, 1986.
Peters, T.M. et al.: Stereotactic Neurosurgery Planning on a PC Based Workstation. Proc. SPIE 89 Medical Imaging III, vol. 1092, pp 556-564, 1989.
Pizer, S.M. et al.: Multiscale Geometric Image Descriptions foor Interactive Object Definition. Mustererkennung 1989, 11. DAGM-Symposium, pp 229-239, Springer 1989.
Ruggieri, P. et al.: Intracranial Circulation: Pulse-Sequence Considerations in Three-Dimensional MR-Angiography. Radiology, vol. 171, pp 785-791, 1989.
Schad, L. et al.: Correction of Spatial Distorsion in MR Imaging: A Prerequisite for Accurate Stereotaxy. J. Comp. Ass. Tom., 11(3), pp 499-505, Raven Press 1987.
Smets, C. et al.: A Knowledge-Based System for the Delineation of Blood Vessels on Subtraction Angiograms. Patt. Rec. Let., 8, pp 113-121, North Holland 1988.
Tofts, P.S. et al.: Nasal Orientation Device to Control Head Movement during CT and MR Studies. J. Comp. Ass. Tom., 14(1), pp 163-164, 1990.

Preliminary Work on the Interpretation of SPECT Images with the Aid of Registered MR Images and an MR Derived 3D Neuro-anatomical Atlas

David J. Hawkes[1], *Derek L.G. Hill*[1], *Eldon D. Lehmann*[1], *Glynn P. Robinson*[2], *Michael N. Maisey*[1], *and Alan C.F. Colchester*[2]

[1]Division of Radiological Sciences and [2]Department of Neurology, U.M.D.S., Guy's Hospital, London Bridge, London, SE1 9RT, UK.

Abstract

This paper describes two methods to aid interpretation and quantification of SPECT or PET images. In the first method 3D SPECT or PET data sets are aligned and scaled to a 3D MRI data set of the same patient using 4 skin markers visible on each modality. Three display schemes have been implemented for viewing the aligned slices. Examples of these displays are provided. The second method uses a labelled 3D MRI reference data set from a volunteer to identify major anatomical structures. The MR reference data set is aligned with the isotope image using the same 4 markers plus a marker on the vertex of the skull. The reference data set is segmented approximately into the major tissue types - cerebrospinal fluid (CSF) and grey and white matter. Major structures are identified via labels in the 3D data set. A linked cursor aids delineation of anatomical regions on the isotope image using the outline of structures on the reference data set as a template. Directions for future research in the generation of complete digital anatomical atlases, which include inter-individual variations, are outlined.

Keywords: 3D Image registration, neuro-anatomical atlases, multi-modality displays, image interpretation, MRI, SPECT, PET.

1. Introduction

Single photon emission computed tomography (SPECT) and positron emission tomography (PET) provide information on tissue perfusion and organ function which cannot be obtained by other means. We propose two alternative approaches to assist in the delineation of anatomical regions of

NATO ASI Series, Vol. F 60
3D Imaging in Medicine
Edited by K. H. Höhne et al.

interest in SPECT or PET images. In the first method we display 3D registered images, derived from 3D MRI and SPECT data sets of the same patient. A linked cursor or colour overlay permits delineation of regions of interest in the SPECT image corresponding to anatomical regions visible in the MRI image.

In the second method we have constructed a segmented and labelled 3D MRI data set from a normal volunteer. This data set is registered with the patient SPECT data set. The two data sets are scaled, aligned and displayed so that the labelled and pre-segmented MRI data set can be used to delineate anatomical structures in the SPECT image. This method could obviate the need for a concurrent MRI investigation and could be used in those centres which do not have access to MRI.

A number of centres have recently demonstrated the combination of PET images with registered CT and MRI data sets from the same patient. This can be done either by using external markers [Chen 1985] and [Schad 1987], by using images of anatomical features such as the skin surface [Levin 1988] and [Pelizzari 1989] or by using the method of moments and principle axes [Gamboa-Aldeco 1986] and [Bajcsy 1989].

It has been reported that, except in gross cases of disease or space occupying lesions, the anatomical image of the brain remains reasonably constant once linear scaling, translation and rotation are performed. Dann and co-workers [Dann 1989] have shown that structures in the brain can be located to within 2-4mm when the skin surface was used for scaling and elastic matching was used between an atlas generated from photographs of cut sections of a cadaver head and CT scans of six neurologically normal patients. This precision is sufficient for our purposes due to the inherently poor spatial resolution of SPECT.

A number of different approaches have been adopted in the generation of neuro-anatomical atlases. Fox and co-workers [Fox 1985] used a stereotactic atlas from text books to facilitate anatomical localisation of structures in PET. Keyserlingk and co-workers [Keyserlingk 1988] generated a statistical brain atlas in order to study and quantify anatomical variations in sulcal patterns between individuals. Bohm and co-workers [Bohm 1989] have created an adjustable computerised brain atlas using digitised photographs of sections from a cryosectioned brain. Their atlas is primarily intended for use with PET but the authors suggest that their approach could also be used in other fields of neuro-imaging, such as stereotactic surgery, CT and MRI. All these methods have used cadaver specimens as original sources of data which have then been manually labelled to provide details of anatomical structure. In order to test the feasibility of our method of labelling and comparison reasonably quickly, we collected a single high resolution MRI data set from a volunteer and used this data set as input to our anatomical atlas. More elaborate and complete atlases could be incorporated at a later stage. The use of MRI derived atlases has the advantage that the data is collected from normal healthy volunteers and thus problems of cadaver specimen preparation, shrinkage, distortion and atrophy have been avoided.

2. Method

2.1 Image acquisition

The MR images were acquired from a 1.5 Tesla whole body MR system (Philips Medical Systems Gyroscan S15). Spin echo images with T_1 weighting and T_2 weighting and inversion recovery images were collected from each patient. The precise repetition times, echo times and slice thicknesses were altered in accordance with normal clinical considerations. For the anatomical atlas 126 transaxial 256x256 pixel 2mm thick contiguous slices were collected using a repetition time (TR) of 350ms and an echo time (TE) of 30ms. The SPECT images were acquired from an IGE Starcam AC400 tomographic gamma camera, using 64 30 second acquisitions, with 360 degree camera rotation. The patient was injected with hexamethylpropyleneamine oxime (HMPAO) labelled with 750MBq of technetium-99m 20 minutes prior to commencing the study.

2.2 Registration of 3D SPECT, MRI and Atlas data sets.

External markers. Registration of the 3D image data sets was accomplished by means of four external fiducial markers attached to the patient skin. The surface of the skin above the mastoid processes of the temporal bones and the zygomatic processes of the frontal bone were chosen as marker positions because of the ease with which they can be localised, and because normal movement of the skin at these points is small. Markers visible in both modalities were constructed and these were left in place on the patient's skin for both scans. This avoided errors introduced by using different markers for the SPECT and MRI acquisitions.

The markers were constructed by moulding silicone rubber around a ball bearing of diameter 3mm. The spherical void created was filled with a solution of sodium pertechnetate ($Na^{99m}TcO_4$) and copper sulphate solution. The activity of the pertechnetate was 0.2MBq per marker, giving a similar signal to the highest uptake expected within the brain for a 750MBq HMPAO SPECT study. The copper sulphate solution had a concentration of 1g per litre, giving the marker a sufficiently short T_1 and T_2 for it to appear with good image contrast in both T_1 weighted and T_2 weighted spin echo MR scans. Copper sulphate is replaced with iodine contrast medium (which gives high contrast in both CT and T_1 and T_2 weighted MR scans) when aligning CT data.

The approximate marker positions were selected manually on the displayed image. Interpolation was used to subtract the background signal from the image in the region surrounding the marker. The marker co-ordinate was defined as the position of the centre of gravity of the residual grey level distribution.

Registration algorithm. The two marker 3D point sets $\mathbf{p}_i$ and $\mathbf{p'}_i$ obtained from the two images were assumed to be related by a 3D translation **T**, a 3D rotation **R**, an isotropic scaling factor **S**, and noise N_i. We have not corrected for geometric distortion or warping in either modality.

$$\mathbf{p}_i' = \mathbf{T}(\mathbf{S}(\mathbf{R}(\mathbf{p}_i))) + N_i$$

The 4x4 matrices **T**, **S** and **R** allow us to express the entire transformation as a single matrix, the product of **T**, **S** and **R**. Using an algorithm derived from Arun et al (1987), the scaling, rotation and translation components of the transformation were decoupled, and solved separately. For registering SPECT and MRI images of the same patient, a single isotropic scaling factor was calculated from the maximum marker separations in the two images. For registration of SPECT data to the MRI atlas data, the three orthogonal cartesian coordinates were scaled separately using the coordinates of the 4 skin markers and the vertex of the skull.

The rotation **R** about the centroid of the marker 3D point set was calculated using a least square fit based on singular value decomposition (SVD), in the way described below.

The centroids **c** and **c'** of the marker 3D point sets $\mathbf{p}_i$ and $\mathbf{p}_i'$ are first calculated using the equation

$$\mathbf{c} = \frac{1}{N}\sum_{i=1}^{N} \mathbf{p_i}$$

where N is the number of markers, and similarly for **c'**.

These values are subtracted from the marker co-ordinates to generate marker 3D point sets $\mathbf{q_i}$ and $\mathbf{q'_i}$ relative to the centroids c and c'. The rotation matrix **R** has now been decoupled from the translation, and can be determined by minimising the sum of squares, Σ^2, given by

$$\Sigma^2 = \sum_{i=1}^{N} || \mathbf{q_i'} - \hat{\mathbf{R}}\mathbf{q_i} ||^2$$

The rotation matrix **R** can be found from the SVD of the 3x3 matrix **H** [Arun 1987].

$$\mathbf{H} = \sum_{i=1}^{N} \mathbf{q_i}\, \mathbf{q_i}'^{t} = \mathbf{U\Lambda V^t}$$

where $\mathbf{q}^t$ represents the transpose of **q**, and $\mathbf{U\Lambda V^t}$ is the SVD of **H**. The rotation transformation can then be found from

$$\mathbf{X} = \mathbf{VU^t}$$

where **R**=**X** provided the determinant of **X** is +1. If the determinant of **X** is −1, then the correct rotation can still be found by changing the sign of the most singular column of **V**, and recalculating **X** and hence **R**.

Once the rotation transformation has been established, the translation **T** can be calculated from the separation of the marker 3D point set centroids:

$$\mathbf{T} = \mathbf{c'} - \mathbf{Rc}$$

Transformation from one image co-ordinate system to the other proceeds as follows. A ray is cast through the untransformed 3D data set at an angle and position corresponding to a single column of the transformed 3D data set. Trilinear interpolation between the 8 nearest neighbours in the untransformed data set is used to compute the transformed data set at each sampling interval along the ray. Once the transformation and interpolation have been performed, the two images have the same dimensions, and equivalent voxels in the two 3D images represent the same volumes in the patient, within the accuracy of the registration.

2.3 Generation and use of labelled, reference MR data set

We used a simple modification of classical region growing techniques to label regions corresponding approximately to the major tissue types within the brain - the CSF, grey matter and white matter. For each 2D slice an edge map was generated using the Canny edge detector with hysteresis [Canny 1983] and a sigma of 3.0. Upper and lower thresholds for each tissue type were set by inspection of the original images. For each tissue type, regions were grown in 3D (27 nearest neighbours) using as seed points all pixels which were between the threshold values for that tissue type. Each region was grown until an edge was encountered, or the voxel value to be grown had a grey level which differed by more than 8% from the average grey level of the current grown region. The region growing was stopped and regions were merged when they encountered a previously grown region of the same tissue type. When all the tissue types had been grown independently, voxels that had been categorised as belonging to more than one tissue type were assigned to one particular tissue type depending on which of the original threshold ranges was closest to the original grey level value of the voxel concerned. Unconnected regions of isolated pixels were removed. In this way an approximate map of CSF, white matter, and grey matter was generated.

In order to label the 3D data set we have developed a scheme which identifies particular anatomical structures by means of points within each 2D slice. The user is left to judge the boundary of the region using visual, anatomical and clinical skills. This is an attempt to simulate the use of pointers in conventional anatomical text books.

Each label allows access to an on-line data base which contains corresponding entries of anatomical, physiological, pathological and clinical information. Thus selection of a particular label and type of information required, leads to display of the relevant text information. Information is displayed in a hierarchy from coarse to fine detail. At each stage access to functional, pathological and clinical information could be incorporated, and this could include images, diagrams and explanations. These schemes therefore provide the basic structure of an interactive training and teaching tool for image interpretation.

2.4 Display

Three techniques for the display of aligned slices have been implemented. The first utilises adjacent displays with separate controls for contrast and brightness in each image and a linked cursor for the delineation of regions of interest. In the second method mixing of the information from each modality in a single display is achieved by assigning alternate pixels in the display with colours from the two different modalities. In the third method a visual transparency effect is achieved by coding the pixel values from one modality to display brightness and the pixel values from the other modality to display chromaticity.

3. Results

The three types of display are shown in figure 1 for a single aligned pair of SPECT and corresponding MRI slices from the same patient on the same day. The scans were reported as being radiologically normal. Figure 1a shows the two slices aligned and scaled with a cursor, movable under mouse control, linking corresponding points on each slice. In the second example, figure 1b, the same data is displayed using alternate pixels to represent grey intensity scale (MRI data) and a saturated rainbow colour scale (SPECT data). The third example, Figure 1c, shows the use of 8 colour scales, each with 31 intensity values to generate transparent overlay. The chromaticity of each scale is determined by the SPECT image intensity and the intensity on the colour scale is determined by the MRI image intensity.

Figure 2 shows an example slice from the MR derived neuro-anatomical atlas displayed aligned and scaled adjacent to the corresponding slice from a SPECT data set. Labels of the major anatomical structures are shown. The red region corresponds approximately to grey matter, the yellow region to white matter and the green region to cerebrospinal fluid.

4. Conclusions

The use of 4 markers, with a mixture of copper sulphate and ^{99m}Tc-HMPAO is effective at registering SPECT and MRI images. There will be errors due to geometric distortion , primarily in the MRI images, and due to uncertainty in marker position. These errors will be small compared with the inherent resolution of the SPECT data. The markers are simple to prepare, easy to position and there is no patient discomfort. They do, however, necessitate the patient having contiguous SPECT and MRI scans.

The linked cursor display allows rapid and straightforward delineation of

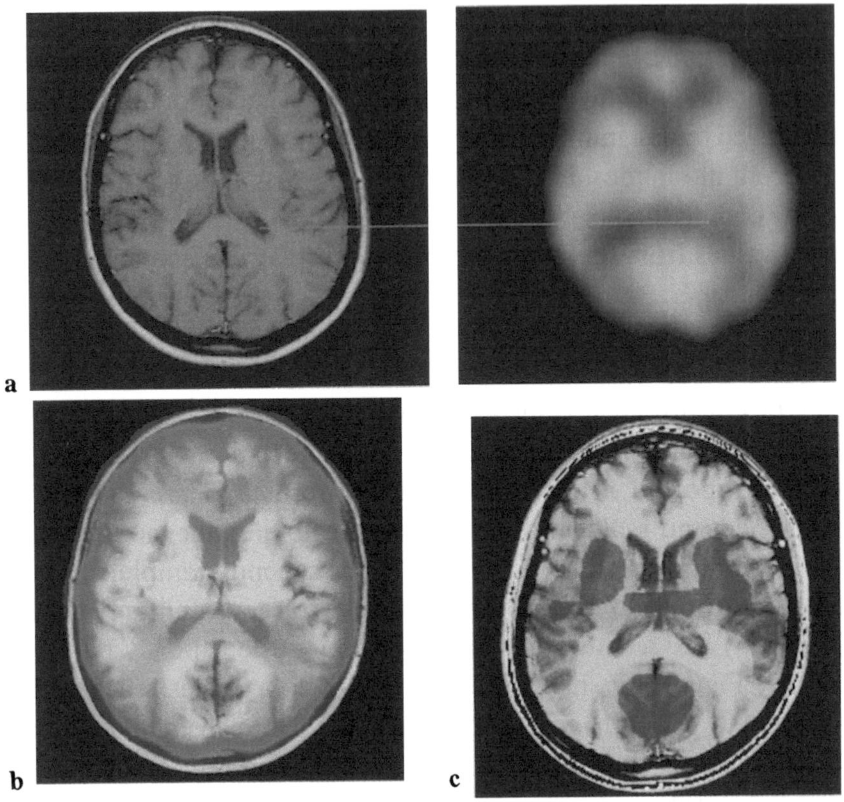

Fig. 1. Three methods of display for a single pair of aligned SPECT and MRI slices from the same patient. **a.** The use of a linked cursor under mouse control. **b.** Alternate pixel overlay. **c.** Transparent colour overlay.

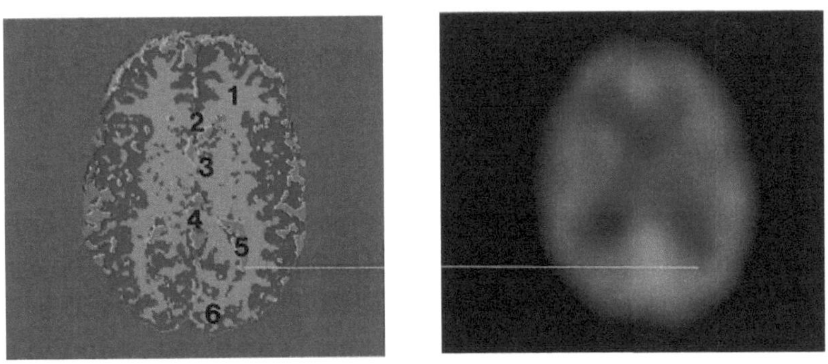

Fig. 2. Example slice from the MR derived neuro-anatomical atlas displayed aligned and scaled adjacent to a corresponding slice from a normal SPECT study. On the atlas, red corresponds approximately to grey matter, yellow to white matter, and green to CSF. **Index:** 1. Frontal lobe; 2. Genu of corpus callosum; 3. Thalamus; 4. Splenium of corpus callosum; 5. Posterior horn of lateral ventricle; 6. Occipital lobe.

corresponding regions in the two images and is the easiest to interpret. The combined colour displays potentially allow direct visualisation of the combined image data. Unfortunately, due to colour mixing, it is difficult to interpret the information in the alternate pixel display despite the visually pleasing effect. The third display suffers from coarse discretization of the SPECT image intensity scale. This is a limit imposed by an 8 bit display system, allowing only 256 colours to be displayed simultaneously. This limitation would be removed by using a 10 or 12 bit display system capable of displaying 1024 or 4096 colours simultaneously.

We registered and scaled the SPECT image data to the atlas data using the same four external markers plus a marker on the skin above the vertex of the skull. Despite the inter-individual consistency of neuro-anatomical structures, there will be errors in the registration of the atlas data to the SPECT data. In our application the SPECT resolution is approximately 10mm, and these errors are probably small by comparison. Keyserlingk and co-workers [Keyserlingk 1988] were able to show greater consistency of the major sulci by registering the data set from each individual brain to the midpoint of the anterior commissural-posterior commissural (AC-PC) line and scaling it to the maximum extent of the brain laterally, superiorly, inferiorly, anteriorly and posteriorly. These features are often visible on the MR scan but not on the SPECT data. Some improvement on location accuracy may be achieved by deriving the rotation and translation transformations from the markers and then scaling from the SPECT data itself. It is difficult to define accurately the cortical surface from the blurred SPECT image.

Work is in progress to check the accuracy and reproducibility of registration of the scaled anatomical atlas with MR data collected on a small series of patients with markers in position.

The use of a labelled reference MRI data set as an anatomical atlas is at an early stage of development and evaluation, yet we have been able to show that in certain circumstances its use would obviate the need for concurrent MR imaging with markers. The atlas permits labelling of SPECT images which have been reformatted in a variety of orientations including double oblique reconstructions. This could be particularly useful in focal epilepsy studies of the temporal lobe when oblique transaxial slices are viewed. The use of the atlas will be invalidated by the presence of mass lesions which lead to gross distortion of anatomical structures. Such gross distortions should, however, be apparent from inspection of the SPECT images.

All the voxels in the atlas have been given a label which corresponds approximately to tissue type but we have not attempted to delineate all structures visible in the reference MRI data set. Instead we have used labelled points to indicate a particular anatomical feature in the 2D slices. The clinician can use the segmented MR image as a guide to shape and size. In this way, the atlas mimics the operation of an anatomical text book. The simple data driven tissue type classification gives good cues of shape for most of the dominant structures in the brain, but significant errors still arise in identification of tissue types. For example the thalamus is classified as white matter when it should be grey matter.

The alternate approach is to interactively delineate all anatomical structures on each slice. This is extremely time consuming and it is difficult for even an expert to define all visible anatomical regions consistently. The labelling system via pointers to regions classified by tissue type is easier to implement but difficulties can still arise for complex convoluted structures.

Future research must address the problems of unambiguous labelling of cerebral structures. We propose that voxel labelling of well defined complete anatomical structures is most appropriate while pointers could be used to indicate parts of these structures. These pointers could be points in 2D or lines in 3D.

In future work the data representation in the digital anatomical atlas should contain information on both the uncertainty in the labelling of a particular voxel and uncertainties in shape and size of a structure. A promising approach which includes a statistical measure of inter-individual variations in anatomy has been described by Keyserlingk and co-workers [Keyserlingk 1988]. They have collected data from more than 80 brains post mortem and from careful measurements of the cortical surface and sulci have generated a statistical brain atlas with mean and spread of position of the sulci and gyri. Extending this work to structures derived from MRI on living individuals and improving the averaging process to include measures of variations in shape and size as well as location would be a valuable avenue of future research.

In conclusion, we have developed a simple practical method for the 3D registration of MR and SPECT images of the same patient. In addition we have shown how a simple neuro-anatomical atlas could be used to aid interpretation of SPECT images. The combination of clinical SPECT data of a particular patient with this atlas linked to an interactive database containing clinical, anatomical, physiological and pathological information would be of great benefit as a teaching and training tool for medical students and clinicians.

Acknowledgements

We are grateful to the Leverhulme Trust for supporting DJH and DLGH and to Alvey project MMI 134 for supporting GPR. We acknowledge many of our colleagues for their help and support in this work. In particular we thank Philippa Graves, Peter Liepins, Charles Bird, Martin Graves, John Lynch, Ron Mistry, and Colin Lazarus. We are also grateful to Prof. Rich Robb at the Mayo clinic for supplying the ANALYZE software package, which was used to assist in this work.

References

Arun KS, Huang TS, Blostein SD. **Least-Squares Fitting of Two 3D Point Sets.** IEEE t-PAMI-9:698-700 (1987)

Bajcsy R, Kovacic S. **Multiresolution Elastic Matching.** Comp. Vis. Graph. Image Proc. 46:1-21 (1989)

Bohm C, Greitz T, Eriksson L, **A Computerised Adjustable Brain Atlas**, Eur J. Nucl. Med 15:687-689 (1989)

Canny JF **Finding Edges and Lines in Images** MIT AI Laboratory Technical Report No 720 (1983)

Chen GTY, Kessler M, Pitluck S, **Structure Transfer Between Sets of Three Dimensional Medical Imaging Data.** Proc Computer Graphics '85 (Dallas) 171-177 (1985)

Dann R, Hoford J, Kovacic S, Reivich M Bajcsy R. **Evaluation of Elastic Matching System for Anatomic (CT, MR) and Functional (PET) Cerebral Images** J CAT 13:603-611 (1989)

Fox PT, Perlmutter JS, Raichle ME. **A stereotactic method of anatomical localisation for positron emission tomography.** J CAT 9:141-153 (1985)

Gamboa-Aldeco A, Fellingham LL, Chen GTY. **Correlation of 3D surfaces from multiple modalites in medical imaging.** Proc SPIE 626:467-473 (1986)

Graf v. Keyserlingk K, Niemann K and Wasel J, **A quantitative approach to spatial variation of human cerebral sulci**, Acta anat. 131:127-131 (1988)

Levin DN, Pelizzari CA, Chen GTY, Chen CT, Cooper MD. **Retrospective Geometric Correlation of MR, CT and PET Images** Radiology 169:817-823 (1988)

Pelizzari CA, Chen GTY, Spelbring DR, Weichselbaum RR, Chen CT. **Accurate Three-Dimensional Registration of CT, PET, and/or MR Images of the Brain.** J CAT 13:20-26 (1989)

Schad LR, Boesecke R, Schlegel W, Hartmann GH, Sturm V, Strauss LG, Lorenz WJ. **Three Dimensional Image Correlation of CT, MR, and PET Studies in Radiotherapy Treatment Planning of Brain Tumors** J CAT 11:948-954 (1987)

Volume Visualization of 3D Tomographies

H. P. Meinzer, U. Engelmann, D. Scheppelmann, R. Schäfer

German Cancer Research Center
Department of Medical and Biological Informatics
(Head: Prof. Dr. C. O. Köhler)
Im Neuenheimer Feld 280, D-6900 Heidelberg, F.R.G.

Abstract

With the aid of computer-supported 3D visualization procedures, such as triangulation [7, 3], Cuberille method [9, 10, 11, 4] or volume rendering approaches [5, 18, 20, 26, 13, 12, 8] 3D images can be generated from spatially associated tomography image series. These 3D images assist the physician in diagnosing, therapy planning and therapy control.

In this paper, a visualizing procedure is introduced which is based on a ray tracing approach by Kajiya [15, 16]. This model was modified and simplified according to medical requirements so that expressive images can be produced at a relatively low expense. Next to the presentation of surfaces, this visualization procedure also permits the presentation of tissues in a translucent form. Thus, organs may be shown in their environment in context.

The visualization of organs in a volume of data implies the association of the voxels with organs. In medical applications, it is useful to apply soft classifiers, e.g., the selflearning, selforganizing topological map [17, 1, 21]. It is, furthermore, necessary to integrate anatomical and morphological knowledge by AI methods.

Keywords

Volume Visualisation; Segmentation; Tomography

1 Volume Visualization

The basis of the 3D visualization method presented is formed by a ray tracing equation by Kajiya [15, 16] for the visualization of volume densities. Here, Kajiya took up an illumination procedure by Blinn [2], which the latter had developed for the purpose of visualizing a model of the Saturn rings, and imbedded it into a ray tracing approach. The ray tracing model is based on the following idea: Light originating from sources to be positioned arbitrarily passes radially through a continous density space $\varrho(x, y, z)$ and, on it's way it is weakened or dispersed, respectively, by the existing matter. Part of the reflected light is emitted in the direction of the observer and is further weakened on its way to the observer.

NATO ASI Series, Vol. F 60
3D Imaging in Medicine
Edited by K. H. Höhne et al.

So that the model may be simulated with a digital computer, one assumes the Low-Albedo case. The Albedo value is a measure for the reflectability of matter; it indicates the share of the emitted light intensity which is reflected. In the Low-Albedo case the Albedo value is chosen so small that with multiple reflections the light intensity of the light ray decreases so rapidly that this reflected light may be neglected. Thus, the computational expense is significantly reduced.

The discrete form of the ray tracing equation by Kajiya is given in equation 1:

$$B = \tau\omega \sum_{i_s \epsilon N_s} [\sum_{j=1}^{n} L_j \varphi(a_{j,r}) \prod_{i_l \epsilon N_j^{i_s}} \exp^{-\tau\varrho(i_l)\Delta t_l}](1 - \exp^{-\tau\varrho(i_s)\Delta t_l})\Delta t_s \prod_{k \epsilon N_{i_s}} \exp^{-\tau\varrho(k)\Delta t_s} \quad (1)$$

B computed intensity of a viewing path
τ Constant that determines the light absorption
ω Albedo value
N_s Sampled points along the viewing path
N_{i_s} Sampled points on the viewing path between the image plain and point i_s
$N_j^{i_s}$ Sampled points on the light ray between the light source j and point i_s
Δt_s Sampling interval along the viewing path
Δt_l Sampling interval along the light ray
n Number of light sources
L_j Intensity of the light source j
φ Phase function
$a_{j,r}$ Phase angle between the direction of the light source j and the direction of the observer at point r
ϱ Density function

In determining equation 1, special attention was paid to the fact that only so much light energy may at most be emitted from an image point as was also admitted [22].

With the aid of equation 1 diffuse objects, objects without precise surfaces, may very well be visualized. However, with anatomical structures, also 'hard' surfaces and tissue transitions always occur. A useful visualization procedure will have to do justice to this fact, i.e. it must be in a position to present diffuse tissue transitions as well as hard surfaces. This can be achieved with equation1 via a corresponding choice of phase function. The strength of a light reflected at a space point in the direction of the observer is determined by the direction and the amount of the density gradient existing there. The phase function is thus calculated via a modified Phong illumination model [19]:

$$\varphi = \frac{1-S}{6} + S(k_d\vec{N}\vec{L} + k_s(2(\frac{\vec{L}+\vec{V}}{\| \vec{L}+\vec{V} \|}\vec{N})^2 - 1)^g) \quad (2)$$

In equation 2, it is implicitly assumed that all terms are normalized.

S Length of the density gradient
$\vec{N}$ Direction of the density gradient
$\vec{L}$ Vector in the direction of the light source
$\vec{V}$ Vector in the direction of the observer
k_d Diffuse reflection constant

k_s Specular reflection constant
g Exponent, that depends on the glossiness of the surface

By this phase function, it is safeguarded that the light reflected from a volume element (voxel) with a large gradient is determined via a surface illumination model. The reflected intensity is depending on the direction. With a small local gradient, the reflected light is emitted diffusely, without directional preferences. The visualization according to the model of equation 1 with free choice of the positions of the light sources and the observer still implies an extremely high computational expense and is thus very time consuming. By purposeful simplifications of the model [22], however, it can be reduced without seriously impairing the expressiveness of the results. The most important simplification lies in the fact that the light sources as well as the observer are positioned in infinity. Thus, the light rays of a light source as well as the view paths of the observer are parallel. A perspective representation is thus prevented, but this can be neglected in medical applications.

2 Knowlegde based Object Segmentation

The result of computer calculations of tomographies are density values for every picture element (pixel). They are standardized from -1000 to +1000 units (Hounsfield-units). In these images it's easy to distinguish air, soft tissue and bones. This separation is of interest and useful in many cases. Very often, however, physicians desire an insight into soft tissue and a more exact differentiation of it. An exact differentiation by the gray values of the computer tomographies alone is not possible. For the differentiation of soft tissues the MRI (magnetic resonance imaging) is better suited which, in turn, separates bones structures badly. In both cases of primary volume data one cannot do without a further preprocessing for the separation of soft tissues. Local, spatial texture measures are useful. These are mainly simple measures such as local median and local variance as well as gray value morphological operators [24, 25].

Direct values (e.g., density) and derived measures (e.g., textural features) of every voxel are now assigned to a certain group (e.g., an organ) with the aid of a classification procedure. Here, the classical multivariate statistical procedures of discriminance analysis and cluster analysis turned out to be somewhat to rigid and not flexible enough. Good experiences in segmentation, however, were made with the selflearning topological map according to T. Kohonen [17], a concept related with neuronal nets and artificial intelligence. This is a 'soft' classifier without a sharp separation of groups which, with the existing blurred biological-medical data, proves to be an advantage. We applied the concept of the topological map to images. Its main feature is the retention of the topology in a measure-space. The basic algorithm is simple and can be split into several steps.

1. To every pixel of the image we assign a number of descriptive, mostly textural, measures, e.g. the gray value itself or its variance. These measures aare of course spatial as we take into account e.g. a 5x5 environment in 3 slices.
2. A topological matrix (a map) of a dimension of e.g. 32x32 and a depth of the number of the textural measures used is randomly initialised.
3. For every image pixel a set of textural measures is computed.
4. For every set of measures the nearest matching neighbour in the map is localised.
5. This point is updated with the following formula:

$$NEW < - OLD - \alpha (OLD - NEW)$$

where NEW is the new map value, OLD is the old map value and α is a learning factor.

6. The adjacent locations in the map are also updated with the described algorithm but with a somewhat lesser learning factor αa.

The steps 3 - 6 are repeated for all image pixels or volume voxels. Pixels with similar textural information are thus collected in connected patches in the map. Two different pixels that differ a lot are distant in the map, and pixels that are texturally close are closer in the map. The topology is enhanced and not destroyed. Connected map areas are then associated with "names" of the application world e.g. "bone", "air", etc.

It is funny to see a map learing. The learning process is massively depending on the learning factor α, which is a function of distance and time. If the learning factor α is not constantly becoming smaller, i.e. if it is basically constant then the map either does not learn (i.e. the classes are not converging to a center or attractor) or the map "forgets" already learned information. This feature can deliberately be used to gain "observant" knowledge bases where new information updates and then overwrites old, not any more used details.

All this learning of a map as it is described here is fully automatic and independent of a primary random initiation of the map. It is of course not independent of the used textural/spatial measures. The clever selection of these can be understood as a piece of a-priori knowledge contributed by an image analyst. We checked a few hundred textural measures only to find that 98% of them strongly correlate and are therefore useless.

A successful segmentation requires in addition the existence of further a-priori knowledge, in our case, e.g., knowledge about anatomy and morphology. This previous knowledge is available in the very formal form of a symbolic description (e.g., 'the eye is about ball-shaped', 'there are two of them', etc.). If the reduction of information on digital images by the topological map is understood as a bottom up process, then the integration of formal a priori knowledge is a top down approach. The connection of both approaches is very important.

3 Examples for Application

Figure 1 illustrates our volume visualization method. Fine structures in volumes such as hair become visible.

Figure 5 shows a human heart reconstructed from a series of MR images. It is opened in the middle with our 'software scalpel'. It shows very fine details of the inner structure of the heart.

In Figure 2 the skull of a girl is to be seen with a bone desease (continuous growth) of the skull. For cosmetic reasons, some bone material was taken away at the lower jaw. Thereby the inner part of the bone (spongiosa) became visible. Thus, Figure 3 is a good example for postoperative control.

Figure 6 is an illustration of the mesogastric region. To gain a better insight, the ribs have been partially removed. A part of the spine as well as the diseased liver can be seen. A complete rotation of the volume data permits insights into and knowledge about healthy and diseased regions of the liver.

In figures 3 and 4 the power of the procedure is finally illustrated. In Figure 3 a visualization of an MR image sequence is to be seen. We recognize a brain with its characteristic folded surface structure and some blood vessels. In Figure 4, this brain is translucent containing an oedema surrounding a tumour which is located between cerebellum and cerebrum.

The location and form of objects becomes particularly comprehensible in objects with a dynamic view, i.e. in a film. From thousands of individual images we composed a film in which the tumour as well as the surrounding oedema, the brain and other parts of the human head are made visible. Making such a film asks for enormous computing power, even supercomputers are still too slow.

4 Implementation

For an image containing e.g. 128 slices of 256 x 256 pixels (that makes 8 million voxels) it takes about one minute on an IBM 3090/150 with an attached vector facility (VF). A realtime realization with 25 images per second demands a computational power of about 15 Giga Flops.

Our programs were developed in the programming language APL2 [14, 23]. This language has a high efficiency in program development. It is suited for the programming of numeric applications as well as for symbolic computations, this allows the integration of image processing and artificial intelligence routines [6].

APL2 makes automaticaly use of the vector facility VF. The vectorizing is done by the programming language and not by the programmer. Our APL functions can be executed on an array processor as well as on a classical von Neumann computer without any changes in the code.

APL has inherent parallel concepts, an implementation on multiprocessor machines is possible. This will give access to virtually unlimited computational power.

5 Prospects

Volume visualization makes unusual insights possible. If today's computer restrictions are solved, a preoperative navigation through the human body will become possible which, for instance, will constitute a valuable assistance of the surgeons. A further possibility for the future will be the presentation of quasi 4-dimensional data, e.g., a view of the beating heart with a chance to detect and treat defective or diseased parts.

In practice, volume visualization will only prove its worth, once the enormous increases in performance have been achieved. All the same, the equipment must have the size of today's workstations because they will have to be located directly in the hospitals next to the image producing devices (e.g., CT, MRI). The use of such powerful hardware in connection with efficient computer programs will assist the physician with diagnosis and therapy planning and contribute to the safety and improvement of medical care.

References

[1] Bertsch, H.: Die selbstlernende topologische Merkmalskarte zur Bildsegmentierung und Klassifikation. Technical Report 23/1988, Deutsches Krebsforschungszentrum Heidelberg, Abteilung MBI.

[2] Blinn, J.F.: Light Reflection Functions for Simulation of Clouds and Dusty Surfaces. Computer Graphics **16** 3(1982) 21–29.

[3] Boissonnat, J.-D.: Shape Reconstruction from Planar Cross Sections. Computer Vision, Graphics, and Image Processing **44** (1988) 1–29.

[4] Chen, L.-S.; Herman, G.T.; Reynolds, R.A.; Udupa, J.K.: Surface Shading in the Cuberille Environment. IEEE CG&A **5** 12(1985) 33–61.

[5] Drebin, R.A.; Carpenter, L.; Hanrahan, P.: Volume Rendering. Computer Graphics **22** 4(1988) 65–74.

[6] Engelmann, U., Gerneth, Th., Meinzer, H.P.: Predicate Logic in APL2. APL Quote Quad (USA) **19** 4 (1989) 124-128.

[7] Fuchs, H; Kedem, Z.M.; Uselton, S.P.: Optimal Surface Reconstruction from Planar Contours. CACM **20** (1977) 639–702.

[8] Glassner, A.S. (Ed.): An Introduction to Ray Tracing. Academic Press, London–San Diego 1989.

[9] Gordon, D.; Udupa, J.K.: Fast Surface Tracking in Three-Dimensional Binary Images. Computer Vision, Graphics, and Image Processing **45** (1989) 196–214.

[10] Herman, G.T.; Webster, D.: Surfaces of Organs in Discrete Three-Dimensional Spaces. In Herman, G.T.; Natterer, F. (Eds): Mathematical Aspects of Computerized Tomography, 204–224, Springer, Berlin–Heidelberg–New York–Tokio 1980.

[11] Herman, G.T.; Udupa, J.K.: Display of 3D–Digital Images: Computational Foundations and Medical Applications. IEEE CG&A **3** 8(1983) 39–46.

[12] Heyers, V.: Raytracing in Grauwertvoxelräumen zur Visualisierung medizinischer Schichtdaten. Technical Report 20/1988, Deutsches Krebsforschungszentrum Heidelberg, Abteilung MBI.

[13] Höhne, K.H.: 3D–Bildverarbeitung und Computer-Grafik in der Medizin. Informatik Spektrum **10** (1987) 192–204.

[14] International Business Machines Corporation, Santa Teresa Laboratory: APL2 Programming Language Reference (SH-20-9227) 1987.

[15] Kajiya, J.T.; Von Herzen, B.P.: Ray Tracing Volume Densities. Computer Graphics **18** (1984) 165–173.

[16] Kajiya, J.T.: The Rendering Equation. Computer Graphics **20** (1986) 143–149.

[17] Kohonen, T.: Clustering, Taxonomy, and Topological Maps of Patterns. Proc. ICPR (1982) 114–128.

[18] Levoy, M.: Display of Surfaces from Volume Data. IEEE CG&A **8** 5(1988) 29–37.

[19] Phong, B.T.: Illumination for Computer Generated Pictures. CACM **18** (1975) 311–318.

[20] Sabella, P.: A Rendering Algorithm for Visualizing 3D–Scalar Fields. Computer Graphics **22** 4(1988) 51–57.

[21] Saurbier, F.: Automatische Segmentierung aus CT– und MR–Bildern mit Hilfe der Topologischen Karte. Technical Report 28/1989, Deutsches Krebsforschungszentrum Heidelberg, Abteilung MBI.

[22] Schäfer, R.: 3D–Visualisierung von Voxelräumen. Technical Report 31/1989, Deutsches Krebsforschungszentrum Heidelberg, Abteilung MBI.

[23] Scheppelmann, D.; Baur, H.J.; Engelmann, U.; Gerneth, Th.; Heyers, V.; Meinzer, H.P.; Saurbier, F.; Schäfer, R.; Wolf, Th.: APLTREE — Bildverarbeitung in APL —. Technical Report 24/1989, Deutsches Krebsforschungszentrum Heidelberg, Abteilung MBI.

[24] Serra, J.: Image Analysis and Mathematical Morphologie. Academic Press, New York–San Francisco–London 1982.
[25] Sternberg, S.R.: Grayscale Morphology. Computer Vision, Graphics and Image Processing 35 (1986) 333-355.
[26] Trousset, Y.; Schmitt, F.: Active-Ray Tracing for 3D–Medical Imaging. In Maréchal, G. (Eds), Eurographics'87, 139–150, North Holland 1987.

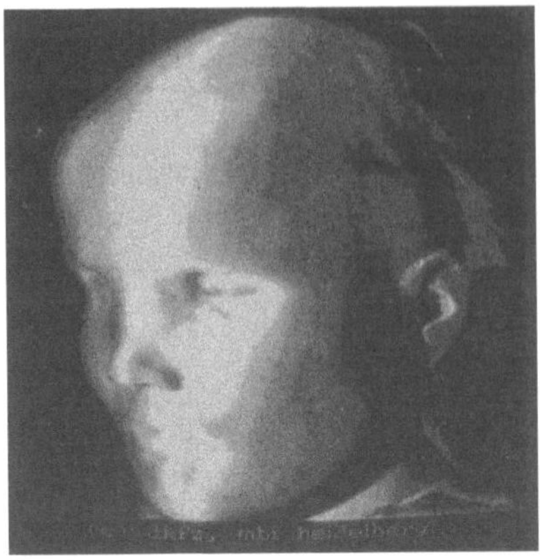

Figure 1: Girl with hair

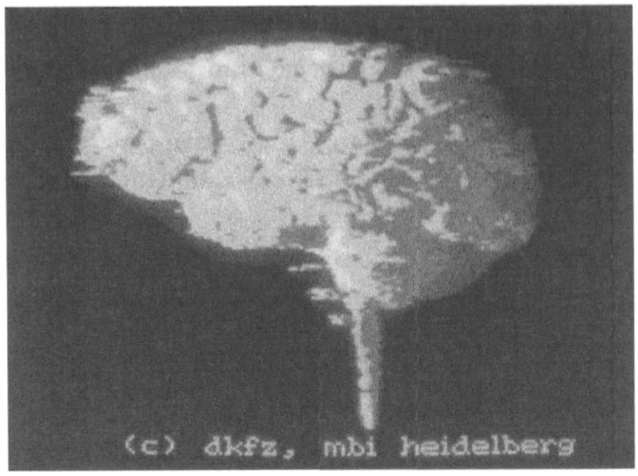

Figure 3: Human brain

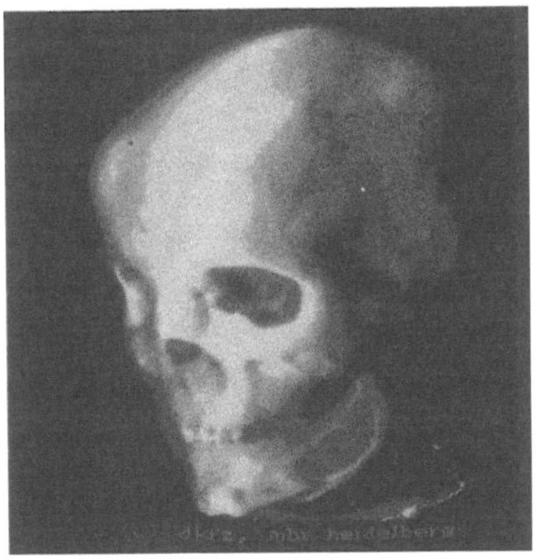

Figure 2: Skull of the girl

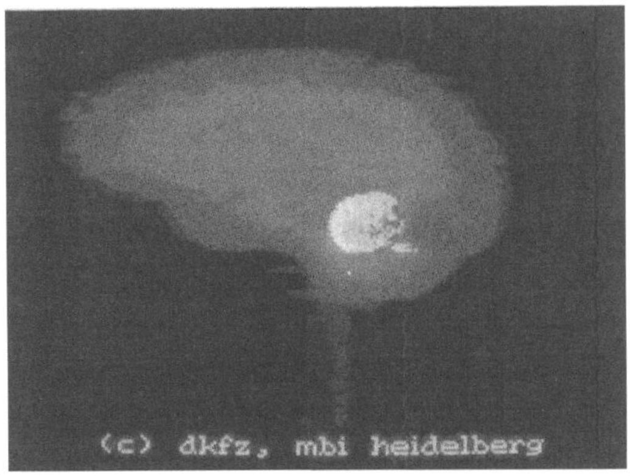

Figure 4: Transparent human brain with a tumour

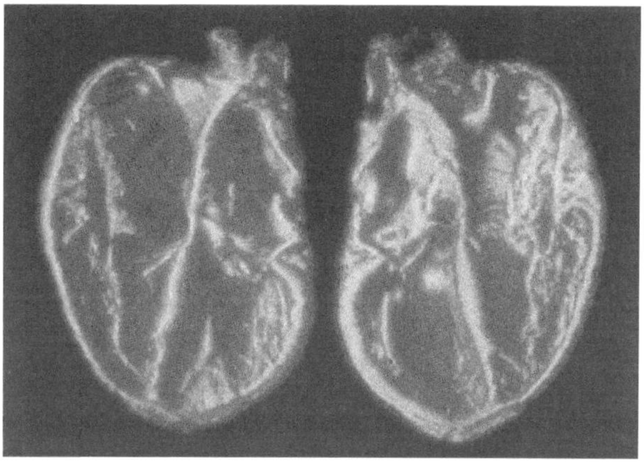

Figure 5: Human heart

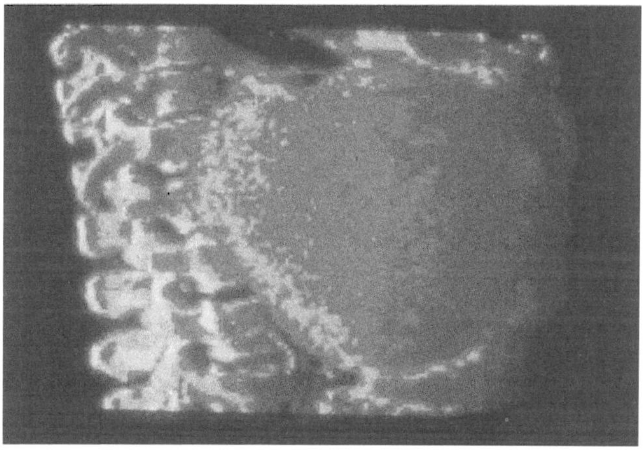

Figure 6: Chest with liver

Echocardiographic Three-Dimensional Visualization of the Heart

Riccardo Pini[1], *Elisabetta Monnini*[2], *Leonardo Masotti*[2], *Kevin L. Novins*[3], *Donald P. Greenberg*[3], *Barbara Greppi*[4], *Marino Cerofolini*[4], *Richard B. Devereux*[5]

[1]Institute of Gerontology and Geriatrics, University of Firenze, Viale Pieraccini 18, 50139 Firenze, Italy

[2]Department of Electronic Engineering, University of Firenze, Via S. Marta 3, 50139 Firenze, Italy

[3]Program of Computer Graphics, Cornell University, 120 Rand Hall, Ithaca, NY 14853-5501

[4]ESAOTE Biomedica, Via di Caciolle 15, 50139 Firenze, Italy

[5]Department of Medicine, The New York Hospital-Cornell Medical Center, 525 East 68th Street, New York, NY 10021, U.S.A.

Abstract

To perform three-dimensional (3-D) reconstruction of the heart by ultrasound, we developed a novel rotating echocardiographic probe which, with computer assistance, allows "real" 3-D reconstruction of the beating heart from 62 standard fan shaped two-dimensional (2-D) images acquired at 2.903 degree increments of rotation around its central axis. To reconstruct 3-D images of the beating heart, an entire cardiac cycle was recorded from each transducer position with electrocardiographic gating; acquisition time is 75 to 123 seconds in normal sinus rhythm. For each frame of the cardiac cycle, the 62 images digitized in cylindrical coordinates were processed by a scan converter algorithm to reconstruct a 3-D cone of information in cartesian coordinates. From the 3-D matrices stored in the computer, 2-D echocardiographic images in any plane at specified times in the cardiac cycle, or throughout the entire cardiac cycle, can be derived and visualized. A computer workstation-based system was developed to create full 3-D perspective projections of the echocardiographic data based on a technique called ray tracing, and adapted for use in visualizing 3-D scalar fields. The 3-D images obtained in normal volunteers demonstrated that our system permits an accurate reconstruction of

NATO ASI Series, Vol. F 60
3D Imaging in Medicine
Edited by K. H. Höhne et al.

the heart with the same spatial and temporal resolution as the original 2-D echocardiograms without cumbersome external reference systems.

Keywords: Echocardiography, Ray-tracing algorithm, 3-D computer graphics.

1. Introduction

Because the heart is a muscular pump whose performance is related to its 3-D cyclic deformation, attempts have been made to derive 3-D heart reconstructions from 2-D X-ray computed tomograms [Feiring 1985], magnetic resonance images [Higgins 1988, Lanzer 1984, Lanzer 1985], and positron emission tomography [Ritchie 1982]. Compared to these imaging techniques, ultrasound presents the advantage of relatively low cost, portability, high temporal resolution, and absence of ionizing radiation; although its lateral resolution is limited, it is sufficient for an accurate estimation of heart dimensions [Pini 1987, Wallerson 1986]. Previously proposed systems for the 3-D reconstruction of the heart by ultrasound from multiple 2-D images with known spatial orientation [Geiser 1982, McCann 1988, Moritz 1983, Raichlen 1986, Skorton 1986] permit only a schematic representation of the heart because the chest wall and lungs allow the visualization of the heart from only limited number of acoustic windows. To overcome these technical limitations of the 3-D reconstruction of the beating heart by ultrasound, we developed a novel rotating echocardiographic probe which, with computer assistance, consents a “real” ultrasound tomography of the heart in any plane with the same spatial and temporal resolution as standard 2-D echocardiograms and permits 3-D visualization of the beating heart.

2. Methods

To perform the 3-D reconstruction of the heart by ultrasound, we developed an echocardiographic system (ESAOTE Biomedica, Firenze, Italy) with a new 3.5 MHz dynamically focused anular array transducer that allows the acquisition from any acoustic window of 63 standard fan shaped 2-D images at 2.903 degree increments of rotation around its central axis (Figure 1) [Pini 1989].

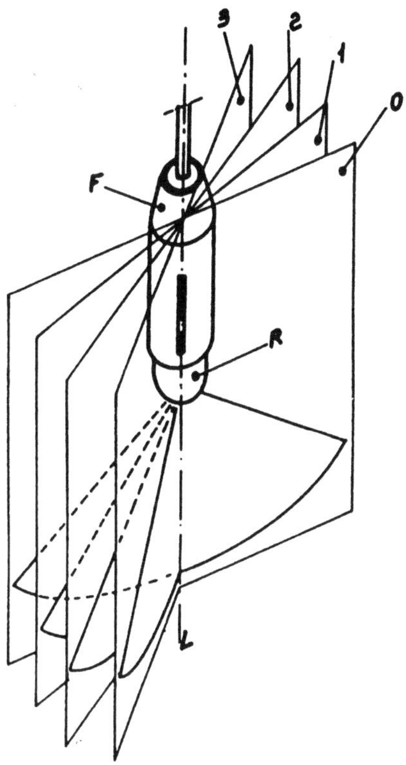

Fig.1 Schematic representation of the new echocardiographic transducer (R) rotating around its central axis (L) and contained within a fixed external housing (F) that represents the reference system. The planes 0-3 represent the spatial orientation of four of the 63 2-D echocardiographic image planes acquired with this transducer with 2.903 degree increments of rotation.

Since the rotation axis is in the center of the 2-D fan, a rotation of the transducer through 180 degrees permits the 3-D reconstruction of a solid cone encompassing the heart through the chest wall. Comparing the 0 (first 2-D scanning plane) and the 180 degree (63rd 2-D scanning plane) images provides an immediate check of the rotation axis stability during the recording, since, if the transducer maintains the same relative position to the heart during the examination, the 0 and 180 degree images are mirror images. To reconstruct 3-D images of the beating heart we gated the echocardiographic acquisition with the electrocardiogram. An entire cardiac cycle (up to 30 echocardiographic frames) was recorded from each transducer position while the subsequent cycle was utilized to rotate the transducer; thus, the system acquires images over 123 cardiac cycles, or 75 to 123 seconds in normal sinus rhythm. The videotaped 2-D echocardiographic images were digitized using a frame grabber (ECS TESAK, Firenze, Italy) with 512 x 512 pixel resolution and 256 gray levels connected with a PDP 11/44 computer (Digital Equipment, Maynard, Massachusetts, U.S.A.). Only 62 of the 63 2-D echocardiographic images were digitized because the 63rd image (180°) was used exclusively to check the rotation axis stability; up to 1860 images were used to reconstruct the 3-D images of the beating heart. For each frame in the entire cardiac cycle, the 62 images acquired in cylindrical coordinates were processed by an algorithm for linear interpolation to reconstruct a 3-D cone of information in cartesian coordinates. The 3-D images were stored in matrices of 512 x 512 x 512 pixels and 256 gray levels with a resolution of 2 to 4 pixels per millimeter. From the 3-D matrices stored in the computer, 2-D echocardiographic images in any plane at specified times in the cardiac cycle, or throughout the cardiac cycle, can be derived and visualized.

In addition to allowing exploration of the 3-D dataset by viewing 2-D slices, a system has been developed to create perspective projections of the data. These images portray full three-dimensionality, making it possible to comprehend the true shapes of cardiac structures. The algorithm used to create the 3-D images [Levoy 1988] is based on a technique called ray tracing [Whitted 1980], and adapted for use in visualizing 3-D scalar fields; the grid of data is given 3-D spatial coordinates, and a light source and camera position (or viewer's location) are specified. A simulation of the light propagation in the environment is then executed, revealing how the scene appears from the camera's point of view [Greenberg 1989]. A crucial step in this visualization process is the assignment of material properties (i.e. reflectiveness, transparency) to each element in the 3-D matrix. Two alternative methods have been used to assign these material properties. The first method (employed in figures included in this manuscript) uses a threshold value to screen out low intensity echoes, in order to visualize a single isovalue surface in the dataset. Using isovalues, a crisp surface may be extracted, at the expense of filtering out much of the information present in the raw data. The second method is to use the echocardiographic intensity of a voxel as a measure of its opacity -the brighter the value on the 2-D echocardiogram, the more opaque the surface.

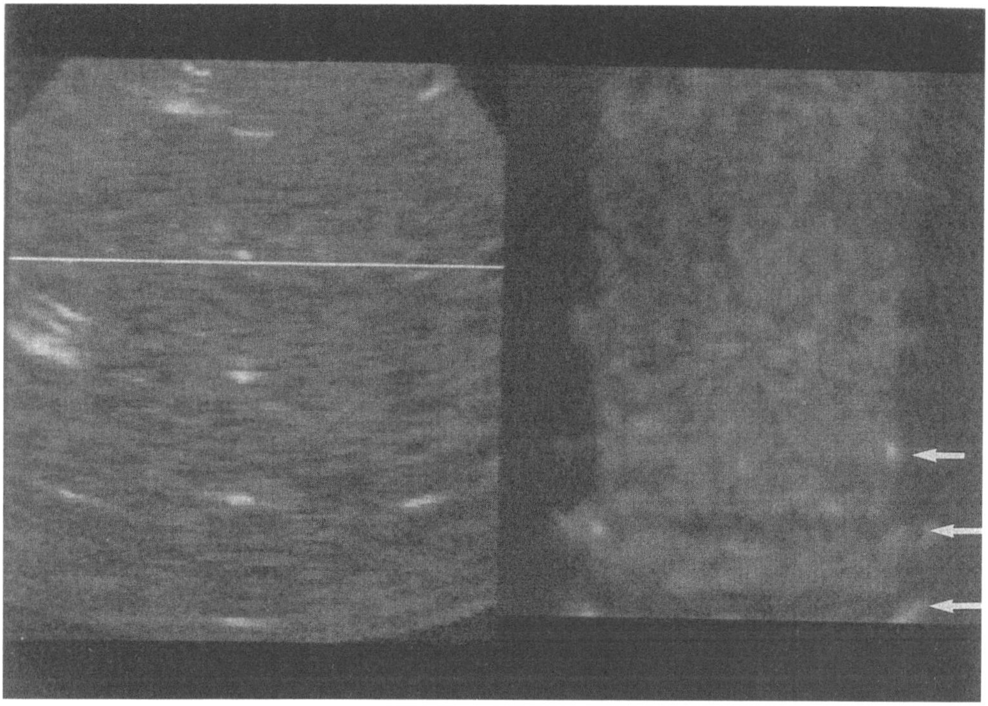

Fig. 2 Original 2-D image of a standard ultrasound phantom (left) with the horizontal line indicating the plane selected to derive the computed tomographic slice (right) from the 3-D matrix. The 2 mm diameter tube (upper arrow) is partially visualized in the tomographic reconstruction even though it is actually not distinguishable in the original 2-D image; the 4 mm and 6 mm diameter tubes (lower arrows) are also visualized.

This technique results in images with fuzzy surfaces, accurately reflecting the noise in the original images. It has the advantage of producing results which are very closely tied to the raw data, but does not readily allow the observation of surface detail. While lighting models make surface orientation readily available to the observer, perspective images of these cloud-like surface suffer from a lack of adequate depth cues. In preliminary studies we found that the kinetic depth effect [Wallach 1953], which can be obtained by producing animations in which the object is rotated in the viewer's field of vision, greatly enhances the 3-D perception of the scene. The computing environment for this visualization work is a 14 MIPS computer graphics workstation (Hewlett Packard, Fort Collins, Colorado, U.S.A.) with 50 Mbytes of main memory and 1 Gbyte of disk storage. With our current generation of algorithms, several minutes are required to obtain a single perspective image. In order to give the

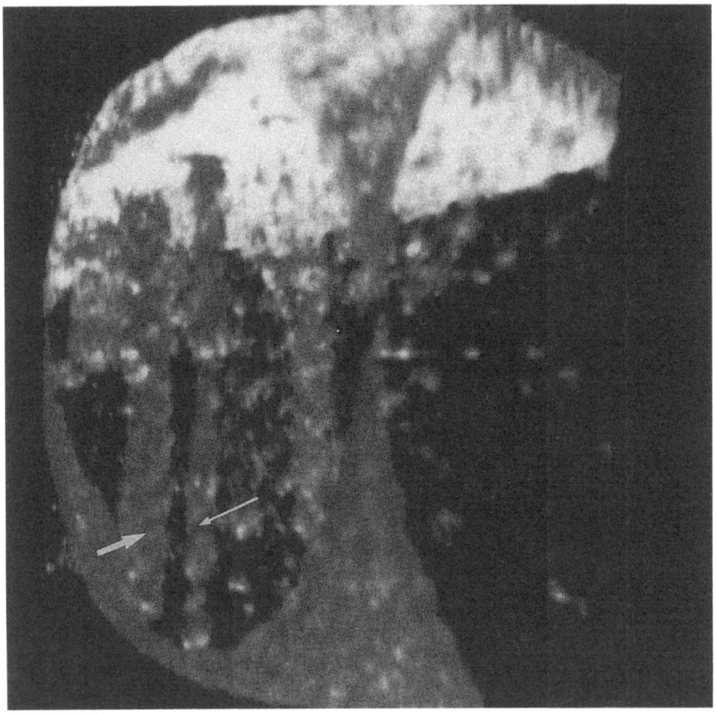

Fig. 3 3-D image of a standard phantom visualizing the 6 mm (thick arrow) and the 4 mm (thin arrow) diameter tubes inside the phantom. Due to the limited memory available in the computer, the image resolution was reduced to 64 x 64 x 64 pixels.

physician some interactive control over the data viewing, a full rotation sequence of frames is computed in approximately 24 hours; the images are then displayed at video rates, with user control of speed and direction of rotation.

3. Results

The accuracy of our tomographic method was tested with a standard phantom (RMI, model 412A, Middleton, Wisconsin, U.S.A.); both computed tomographic 2-D images derived from the 3-D matrices and the 3-D images demonstrated that wires and tubes were accurately reconstructed at any depth in the visualized solid cone. As shown in Figure 2, the 2 mm diameter tube is partially visualized in the tomographic reconstruction even though it is actually not distinguishable in the original 2-D images. In clinical studies performed in normal volunteers and in patients with valvular diseases, our system permitted precise reconstruction of the heart with good spatial and temporal resolution as demonstrated by clear definition of endocardial border as well as mitral valve apparatus and aortic valve cusps. In particular, the system allowed accurate reconstruction of tomographic slices near the edge of the original acquired echocardiographic images. Moreover, the ability to derive tomographic images with any spatial orientation permitted 1) visualization of structures, like the pulmonic valve, that usually are not completely visualized by the standard 2-D echocardiograms because the lung and ribs prevent the correct orientation of the transducer and 2) the reconstruction of tomographic slices in the frontal plane that could be obtained only with the magnetic resonance.

The 3-D images demonstrated accurate visualization of both the ultrasound phantom (Figure 3) and of the human heart (Figure 4); perception of depth in these still frames is dramatically reduced compared to visualization of the 3-D images on a monitor due to loss of the kinetic depth effect.

4. Discussion

The heart is a 3-D structure which undergoes cyclic deformation closely related to its performance. As such, the ultimate goal of any cardiac imaging technique is the generation of a 3-D representation of cardiac morphology throughout the heart cycle. Recently, efforts have been made to derive 3-D heart reconstructions from 2-D X-ray computed tomograms [Feiring 1985], magnetic resonance imaging [Higgins 1988, Lanzer 1984, Lanzer 1985], and positron emission tomography [Ritchie 1982]. Compared to these imaging techniques, ultrasound has the advantage of a higher temporal resolution, portability, relatively low cost, and absence of ionizing radiation. However, the limited ability of ultrasound to pass through bones or anatomical structures containing air, like the lungs, prevents the acquisition of a sufficient number of parallel 2-D images as it is possible with other tomographic techniques. To

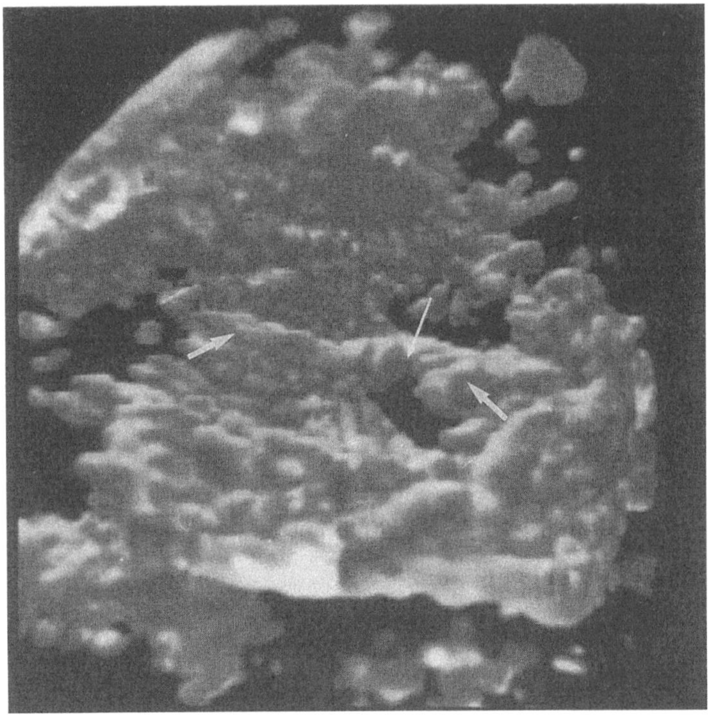

Fig. 4 3-D reconstruction of the left ventricle at end-diastole in a normal subject. The left ventricular cavity is viewed from the apex. The two papillary muscles (thick arrows) and the anterior mitral leaflet (thin arrow) are clearly visualized; depth perception in still frame is dramatically reduced compared to visualization of the 3-D images on a monitor. Due to the limited memory available in the computer, the image resolution was reduced to 64 x 64 x 64 pixels.

overcome this limitation, the acquisition of nonparallel 2-D images with known spatial orientation has been proposed by many authors [Geiser 1982, Linker 1986, Matsumoto 1981, McCann 1988, Moritz 1983, Raichlen 1986, Skorton 1986]. Transducer locating systems have included a mechanical arm with position sensors or an acoustic ranging system with three fixed microphones and spark gaps affixed to a freely movable ultrasound transducer. Both of these types of systems have been used to obtain accurate measurements of the volume of heart chambers. However, these methods have limited clinical applicability because they reduce the echocardiograph portability and, especially, because they do not permit the acquisition of a sufficient number of 2-D images for "real" echocardiographic 3-D reconstruction. Moreover. manual identification and tracing of endocardial borders is needed in each 2-D image, a tedious process that has precluded routine use of this technique. Finally, 3-D representation produced using this methodology, independent from the transducer location system adopted, are wire-frame models of the heart cavities or valves rather than realistic images. Recently, the in vitro accuracy of the 3-D reconstruction of the heart using a rotating transducer similar to ours has been demonstrated [McCann 1989], but only limited data derived from in vivo studies have been presented. Moreover, they suggested the use of only the apical window that is more sensitive to the changes of thoracic shape during normal breathing; to avoid artifacts due to the respiratory movements, McCann et al proposed the synchronization of the acquisition process with the respiratory cycle increasing the time needed for the entire examination. Finally, they presented only schematic 3-D representation of the left ventricle derived from manually traced endocardial silhouettes.

Our apparatus, based on a 2-D echocardiographic probe rotating around its central axis, overcomes the limitations of the previous systems and allows a "real" echocardiographic tomography of the heart. In fact, because the dynamically focused anular array transducer as an uniform lateral resolution along the entire depth used to visualize the heart and the 62 2-D images are separated by angles narrower than the lateral resolution of the transducer, from the 3-D matrices we can derive tomographic 2-D images with any spatial orientation with the same resolution as the original 2-D echocardiograms. Moreover, 2-D images with the correct spatial orientation of standard imaging planes can be derived from the 3-D matrices even if the heart is visualized from a nonstandard acoustic window. From the same matrices, using algorithms for 3-D reconstruction of anatomic structures, it is possible to obtain 3-D images maintaining the same resolution as the original 2-D tomographic slices.

4.1 Clinical Applicability

Preliminary results obtained in normal volunteers and in patients with valvular diseases demonstrated the clinical feasibility of our method. The absence of any mechanical or acoustic system connected to the transducer to determine its spatial location provides an easy orientation of the ultrasound probe. The acquisition time for each window ranges from 75 to 123 seconds in patients with normal sinus rhythm; the entire examination time, therefore, is reduced compared to the time needed for a standard 2-D echocardiogram. The ability to derive tomographic images with any spatial orientation maintaining the same resolution as the original 2-D images from the 3-D matrices, constructed using our rotating transducer, makes it possible to visualize standard 2-D echocardiographic planes with correct orientation even in technically difficult patients with nonstandard acoustic windows, potentially decreasing the number of subjects with limited studies. From the 3-D matrices, 2-D echocardiographic images in any plane throughout the cardiac cycle can be derived and used to calculate both qualitative and quantitative indices. This method may permit accurate estimation of chamber sizes and myocardial mass, by methods free of assumptions about cardiac geometry, in patients with distorted ventricular shape. As previously described using our rotating echocardiographic transducer [Katz 1990], the assumption of orthogonality between the long- and short-axis views in the parasternal window and between the four- and two-chamber views in the apical window is correct on average but there is wide deviation from this principle in individual patients; these deviation limit the accuracy of 2-D echocardiographic measurements of chamber volumes and mass by methods that assume orthogonality of projections. In addition, the 3-D visualization of the entire beating heart permits an easier evaluation of both normal and abnormal heart kinetics.

The major limitation of our system are common to all echocardiographic techniques: 1) limited lateral resolution of the transducers currently available compared to X-ray and magnetic resonance imaging systems, and 2) difficulty in visualizing the heart of patients with very limited acoustic windows. As any imaging system gated to cardiac activity, echocardiograms acquired with our method could present artifacts in patients with arrhythmias (e.g.: atrial fibrillation, frequent premature contractions).

4.2 Future developments

Our system is capable of reconstructing, 3-dimensionally, all the anatomic structures included in the cone of ultrasonic information encompassing the heart through the chest wall. Usually, the heart is only partially visualized from each single acoustic window; the 3-D visualization of the entire heart throughout the cardiac cycle could be attempted combining 3-D matrices acquired from different windows. Applying our rotating technique to Doppler echocardiographic color flow images instead of 2-D echocardiographic images, the blood velocity jets could be reconstructed 3-dimensionally and their actual spatial orientation correctly defined. Rotation of the imaging plane by electronic phased array technology might make it possible in the future to obtain a complete cone of echocardiographic information on 62 rather than 123 cardiac cycle, thus both speeding image acquisition and making it easier to sustain a stable axis of rotation for the entire period needed. We are currently researching new algorithms and hardware technologies to allow the generation of useful perspective images for reviewing the data at interactive speeds within minutes of the initial 2-D echocardiographic scan. At the same time, we are exploring more accurate lighting models to produce highly detailed images as an overnight process. Within a few years it may be possible to package a computer with an echocardiograph to achieve these functions without significant extra cost or decreased portability. Our system was developed for tomography of the heart but it may also be used to study other anatomical structures that can be visualized by ultrasound (e.g.: kidneys, great vessels).

In conclusion, we developed a novel rotating echocardiographic transducer which, with computer assistance, allows "real" ultrasonic 3-D reconstruction of the heart with the same spatial and temporal resolution as the original 2-D images without cumbersome external reference systems, expensive equipment, or ionizing radiation exposure.

5. References

Feiring A., Rumberger J.A., Reiter S.J., Skorton D.J., Collins S.M., Higgins C.B., Lipton M.J., Ell S., Marcus M.L.: Determination of left ventricular mass in dogs with rapid-acquisition cardiac CT scanning. Circulation 72, 1355-1364 (1985)

Geiser E.A., Christie L.G., Conetta D.A., Conti R., Grossman G.S.: A mechanical arm for spatial registration of two-dimensional echocardiographic sections. Cathet. Cardiovasc. Diagn. 8, 89-101 (1982)

Greenberg D.P.: Light reflection models for computer graphics. Science 244, 166-173 (1989)

Higgins C.B., Holt W., Pflugfelder P., Sechtem V.: Functional evaluation of heart with magnetic resonance imaging. Magn. Reson. Med. 6, 121-139 (1988)

Katz A.S., Wallerson D.C., Pini R., Devereux R.B.: Visually determined long-axis and short-axis parasternal and four-chamber and two-chamber apical views do not represent paired orthogonal projections. J. Am. Coll. Cardiol. 13, 94A (1990)

Lanzer P., Botvinick E.H., Schiller N.B., Crooks L.E., Arakawa M., Kaufman L., Davis P.L., Lipton M.J., Higgins C.B.: Cardiac imaging using gated magnetic resonance. Radiology 150, 121-127 (1984)

Lanzer P., Barta C., Botvinick E.H., Weisindinger H.V., Modin G., Higgins C.B.: ECG synchronized cardiac MR imaging: Method and evaluation. Radiology 155, 681-686 (1985)

Levoy M.: Display of surfaces from volume data. IEEE Computer Graphics and Applications 8, 29-37 (1988)

Linker D.T., Moritz W.E., Pearlman A.S.: A new three-dimensional echocardiographic method of right ventricular volume measurement: in-vitro validation. J. Am. Coll. Cardiol. 8, 101-106 (1986)

Matsumoto M., Inoue M., Tamura S., Tanaka K., Abe H.: Three-dimensional echocardiography for spatial visualization and volume calculation of cardiac structures. J. Clin. Ultrasound 9, 157-165 (1981)

McCann H.A., Sharp J.C., Kinter T.M., McEwan C.N., Barillot C., Greenleaf J.F.: Multidimensional ultrasonic imaging for cardiology. Proc. IEEE, pp 1063-1073 (1988)

Moritz W.E., Pearlman A.S., McCabe D.H., Medema D.K., Ainsworth M.E., Boles M.S.: An ultrasonic technique for imaging the ventricle in three dimensions and calculating its volume. IEEE Trans. Biomed. Eng. 30, 482-491 (1983)

Pini R., Ferrucci L., DiBari M., Greppi B., Cerofolini M., Masotti L., Devereux R.B.: Two-dimensional echocardiographic imaging: in-vitro comparison of conventional and dynamically focused annular array transducers. Ultrasound in Med. and Biol. 13, 643-650 (1987)

Pini R., Monnini E., Masotti L., Greppi B., Cerofolini M., Devereux R.B.: Echocardiographic computed tomography of the heart: preliminary results. J. Am. Coll. Cardiol. 13, 224A (1989)

Raichlen J.S., Trivedi S.S., Herman G.T., Sutton M.G.J., Reichek N.: Dynamic three-dimensional reconstruction of the left ventricle from two-dimensional echocardiograms. J. Am. Coll. Cardiol. 8, 364-370 (1986)

Ritchie J.L., Williams D.L., Harp G., Statton J.L., Caldwell J.H.: Transaxial tomography with Thallium-201 for detecting remote myocardial infarction: Comparison with planar imaging. Am. J. Cardiol. 50, 1236-1241 (1982)

Skorton D.J., Collins S.M., Kerber R.E.: Digital image processing and analysis in echocardiography. In: Cardiac imaging and image processing (Collins S.M. and Skorton D.J., eds.) pp 171-205. McGraw-Hill, New York: 1986

Wallach H., O'Connell D.N.: The kinetic depth effect. Expmtl. Psych. 45, 205-217 (1953)

Wallerson D.C., Devereux R.B.: Reproducibility of quantitative echocardiography: Factors affecting variability of imaging and Doppler measurements. Echocardiography 3, 219-235 (1986)

Whitted T.: An improved illumination model for shaped display. Communications of the ACM 23, 343-349 (1980)

Object Manipulation and Interaction

Surface Modeling With Medical Imagery

H. Harlyn Baker

Artificial Intelligence Center, SRI International
Ravenswood Avenue, Menlo Park, CA 94025

Abstract

3D biomedical data obtained through tomography provide exceptional views of the interior structure of biological material. While visualization is one of the primary purposes for obtaining these data, other more quantitative and analytic uses are possible. These include modeling of tissue properties and interrelationships, simulation of physical processes, interactive surgical investigation, and analysis of dynamics. In large part these uses require the representation of tissue geometry, and this suggests the use of descriptive formalisms that make this geometric structure explicit. The descriptive formalism we are developing models tissue imaged from tomography as two-dimensional manifolds in three space. The uses demonstrated here include simple versions of surgical simulation, kinematic modeling, and kinematic analysis.

Keywords: Surface modeling, tissue models, parallel methods, interactive 3D manipulation. surgical simulation, higher-dimensional analysis.

1. The Modeling Issue

From a computational viewpoint, modeling is the representation of some collection of data as a coherent entity. It involves defining a means to associate together related data elements – related semantically, physically, or in some other useful sense – distinguishing them from those with which they have no such connection. The modeling, then, groups together the relevant components of some entity and enables their treatment as a coherent structure. Although these structures can be displayed, display is a secondary concern in modeling, with the main interest being in *use* of the models. The uses of interest to us comprise both interactive physically-based simulation and higher-dimensional analysis.

NATO ASI Series, Vol. F 60
3D Imaging in Medicine
Edited by K. H. Höhne et al.

1.1 Modeling for assessment and interaction

One of the principal purposes of 3D display for a surgeon is to enable noninvasive assessment of patient morphology. This must involve more than just viewing, and the representation used should enable him a variety of investigative *operations* including tissue preparation, instrument selection, incision and tissue restructuring, use of cephalometric tools, evaluation of potential morbidity conditions (connective tissue, nerves, and blood supply), and site repair. More futuristic elements include advanced modeling where a severed vein bleeds, stimulated muscle moves its connecting tissue, forces exerted against tissue by the surgeon are reflected through his tools, and tissue healing and growth can be extrapolated to suggest postoperative results. Such a scenario calls for not just sophisticated visualization and representation capabilities, but for detailed modeling of the physics throughout.

With their lack of geometry, voxel-display representations provide little of the capability required for treatment of the data in a manner consistent with their physical properties. Surface models, on the other hand, do have the capability of representing this geometric structure. This power allows surface models to facilitate such varied tasks as: fabricating prostheses [Woolson 1986] and facsimiles [Marsh 1985]; inspecting occluded joints [Brewster 1984]; planning and rehearsing surgical operations [Cutting 1986]; evaluating postoperative results [Brewster 1984]; and modeling of articulation kinematics [Thompson 1988].

1.2 Modeling for computational analysis

A second major purpose of modeling is to provide for such computation as detection of cephalometric landmarks in 3D data, registration of data sets from different scanner modalities, tracking of 2D and 3D structures over time for inference of dynamic characteristics (such as TMJ constraints, subtalar axis in the ankle, evolution of chromosomes during mitosis), and analysis of electrical activity over the heart. Very little has been done in medical imaging that relates to such analysis issues. We raise these points to suggest that the structural interpretations prerequisite to these sorts of analyses are critically dependent upon the modeling. These are activities applying to discrete (although interrelated) physical structures, and to attribute characteristics to the actions likely requires attributing the actions to physical structures – and this brings in the modeling. Furthermore, the representations we have developed for three-dimensional (and four-dimensional) data sets can themselves facilitate analysis and inference of these characteristics. Modeling the structure of the data is the first step in modeling their processes.

2. Representational Issues

Surface modeling entails first identifying the frontier of a desired tissue in the volume, and then constructing a description of this frontier as a surface in space. There are several styles of surface modeling in use, with varying levels of geometric sophistication. The most simple is one where the surface is a collection of 2D contours, each one slice thick [Cutting 1986]. An improvement is obtained by explicitly linking the contours in the vertical direction [Fuchs 1977], but this reconstruction is inherently ill-posed. The best modeling of 3D surfaces results from treating all three dimensions of the data isometrically ([Artzy 1981], [Lorensen 1987],[Baker 1989b]).

2.1 Surface Definition

We use a simple initial definition for grouping elements together into models: They are all on the surface of an object – a 2D manifold in three space. Surfaces are defined by either threshold intervals on the sensor output or differential measures on these values. A simple and direct mapping of surface elements can be made from an eight-bit signature computed from the values in a $2 \times 2 \times 2$ neighborhood on the 3D data volume. The values determine *inside* versus *outside*, and these are represented at each voxel by a computed sign. If absolute density measures are being used for surface definition, the signs are those of the voxel values with respect to the threshold; if differential methods are to be used (for example, the Laplacian of a Gaussian) then their output will provide the polarity. In a $2 \times 2 \times 2$ window of the data there will be 256 combinations of signs possible, and thus 256 ways for the window to be intersected by tissue surfaces. This simple mapping, applied everywhere in the data set, will result in specification of all contained surfaces (see [Baker 1989b]).

2.2 The Computation

The surface builder operates on images sequentially in the order of acquisition, knitting together a connected representation of tissue evolution over the third dimension. It maintains the continuity of structures over the volume, even allowing varying slice angle and separation. The processor acts as a *loom* during surface construction, with a wall of accumulators meeting each image in sequence and weaving its elements into the mesh of surfaces prepared behind (see Figure 1). Simultaneous with this surface creation is the mapping of the observed values into a Cartesian frame, and separation of the data into distinct surfaces. Although currently implemented to work sequentially within the first two dimensions (as it must on a sequential machine), the process could be recoded to operate in parallel over the images. The sequential implementation runs at about 1000 voxels per second on a 1-MIPS Symbolics Lisp Machine. Real time operation is not an obvious requirement for medical applications, however this system was developed for motion analysis in computer vision [Baker 1989a], and here speed is vital. Operating in a pixel-parallel mode on a dedicated processor could result in a rate of better than 1000 slices per second; or, with scan-line parallelism, upwards of 30-Hz. Either approach to parallelism would be plenty fast for any conceivable real-time requirement.

3. Uses of 3D Models

The utility required for interactive model manipulation will serve a variety of purposes. Cell structure studies [Agard 1980] will benefit from the ability to segment individual chromosomes, unravel them to identify their structure, and investigate the relationships between their structure and function. Models that can encode growth patterns and develop by them will provide a critical tool for statistical growth studies and extrapolations of surgical results. A tissue representation that enables propagation of electrical current and variation of potential will be a powerful tool for an electrophysiologist. Quantifying studies of the dynamics of the musculoskeletal system have generally been done with vivisection or postmortem dissection [Williams 1977]. 3D tools can be used to specify tissue composition and relationships – muscle and connective tissue could be created and attached to specific bone sites and given properties that will allow them to act.

Each of these applications involves the use of graphics and modeling tools to facilitate a researcher's investigations and experimentation with his data. From a data point of view these are all quite passive, with the action being almost exclusively that of the operator. Each of these applications could be viewed in the reverse – of analysis rather than synthesis. 4D observations of chromosomes during mitosis could be used in determining something of the biological action taking place; the fossil record or other staged samples could be used by the computer in developing models of growth or dynamics (such as the articulation at the TMJ); scanners measuring field characteristics could provide data for modeling electrophysiology; tissue functional relationships could be determined automatically from anatomical data.

Some of the above analysis targets have been motivations for the modeling and tool-development research I have been involved in for the past few years. Although this work is still in its early stages, in this section I present some of the results of these preliminary analyses.

3.1 Modeling structure

Figure 1 shows the evolution of surfaces judged to be 'bone' in a 70×30 window of a 50-image data set.[1] As depicted through the staged display, these surfaces are built in parallel in the order of image acquisition. Figure 2 shows a rendered stereo pair of this spine.

These spine data were processed by selecting surfaces on the basis of their CT sampling measures. This is adequate for some sensing modalities but not for MRI. Figure 3 shows a reconstruction of a hand from a 125 $\times$ 250 window of a 35-image MRI data set.[2] Here the surface is defined by a differential technique using the zeros of a Laplacian of a Gaussian operator – the Gaussian sets the resolution and the Laplacian selects the surface. The hand was secured between two foam pads and suffered some distortion and crinkling because of this – other blemishes are artifacts of the filter size and the difficulty in distinguishing tissue types in these data. Proper segmentation of MRI data is an open research problem ([Chen 1988], [Höhne 1988]).

Figure 4 shows several views of the skin, skull and brain from a 40-slice CT data set.[3] The different tissue were selected using threshold (the skin and skull) and interval measures (the brain) from the raw density data. Figure 5 shows a model of *Drosophila melanogaster* chromosomes imaged at prophase using optical-sectioning microscopy, and colored by the DNA's absorption of fluorescein marker – such a coding could be used for initial labeling of the chromosomes' gene structure. Again, density measures were the basis for surface definition.

[1]Data specifications: in vivo, .49mm by 0.49mm pixels with 5mm slice separation.
[2]Data specifications: in vivo, 0.91mm by 0.91mm pixels with 0.93mm slice separation.
[3]Data specifications: in vivo, 1.56mm by 1.56mm pixels with 2mm slice separation.

Although the display aspects of this technique are adequate, remember that the primary representation is a surface model, with all the detail appropriate for full model-based analysis. These are *geometric models* built directly from the data. This is an important point. Ours is not an attempt to improve on rendering algorithms: What is presented is a mechanism and a representation for specifying the geometric structure of tissue from tomographic data. This importance of this distinction is shown below.

3.2 Interactive manipulation

Obtaining a static model is a major part of the tissue-modeling task – refining and modifying the models is also a critical component. Figure 6 shows a simplified simulation of jaw kinematics from the data of which Figure 2 showed the spine. This simple depiction, while having no structural model of jaw motion, no notion of constraints, no knowledge of dynamics, no knowledge of muscle attachments, and no ability to consider collision possibilities (all of which must be developed), demonstrates the principle behind the modeling of kinematics and dynamics. We wish to be able to define the properties of parts and their functional relationships to enable this type of simulation: With such representations we could build working models of the musculoskeletal system [Hoffmann 1987].

Figure 7 displays frames from an interactive operation to cut the flesh and remove it from the skull. The reverse of such an operation could enable plastic surgeons to visualize the cosmetic results from a reconstruction of the underlying bone. These simulated surgical actions were performed using a 3D display with a 3D mouse playing the role of a scalpel. Real-time dynamic interaction on wire-frame displays allowed free choice of tissue, cutting path, and further action on the tissue. Although the display is simple, for example not including subcutaneous material or cartilage, not modeling the geometry of the scalpel, et cetera, it does represent the connectivity of the tissue and demonstrates some of the initial basics such as an approximation to skin elasticity.

3.3 Kinematic Analysis

Figure 8 shows a display of bones in a human foot (with toes omitted).[4] The top left shows the full set of bones; to its right we color these by surface connectivity – surfaces joined in the data are the same color.

[4]Data specifications: 29 CT slices, in vivo, 0.94mm by 0.94mm pixels with 2.5mm slice separation.

Further partitioning could be done with the 3D editing tools as used for Figure 7. The middle row shows the calcaneus and talus isolated, with the talus rotating about an estimated axis. The bottom left depicts this subtalar axis with a red arrow. To its right is a view of the calcaneus and axis alone, with the calcaneus' articulation facets shaded in yellow. These were indicated on the bone, again, with a 3D surface-marking tool. There is considerable variation in the geometry of these articulation facets [Bruckner 1987], and our interest in these data is to develop computational tools for analyzing individual specimen kinematics. We can compute on these representations, and the next image shows these bones overlaid with their bounding volumes, oriented along their principal axes. Further calculation will enable definition of motion constraints and derivation of the range of axis angles implied by the tissue. We have also affixed 'tendon' and 'muscle', and could simulate response to forces.

4. Modeling in the Fourth Dimension

The surface modeling presented to this point has been three dimensional – more specifically, representing tissue as 2D manifolds in three space. There has been relatively little development in 4D modeling in the medical imaging field, probably because such data have been relatively rare and computation on them could be exceedingly expensive.

4.1 4D Studies

Clinical studies of 3D motility using cadavers give insight into kinematic range and the effects of surgical procedures, but their analysis is manual and time consuming. Dynamic studies in vivo are complicated by movement, gating, and registration problems. The University of Pennsylvania has developed an algorithm for constructing surfaces in higher dimensions (including 4D), but has not presented any developed uses for the surfaces [Udupa 1982]. The MAYO Clinic collects CT and ultrasound data in 4D and uses it in cine-mode to assess time-varying behavior of tissue [Robb 1988], but they do not yet perform any computation that exploits the higher dimension. Rather, they use it as an index with 3D visualization and measurement tools. Some 3D measurements such as ventricular volume determination and flow turbulence can be usefully studied across time [Hoffman 1985]. Much as 3D tomographic data have often been treated as a 2D image sequence, 4D data are treated as a sequence of 3D images.

4.2 Uses of the Fourth Dimension

There are several ways that higher-dimensionality can enter the modeling issue. The first is that manipulation and alteration of 3D structures is a 4D process – three-dimensional objects being manipulated in time. The latter analysis examples suggested this bridge to the fourth dimension with the introduction of data interaction and kinematic simulation and analysis. More critically, data that is intrinsically four dimensional is now being collected. These include 4D CT, MRI and ultrasound tomography, and in vivo optical-sectioning and confocal microscopy – all involving temporal imaging of 3D dynamic structure. Investigation, manipulation and analysis of the resulting massive amounts of data will require tools for interaction, experimentation and display, tailored for the higher dimension.

The semantics of other dimensions also must be considered. Tissue has thickness, while the models we have displayed are just tissue shells. Accurate and realistic tissue modeling will require analysis of this neglected dimension. To incorporate interior necessitates representation of tissue as 3D manifolds in three space – each piece having connections to neighbors on its layer and to adjacent layers. The same 3D manifold representation in four space can be used to model the dynamics of 3D tissue over time; here thickness relates to local velocity. Preliminary studies on 2D imagery over a third dimension of scale for optimizing feature resolution [Baker 1989b] suggest that, for our surfaces, both positioning and segmentation can gain by analysis of their behavior over a fourth dimension of scale.

5. Conclusions

While volume display provides good information for clinical evaluation, more advanced capabilities such as structural manipulation of the data or simulation and analysis of dynamic function and behavior will require the development of both higher-dimensional modeling formalisms and analysis tools appropriate to these dimensions. We are producing surface-based models comprising 2D and 3D manifolds for representation and analysis of these data, and are developing interactive three-dimensional tools for their directed investigation and analysis.

6. Acknowledgements

This research has been supported by a research grant from SRI International. Early developments were motivated by work done for DARPA under contracts MDA 903-86-C-0084 and DACA 76-85-C-0004. 3D tomographic data were provided by CEMAX Corp., Santa Clara, California, Philips Medical Systems Inc., Shelton, Connecticut, the Palo Alto Veteran Administration Medical Center, the Department of Biochemistry and Biophysics at the University of California at San Francisco, and the Indiana University School of Medicine.

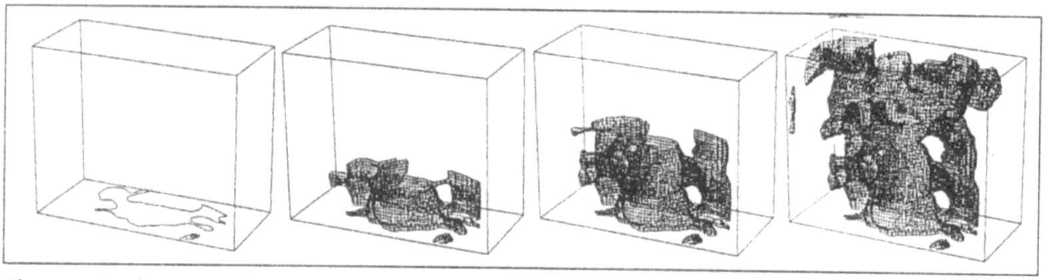

Fig. 1. Evolution of Spine Model from a Sequence of CT Images.

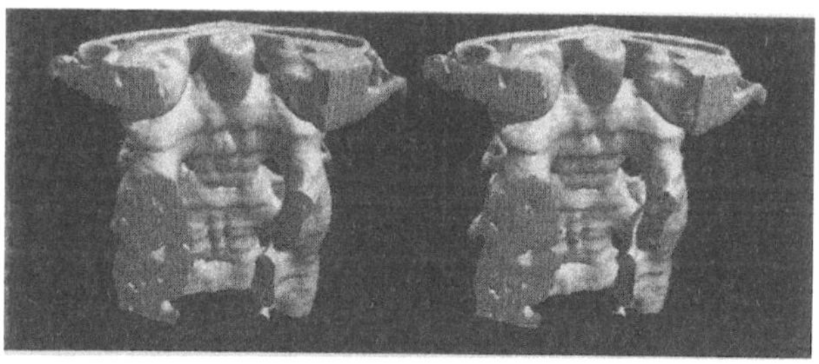

Fig. 2. Stereo View of the Spine.

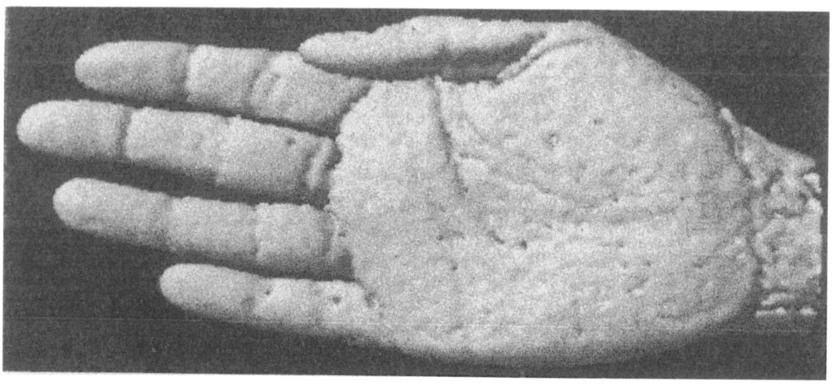

Fig. 3. Hand from MRI Data Set: Surface positioned at Laplacian zeros.

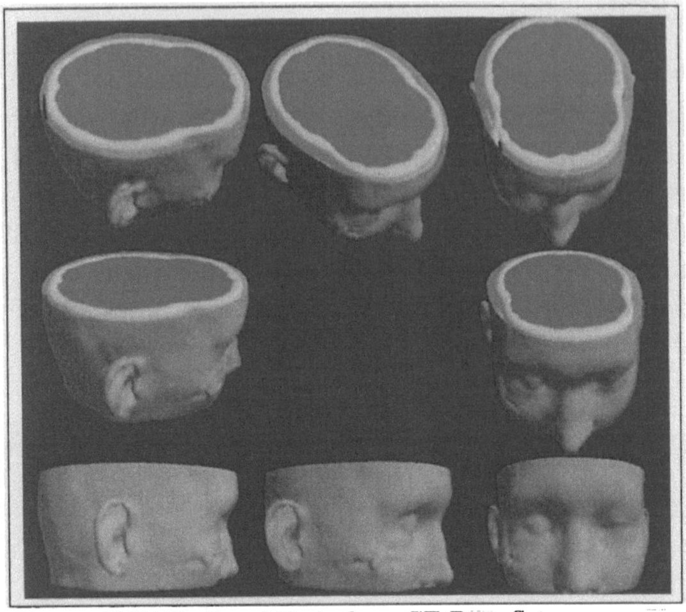

Fig. 4. Skin, Skull and Brain from CT Data Set:
Skin and Skull selected by single thresholds; Brain by interval.

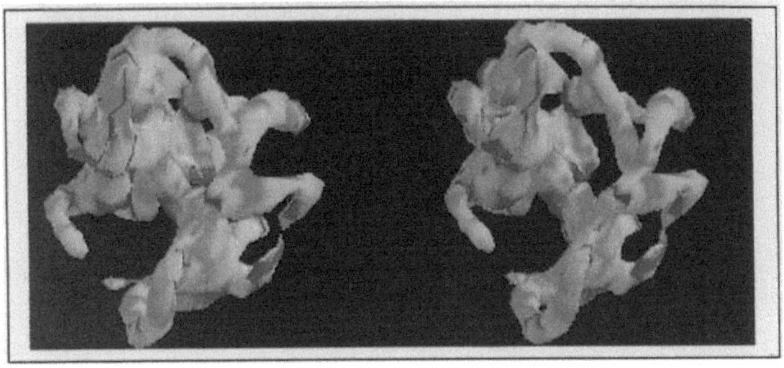

Fig. 5. Chromosomes Color Coded by DNA Marker Absorption.

Fig. 6. Rough Simulation of Jaw Kinematics.

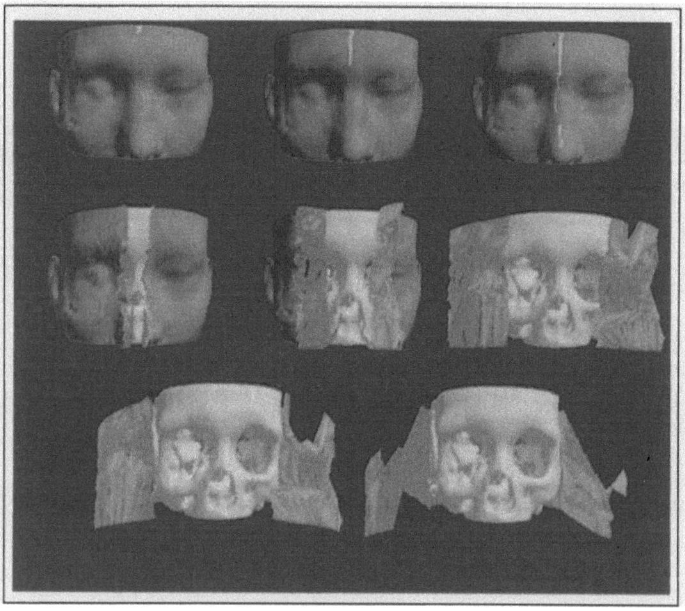

Fig. 7. Simulation of Skin Peeling:
The skin is cut with 3D tools, then pulled to the sides and away to the rear.

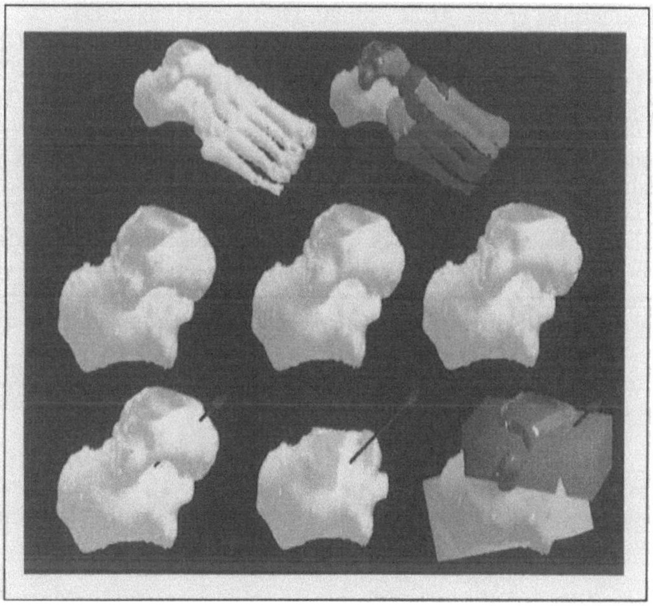

Fig. 8. Models of Human Foot Bones:
Top: Full set; Connected tissue. **Middle:** Talus rotation.
Bottom: Subtalar Axis (est); Joint Facets; Bounding Rectangles.
Our interest is in modeling the joint kinematics.

7. References

Agard, D.A., and J.W. Sedat: Three-Dimensional Analysis of Biological Specimens Utilizing Image Processing Techniques, *Proc. Soc. Optical and Instrumentation Engineers – Applications of Digital Image Processing to Astronomy,* Vol.264, 110–117, 1980.

Artzy, E., G. Frieder, and G.T. Herman: The Theory, Design, Implementation, and Evaluation of a Three-Dimensional Surface Detection Algorithm, *Computer Graphics and Image Processing,* 15, 1–24, 1981.

Baker, H.H., and R.C. Bolles: Generalizing Epipolar-Plane Image Analysis on the Spatiotemporal Surface, *International Journal of Computer Vision,* Vol.3, No.1, 33–49, 1989.

Baker, H.H.: Building Surfaces of Evolution: The *Weaving Wall, International Journal of Computer Vision,* Vol.3, No.1, 51–71, 1989.

Brewster, L.J., S.S. Trivedi, H.K. Tuy, and J.K. Udupa: Interactive Surgical Planning, *Computer Graphics and Applications,* Vol.4, No.2, 31–40, 1984.

Bruckner, J.S.: Variations in the Human Subtalar Joint, *Jour. Orthopedic and Sports Physical Therapy,* Vol.8, 489–494, 1987.

Chen, J., J. Chou, W. Lin, and C.A. Pelizarri: Edge and Surface Searching in Medical Images, *Proc. Soc. Photo. Instr. Engineers,* Vol.914, Medical Imaging II, 594–599, 1988.

Cutting, C., F.L. Bookstein, B. Grayson, L. Fellingham, and J.G. McCarthy: Three-Dimensional Computer-Assisted Design of Craniofacial Surgical Procedures: Optimization and Interaction with Cephalometic and CT-Based Models, *Plastic and Reconstructive Surgery,* Vol.77, No.6, 877–885, 1986.

Fuchs, H., Z.M. Kedem, and S.P. Uselton: Optimal Surface Reconstruction from Planar Contours, *Communications of the ACM,* Vol.20, No.10, 693–702, 1977.

Hoffmann, C.M., and J.E. Hopcroft: Simulation of Physical Systems from Geometric Models, *IEEE Journal of Robotics and Automation,* Vol.RA-3, No.3, 194–206, 1987.

Hoffman, E., and R.A. Robb: CT Heart Volume Measurement: Importance of Thickness and Orientation, *Radiology,* 155, 739–744, 1985.

Höhne, K.H., M. Bomans, U. Tiede, and M. Riemer: Display of Multiple 3D-Objects Using the Generalized Voxel-Model, *Proc. Soc. Photo. Instr. Engineers,* Vol.914, Medical Imaging II, 850–854, 1988.

Lorensen, W.E., and H.E. Cline: Marching Cubes: A High Resolution 3D Surface Construction Algorithm, *Computer Graphics,* Vol.21, No.4, 163–168, 1987.

Marsh, J.L., M.W. Vannier, W.G. Stevens, J.O. Warren, D. Gayou, and D.M. Dye: Computerized Imaging for Soft Tissue and Osseous Reconstruction in the Head and Neck, *Clinics in Plastic Surgery,* Vol.12, No.2, 279–291, 1985.

Robb, R.A.: Multidimensional Biomedical Image Display and Analysis in the Biotechnology Computer Resource at the Mayo Clinic, *Machine Vision and Applications,* 1, 75–96, 1988.

Thompson, D.E, W.L. Buford, L.M. Myers, D.J. Giurintano, and J. Brewer III: A Hand Biomechanics Workstation, *Computer Graphics,* Vol.22, No.4, 335–343, 1988.

Udupa, J.K., S.N. Srihari, and G.T. Herman: Boundary Detection in Multidimensions, *IEEE Transactions on Pattern Analysis and Machine Intelligence,* Vol. PAMI-4, No.1, 41–50, 1982.

Williams, R., and A.A. Seireg: Interactive Computer Modeling of the Musculoskeletal System, *IEEE Transactions on Biomedical Engineering,* Vol. BME-24, No.3, 213–219, 1977.

Woolson, S.T., P. Dev, L.L. Fellingham, and A. Vassiliadis: Three-Dimensional Imaging of Bone from Computerized Tomography, *Clinical Orthopaedics,* Vol.202, 239–248, 1986.

MANIPULATION OF VOLUME DATA FOR SURGICAL SIMULATION

Simon R. Arridge

Dept. of Medical Physics and BioEngineering, University College London
1st Floor, Shropshire House, 11-20 Capper Street, London WC1E 6JA, UK

Abstract

Many systems exist for the 3D visualisation of volume data. The manipulation (translation, rotation, Boolean operations) of such data is less well developed. In surgical simulation, the requirement is to have multiple object fragments that can be independently manipulated and merged. This paper discusses an approach using an interactively specified volume mask to dissect the data, and subsequent operations performed on Boolean expressions of masks and data.

Keywords : Manipulation, Surgical simulation, Transformation, Dissection

1 Introduction

The application of 3D medical graphics to craniofacial surgical planning was one of the earliest ones, and its importance is well established [Vannier 1984, Hemmy 1983]. It was quickly realised that this application needed a degree of manipulation similar to CAD/CAM design packages. Indeed one of the first systems [Vannier 1983], was actually developed in conjunction with a commercial CAD/CAM company (McDonnell Douglas). This was, however, limited to simple operations on 1D primitives such as mirroring to correct unilateral facial assymmetries.

This paper will deal with the implementation of more sophisticated techniques. The discussion will be from the point of view of a computer scientist. Several references from the clinical point of view exist [Hemmy 1983, Cutting 1986, Tessier 1986, Vannier 1983].

NATO ASI Series, Vol. F 60
3D Imaging in Medicine
Edited by K. H. Höhne et al.

2 Other Simulation Systems

In orthognathic surgical planning it is common to take patient photographs and radiographs, and to plan an operation just by cutting these up with scissors and moving them around [Henderson 1974]. Obviously the insight gained by doing this is very limited. The UCL3D system can be thought of as a three-dimensional version of this. It allows the user to cut up volumes of data, and reposition them an arbitrary number of times, making quantitative measurements where desired. There are many more sophisticated approaches that await future development. Even so, the fairly (conceptually) simple methods developed so far represent an enormous clinical improvement over previous methods [Arridge 1989].

The simulation of osteotomy, and reconstruction are the most challenging requirements, and identified by most authors [e.g. Udupa 1986] as the most lacking in current systems. It is this capability that is readily possible in UCL3D. Most other approaches have involved going back to original slices, modifying regions of interest therein, and reconstructing a separate object. Usually, regions drawn on the shaded display are combined with a specified depth to describe a set of cutting planes. The intersection of the planes with the contours of the original slices modifies the borders tracked by the surface detection algorithm which can then construct a new object [Brewster 1984, Weisburn 1986].

Many applications might require multiple osteotomies and trial moves. This was very difficult to achieve in those systems described. Intersection of objects is required for two reasons. Firstly for collision detection, and secondly, if some bone must be removed, it is possible to show how much by moving a piece into overlap and showing the intersection.

Interference detection is trivial to detect in a volume based method. Although suggested by early authors [e.g. Chen 1984], most systems do not provide this because they do not operate on a true volume model. Trivedi identified Boolean operations as a possible method of simulation, and demonstrated simple primitive "masks" in a binary digital array representation [Trivedi 1986]. The segment-endpoint (SE) representation has also been suggested [Trivedi 1985]. However, no clinical implementations have been reported.

The requirements are to demarcate an arbitrary subregion (a VOI) of an object, to segment a new object within such a VOI, and to independently transform the "dissected" objects and merge them with each other. Here *transform* means the process of coordinate transform. If visualisation is the only consideration, this is usually more efficient in Image-Space, although some approaches use an Object-Space transform to ensure the viewing direction is always orthogonal to the Object-Space coordinate axes [Tiede 1987]. When the scene to view consists of multiple objects, which can be independently transformed, Object-Space transforms are required. This is essential in the type of application which is the goal of UCL3D.

3 Dissection

Object Dissection means the interactive process required in a surgical simulation where the need is to visualise cutting of the object and thus to dissect it into one or more sub-objects. Some approaches have been used before. These may be classified as *object-modification* methods, and *object-masking* methods.

3.1 Object Modification

In these methods the data object itself is reconstructed with the required pieces removed. Obviously this requires the overhead of reconstructing the object. Using the display of an object in a given view, a closed contour is drawn on the screen. This contour is projected as a *cylindrical volume* (CV) into Image-Space to a depth specified by the user. The CV is transformed to Object-Space, and all parts of the object falling within this cylinder are deleted from the set of voxels comprising the object. Effectively a modified scene is produced, where the deleted voxels have their density function set to zero. A new object is then constructed from this modified scene.

The technique was first implemented in the 3D MIPG (Medical Image Processing Group) package, by operating in the Cuberille representation, using either 2D or 3D boundary methods [Brewster 1984, Chen 1984]. A number of possibilities were tried. In 2D, either the slices, or the Directed Contour representation of their boundaries were modified. In the former, the 3D boundary tracking algorithm could be applied to the modified slices. The latter case was faster but more limited because rotations out of the plane of the slices could not be accommodated. In 3D, a method using a full binary array was tried, where here the projected cylindrical volume "scooped out" all voxels within it, between minimum and maximum depths from the screen [Chen 1984]. Trivedi suggested using the depth-buffer shading as a visual cue for choosing these depth limits [Trivedi 1986].

The CEMAX approach was very similar, but allowed a more interactive editing session based in a windows environment [Weisburn 1986]. In principle the technique could equally well be applied to boundary representations, although it does not appear to have been done.

3.2 Object Masking

The alternative is to delay the clipping of the object to the display procedures. These algorithms can only operate on volumetric data bases. Alternatives previously suggested are clipping by half spaces, or clipping by a cylindrical volume.

Clipping by half spaces is achieved by testing each voxel to be displayed against a half space. This is efficient but grows linearly in the number of half spaces. If the half spaces are orthogonal to the object space axes the technique is simpler. This method was used by several groups [Meagher 1984a, Goldwasser 1984, Höhne 1987, Cline 1988, Robb 1989]. A major disadvantage is that only *convex* volumes may be delineated in this way.

Alternatively, if the method of 3.1 is used to describe a cylindrical volume, it would be possible not to display voxels within that volume, during traversal of a voxel display algorithm. This method was suggested by Freider *et al* [Freider 1985] although the voxels in the cylindrical volume were then deleted from the input file. However this is still somewhat limited in that a cylindrical volume is a restricted volume, even if concave, since it is (by definition) of identical cross section in any plane perpendicular to its axis.

3.3 The Generalised Volume Mask

In our system an unrestricted volume mask is implemented that does not affect the object in question. A *Volume of Interest* (VOI) of arbitrary complexity may be constructed and a Boolean expression of the mask and object may be viewed. Thus the VOI is like a 3D bitmask - a straightforward extension of

2D Regions of Interest. Then subsequent display and manipulations algorithms operate on Boolean expressions representing, for example, the object inside or outside the VOI.

The VOI may also be useful in the segmentation stage. In 2D it is quite common to manually define a Region of Interest (ROI) on an image, and use it as a mask, within which a segmentation operation is performed.

4 Generation of Arbitrary Volume Masks

This section presents a method for constructing an arbitrary 3D volume mask, by interactively drawing 2D ROIs in multiple views and forming the Boolean intersection of their inverse projections. It is simple to construct any concave simply connected object in this way. With an extension it is also possible to include objects with interior holes. The MIPG suggested the use of volume masks, but only implemented 3D primitives like triangular prisms, and ellipsoids [Brewster 1984, Trivedi 1986]. Also implemented in UCL3D is a method of generating a VOI by combining a set of 2D ROIs generated in the original 2D slices of the data.

The advantage of a volume mask is that it may be stored in a compact form and subsequent operations such as viewing, translation and rotation can be performed on a Boolean expression of masks and data, rather than creating an intermediate dataset.

In UCL3D, where an octree encoding scheme is used for the dataset and VOIs, a high degree of compression is achievable for the usually simply connected masks. The masks are stored as binary trees and the data as greyscale trees. In principle other 3D structures, or hybrid structures would be possible.

4.1 Volume Mask Construction Algorithms

In the Surgical Simulation application, the user most often wishes to view a dataset and interactively demarcate sections to be modified. Thus it is useful to draw the required dissection and observe it in multiple views, editing as required. For other applications it is desirable to build a VOI from the set of ROIs in 2D planes.

The idea of constructing the 3D data set of an object from its projections in multiple views is similar to some Computer Vision applications [Potmesil 1987]. The principle difference here is that instead of having available a number of images and information concerning the viewing conditions used to obtain them, an object is considered as being viewed in successive directions and edited until the required state is achieved.

The process starts with the object Q and displays it in a number of views to obtain a set of displayed images $\{Dy_i\}$. For each view a number of arbitrarily complex *regions* $\{R_i\}$ are drawn on these images, where each R_i may have separately connected areas. Then the inverse projections are derived to form each CV and an iterative method of forming the VOI is allowed by considering at any stage the partial intersection set :

$$(VOI)_k = CV_1 \cap CV_2 \cap \ldots \cap CV_k \tag{1}$$

and creating each subsequent display as one of

$$Dy_{k+1} = \textbf{Display}(Q) \text{ under transform } T_{k+1} \tag{2a}$$

$$Dy_{k+1} = \textbf{Display\_Boolean}(Q \cap VOI_k) \text{ under transform } T_{k+1} \tag{2b}$$

$$Dy_{k+1} = \textbf{Display\_Boolean}(Q - VOI_k) \text{ under transform } T_{k+1} \tag{2c}$$

Thus some flexibility is allowed in creating the VOI. The basic philosophy is to use as many images obtained by equation (2) as are useful to provide the visual cues for creating the VOI.

The actual CV_i's are intermediate and of no interest. Instead the VOI is built by editing the volume described by equation (1) at each stage.

5 Example Construction of an Arbitrary Object

In order to demonstrate the features of the process an artificial object was created. Figure 1 shows the screen image at stages in this editing process. The left hand image shows the sphere in anterior view with R1 drawn on it. The right hand image shows the left view of the expression sphere $\cap$ VOI_1, with R2-R4 drawn on it. In figure 2, the cylindrical volumes of each region are shown explicitly, with their intersection in figure 2c, and difference in 2d. Normally the intermediate sets $\{CV_i\}$ are not generated, but here they have been used to clarify the method. Figure 3 shows two general views of the objects formed from the intersection and difference of the sphere with the complete volume mask.

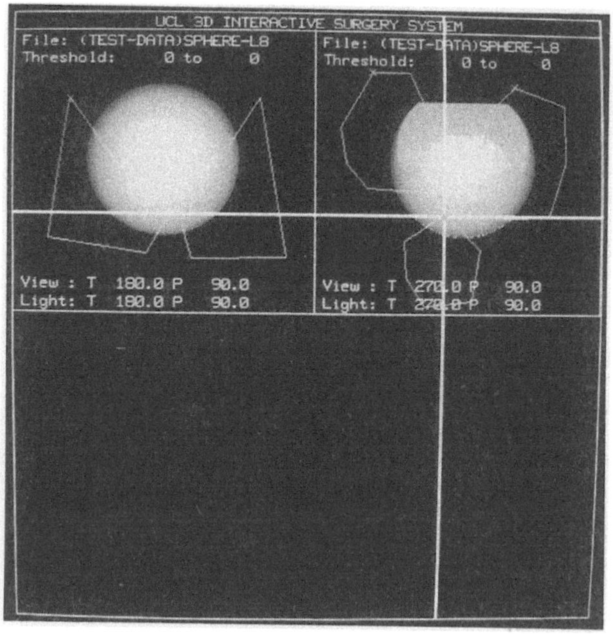

Fig. **1** : Regions R1 to R4 drawn on the shaded display of a sphere in multiple views.

Fig. **2** : The cylindrical volumes generated from the regions of figure 1.

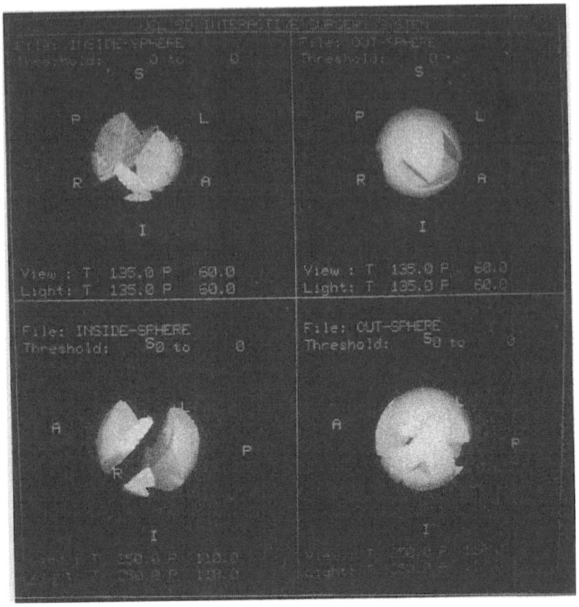

Fig. **3** : The resulting objects from intersection and difference of the sphere and the volume mask delineated in figure 1.

6 Explicit Boolean Operations

Explicit Boolean operations are required actually to create merged and split objects. When **Intersect**ing or **Union**ing two grey objects, one is given priority, for the case where two nodes are Full, so that the colour of the resultant is the colour of the higher priority node. Also, some meaning must be attached to the colour of the complement of an Empty grey node, although this operation has little practical use in the medical field. For the **Difference** operation, the tree that is not being complemented is given priority (i.e A is higher priority in the expression A - B). Intersection is useful for finding interference volumes if bone pieces are moved into each other.

Since the application of these algorithms is concerned with masking an object inside or outside a VOI, the algorithms have to be able to operate on a grey tree and a binary tree. The binary tree is formally treated as a grey tree with a uniform colour (usually arbitrary), and given lower priority.

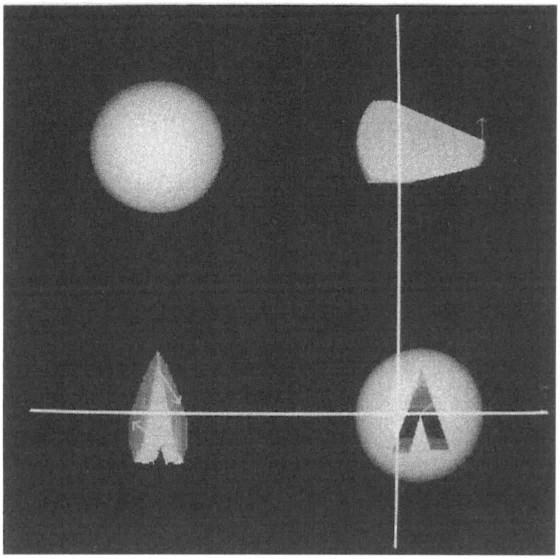

Fig. **4** : A slice dissected from the binary digital sphere, and a rotation specified.

7 Bilinear Transformations

Rotation, or general rotation and translation, procedures are very important for fully flexible simulations, but are lengthy to perform. Several authors [Udupa 1986, Trivedi 1986b] have stressed the importance whilst describing the large computational overhead.

Figure 4 shows the binary digital sphere with a piece dissected out from it and a rotation specified. Figure 5 shows the slice rotated and **Union**ed back with the sphere. The rotation algorithm took 15 minutes on a SUN 3/60.

A clinical example is shown in figure 6. Here a patient with Treacher Collins' Syndrome had a simulation carried out to separate the mandible, rotate and relocate it, as if for a costochondral graft.

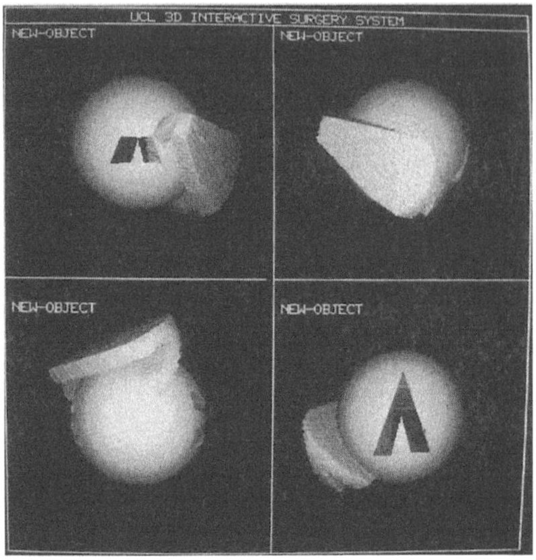

Fig. **5** : The result of dissecting, rotating and merging an arbitrary slice from a sphere.

8 Conclusions

The need for a sophisticated surgical planning package includes the ability to dissect, transform and merge pieces of object. In the UCL3D system this is achieved using hierarchical volume descriptions of objects and masks, and standard octree algorithms for the manipulation. This volumetric approach combines the advantage of complete access to the data, with a space planning ability similar to CAD/CAM applications.

Here the simulated "cutting" of an object by masking inside or outside a Volume of Interest (VOI) has been discussed, together with a simple rigid body transformation and merging by Boolean operations. A binary volume mask created in the way described in this paper may be used on several objects derived from the same data set. For greyscale representations, the visualisation of soft tissue ranges, masked by a VOI created using hard tissue displays, is achieved simply by using a different greyscale range in the **Display_Boolean** process. Obviously the use of greyscale structures saves the expense of a whole object generation process.

The current system has order of minutes response to these operations and

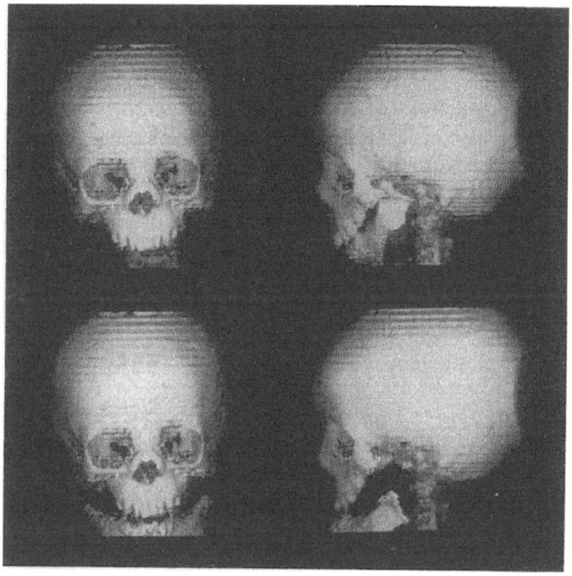

Fig. **6** : The simulation of a rotated mandible compared with the original patient data.

does not approach real-time interaction. With the current rate of increase in hardware performance, such a capability may be in reach in a few years. Instead the intention has been to demonstrate the type of functionality that will be required by a sophisticated surgical planning package. Although this approach allows a very flexible technique for simulation, an acceptable interface will probably need to reduce the options that confront the surgeon, and to suggest procedures based on standard methods.

9 Acknowledgements

This work was funded by the Information Technology Division of the Department of Health and Social Security grant number CTJ/506, as part of the development of a complete package for craniofacial surgical planning. The author wishes to thank Professor Jim Moss and Mr. David James for permission to use clinical data, and Mr. Sherif Gayed for discussion of the clinical requirements.

10 References

Arridge, S.R., *Data Structures and Algorithms for Manipulation and Display in Computer Simulated Surgery*, PhD thesis, University of London, December 1989.

Brewster, L.J., Trivedi, S.S., Tuy, H.K., Udupa. J.K., *Interactive Surgical Planning*. IEEE Computer Graphics and Applications. **4(3)** 31-40, 1984.

Chen, L-S, Hung, H.M., Levkowitz, H., Herman, G.T., Trivedi, S.S., Udupa, J.K., *Interactive manipulation of 3D data via a display device*, Proc SPIE, **507**, 25-37, 1984.

Cline, H.E., Lorensen, W.E., Ludke, S., Crawford, C.R., and Teeter, B.L., *Two algorithms for three dimensional reconstruction of tomograms*. Medical Physics **15(3)** 320-327, 1988.

Cutting, C., Grayson, B., Bookstein, F., Fellingham, L., McCarthy, J.G., *Computer aided planning and evaluation of facial and othognathic surgery*, Computers in Plastic Surgery, **13(3)**, 449-462, 1986.

Freider, G., Gordon, D., Reynolds, R. *Back-To-Front Display of Voxel-Based Objects*. IEEE Computer Graphics and Applications, **5(1)**, 52-59, 1985.

Goldwasser, S.M., *A generalised object display processor architecture*, IEEE Computer Graphics and Applications, **4(10)**, 43-55, 1984.

Henderson, D., *Assessment of bony deformities of the midline and lower face*, British Journal of Plastic Surgery, **27**, 287-296, 1974.

Hemmy, D.C., David, D.J., Herman, G.T., *Three-dimensional reconstruction of craniofacial deformity using computed tomography*, Neurosurgery, **13(5)**, 534-541, 1983.

Höhne KH *3D-Bildverarbeitung und Computer-Graphik in der Medezin*. Informatik-Spectrum, Springer-Verlag (Berlin), **10** 192-204, 1987.

Meagher, D.J., *Interactive Solid Modelling*. Proceedings of CADCON West, February, 1984.

Potmesil, M., *Generating Octree Models of 3D Objects from their Silhouettes in a Sequence of Images*. Computer Vision, Graphics, and Image Processing **40** 1-29, 1987.

Robb, R.A., Barillot, C., *Interactive display and analysis of 3-D medical images*, IEEE Transactions on Medical Imaging **8(3)**, 217-226, 1989.

Tessier, P., Hemmy, D., *Three-dimensional imaging in medicine : a critique by surgeons*, Scand J Plast Reconstr Surgery, **20**, 3-11, 1986.

Tiede, U., Höhne, K.H., Riemer, M., *Comparison of surface rendering techniques for 3-D-tomographic objects*, Computer Assisted Radiology 87 (eds. Lemke HU, Rhodes ML, Jaffe CC, Felix R) Springer-Verlag (Berlin), 610-614, 1987.

Trivedi, S.S., *Representation of three dimensional binary scenes*, Proceedings of 6th Annual Conference NCGA, Technical session **III**, Dallas, 132-144, 1985.

Trivedi, S.S., *Interactive manipulation of three dimensional binary scenes*, The visual computer **2**, 209-218, 1986.

Udupa, J.K., *Computerised surgical planning : current capabilities and medical needs*, Proceedings SPIE **626**, 474-482, 1986.

Vannier, M.W., Marsh, J.L., Warren, J.O., *Three-Dimensional Computer Graphics for Cranio-Facial Surgical Planning and Evaluation*. Computer Graphics **17(3)** 263-272, 1983.

Vannier, M.W., Marsh, J.L., Warren, J.O., *Three-Dimensional CT Reconstruction Images for CranioFacial Surgical Planning and Evaluation*. Radiology **150** 179-184, 1984.

Weisburn, B.A., Patnaik, S., Fellingham, L.L., *An interactive graphics editor for 3D surgical simulation*. Proc SPIE **626**, 483-490, 1986.

COMPUTER ASSISTED MEDICAL INTERVENTIONS

Stéphane Lavallée, Philippe Cinquin

TIMB, Faculté de Médecine de Grenoble
Domaine de la Merci, 38 700 La Tronche, FRANCE
Tel +33 76 51 80 00 Fax +33 76 51 86 67 email cinquin@timb.imag.fr

Abstract

Existing imaging devices can be used to plan complex medical and surgical interventions. Recent advances in Robotics provide the opportunity of assisting the physician or the surgeon in performing the intervention. Assisting both planning and performing of interventions first raises problems of matching of various multimodality data. Then the performance of an intervention with a partially autonomous system puts specific problems which are discussed. A general methodology for Computer Assisted Medical Interventions is proposed, which turns out to be a particular case of the classical loop of Perception - Decision - Action. Stereotactic Neurosurgery is a domain in which this methodology is readily applicable. Results are discussed.

Keywords : Intervention Planning - Intervention Performing - Robotics - Multimodality Data Matching - Computer Assisted Stereotactic Neurosurgery

1. Introduction

The aim of Computer Assisted Medical Interventions (CAMI) is to **help surgeons and physicians to use in a rational and quantitative way multimodality data, in order to plan and to perform medical interventions.** Recent advances in imaging systems such as CT or NMR have induced research about the interpretation of medical images. Nevertheless, very few systems allow for an efficient therapeutic use of the wealth these images contain, which is CAMI's twofold objective :

- to define an operative strategy that takes advantage of the localizing capabilities of imaging, and to make this strategy available in an operative

NATO ASI Series, Vol. F 60
3D Imaging in Medicine
Edited by K. H. Höhne et al.

reference system. This implies to model basic data, to handle these models, and to process them, in conjunction with a priori knowledge, to define an optimal strategy.

- to perform the previously defined operative strategy, thanks to guiding systems, under the supervision of imaging. This performance may in some cases be very difficult for various reasons : surgical tools such as probes most often have to be very precisely positioned, complex interventions may happen to be necessary on organs on which a human operator will only get poor or no visibility, very accurate gestures may be required (microsurgery), and some interventions may be dangerous for the medical staff (in case of contamination or irradiation). In all these instances, guiding systems are required. They may be optical, mechanical or even robotic.

The ultimate developments of this research will deal with complex mechanical systems provided with various sensors, meant to perform "non anthropomorphic" interventions (i.e. interventions where constant accuracy with no fatigue is required, where the target lies in parts of the human body usually difficult to reach, where integration of specific multi-sensors data may be necessary ...).

All the aforesaid objectives aim at improving the quality of the interventions, since they result in :

- making it easier for the surgeon or the physician to define in a rigorous and complete way the objectives, to take consequences of the strategy into account, and to simulate a part of the intervention,
- increasing the accuracy of gestures, thus reducing the gap between results and objectives and improving safety,
- decreasing the intervention length (at the price of a potentially longer and more rigorous preparation),
- possibility to consider new interventions which were unrealizable up to now,
- possibility to validate a protocol of therapeutic research, as the necessary tools allow precise recording of the pre-operatively chosen strategies and comparison with the obtained clinical results.

Obviously, this is a long term research, with many potential clinical applications. Yet, a general methodology can be applied to various clinical situations. This is what we shall present in a first section. The results of a typical implementation of this methodology in Stereotactic Neurosurgery will be presented in a second part.

2. General Methodology

Computer Assisted Medical Interventions are performed in four non sequential steps :

at level of Perception :

- Step 1 : Acquisition and modelling of multi-modality information

at level of Decision :

- Step 2 : Definition of a surgical strategy
- Step 3 : Matching of multi-modality information

at level of Action :

- Step 4 : Performing the surgical strategy with guiding systems (robots)

Steps 1 and 2 will be shortly described, since these topics refer to well known areas (see for instance [Höhne 84]). Steps 3 and 4 are more specific of CAMI and will be emphasized and detailed.

Step 1 : Acquisition and modelling of multi-modality information
At that stage, information is acquired mostly with medical imaging devices (CT, NMR, DSA, US, TEP, ...) , but also with multimodality sensors (coming from Computer Vision, or devised to analyze various signals : pressure, Doppler, "tactile" feelings,...). Geometrical and anatomical rules and models (such as Atlases) have to be taken into account. These data must then be modelled. One of the important steps in this modelling is the representation of the surfaces of anatomical structures after 3D segmentation [Marque 89].

Step 2 : Definition of a surgical strategy
A strategy now has to be defined, using all the available information. A medical objective is assigned, and simulation of the intervention makes it possible to anticipate the morphological and functional consequences of the intervention. This definition of the strategy is first based on pre-operative data (3D images and anatomical models). It is then completed and updated by intra-operative data that come from various sensors (X-rays, Ultrasound, ...), and that are acquired at different stages of the intervention. The uncertainty due to the whole chain noise is taken into account by providing the surgeon with geometrical envelopes of the surgical tools.

Step 3 : Matching multimodality information

Intra-operatively using the whole set of multimodality information implies to "match" all the involved reference systems (one for each imaging device, one for the operating table, one for each "Atlas" describing some a priori knowledge, one for the surgical tools, and finally one for the patient's internal structure).The objective of this section is therefore to propose a general frame for these matching operations. The purpose is to match :

1) pre-operative images (CT, NMR, TEP,...) with intra-operative images (X-ray projections, US images,...) or signals (tactile sensors,...),
2) several pre-operative images together (CT with NMR for instance),
3) 3D images with the anatomical information which is contained in an Atlas,
4) various images and signals with the operative systems (robots).

All the geometrical matching problems can be expressed in a common way : if we associate each modality m with a reference system Rm, the matching between the modalities m1 and m2 is equivalent to the search of a function u: $R^3 \to R^3$ that associates each point of Rm1 with a point of Rm2. The matching process will maximize a similarity function S, while the function u is constrained by a model of transformation (rigid or elastic transformation) [Lavallée 89b].

In cases 1) and 2), which represent the most important part of the global matching process, a classical approach consists in using *material landmarks* in each reference system, and in looking for the 6 or 7-parameters rigid transformation (rotation + translation + scaling if unknown) that matches these material landmarks. Stereotactic neurosurgery is a typical case : some methods consist in fixing a stereotactic frame on the head of the patient during preoperative and intraoperative examinations. A more flexible solution uses some spheres pasted on the skin of the patient during preoperative image acquisitions and during the intervention [Mosges 89] : in this case, the 3D positions of the spheres are known in an operative reference system thanks to a mechanical pointing device. A similar solution uses pins implanted in a bone [Taylor 89]. All these methods raise some problems of "discomfort" for the patient, and of organization between different medical departments. They can remain inaccurate and lack flexibility. A more convenient way consists in *matching anatomical structures* . This can be done by matching two 3D data sets, or by matching one 3D data set with several 2D data sets.

3D/3D matching : An interactive matching of particular points is the first solution [Toennies 89], [Schiers 89]. But the process becomes inaccurate and unreliable if a few points are selected and it is fastidious with a lot of points. More automatic methods are studied [Ayache 89], [Dalton 88], [Pelizzari 89]. Global methods have also been developed (e.g. alignment of the inertia centers and axes of a same structure [Gamboa 86], [Bajcsy 89]).

3D/2D matching : This is necessary when intraoperative sensors are X-ray image acquisition devices, whose images are to be matched with 3D preoperative data. This problem is still unsolved, but work is in progress, inspired by similar problems that rise in classical Computer Vision [Lavallée 89 c]. Here, the method is to minimize a distance between real 2D intra-operatively acquired and synthetized projections of the same 3D structure, with respect to the parameters of the transformation involved between the two reference systems.

Finally, the problem of a dynamic matching between a preoperative model of a structure and its intra-operative deformation remains unsolved.

Case 3), where the purpose is to match an anatomical model (Atlas) with real 3D data, will be discussed in the case of stereotactic neurosurgery.

The geometrical matching between a robot (or any guiding system) and an intraoperative sensor (case 4) is equivalent to the extrinsic calibration of the robot in the operative reference system. This problem is well known in Computer Vision applied to Robotics. [Lavallée 89a] describes a method where a robot carries a calibration cage made of 4 plexiglass planes containing metallic balls in the neighbouring of the patient. A pair of X-ray images of this calibration cage allows then both to calibrate the radiological sensor (two-planes method) and to link the robot with the X-ray device.

Step 4 : Action performing

In this final stage, the physician or the surgeon can be assisted to perform the intervention itself. "Guiding systems" can be divided into three categories, depending on the autonomy that is left to the physician or to the surgeon :

Passive systems : the assistance consists in making possible a comparison between the simulation of the strategy and its actual implementation. Intra-operative sensors such as video cameras or 6-joints articulated arms (see the pointing device described in [Mosges 89]) allow the surgeon to control his action with respect to the whole information, including pre-operative data.

Semi-active systems : the action is "guided", which means that the intervention is performed with respect to the previously defined strategy, but that its final control depends on the surgeon. This is the typical case of linear motions that the surgeon performs by introducing a probe through a guide (see the instance of Stereotactic Neurosurgery, § 3).

Active systems : some subtasks of the strategy are performed with the help of an autonomous robotic system, supervised by the surgeon, and controlled by redundant sensors. This is the case where complex actions are performed, according to a complex strategy, interactively defined on the basis of multi-modality data. This category also includes Advanced Teleoperation where multi-modality data are taken into account for microsurgery. The challenge is here to make a cooperation between the operator and the robotic system possible.

All these systems meet common problems :

a) **matching** between the reference system of the guiding system and the reference system of the imaging devices. This is an "external" calibration problem.

b) **intrinsic calibration** : each system must be specifically calibrated.

c) **security issues** : this is a key point, all the more so because no security norms for the use of robots in medical or surgical conditions exist. Two kinds of problems can occur and account for a difference between the theoretical and the actual intervention :

- *informatics autonomy* : at any time, the coherence of the set of available information has to be checked. This is a Perception problem (which implies the use of *redundant sensors*), a Decision problem (the incoherence has to be detected), and an Action problem (prevent the error from propagating).
- *robotics autonomy* : the intervention is performed with non human strenghts. To limit the consequences of an error, these errors must be unlikely to happen (which raises fiability problems), and the speeds and potential strenghts should be limited, to enable control systems to detect the problem and to "react" quickly enough.

d) **psychological acceptance** : these systems must be considered as tools that help the physician or the surgeon. In our experience, the system really improved a surgical intervention, and was readily accepted by the medical community and by the patients.

e) **adaptation to medical and surgical constraints** : such problems as sterilization, interfaces, limited room in surgical theatres, must be considered.

The following diagram (Fig. 1) sums up these four steps, which are found to be a particular case of the classical loop Perception - Decision - Action.

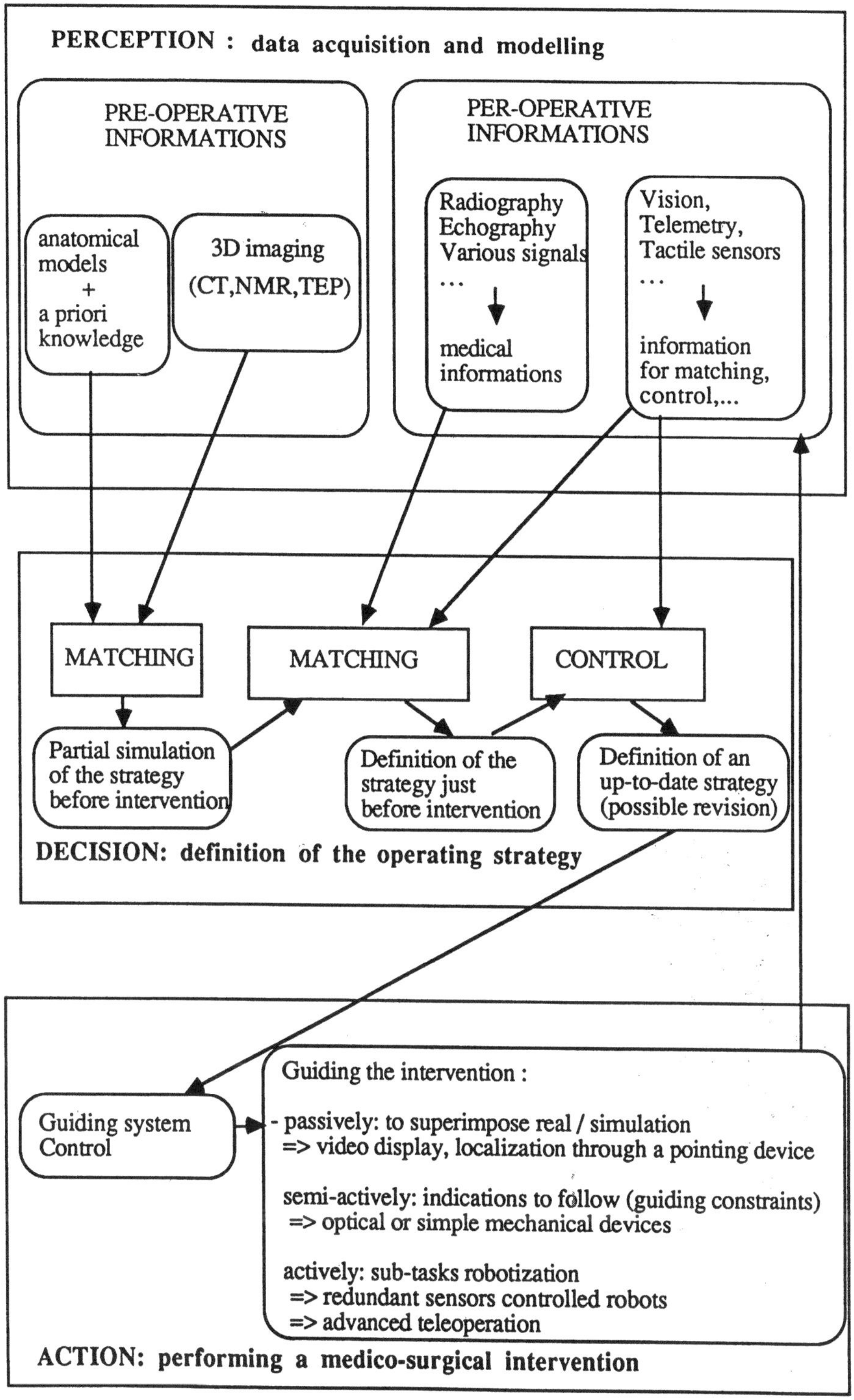

PERCEPTION : data acquisition and modelling
PRE-OPERATIVE INFORMATIONS
anatomical models + a priori knowledge
3D imaging (CT,NMR,TEP)
PER-OPERATIVE INFORMATIONS
Radiography Echography Various signals ...
medical informations
Vision, Telemetry, Tactile sensors ...
information for matching, control,...
MATCHING
MATCHING
CONTROL
Partial simulation of the strategy before intervention
Definition of the strategy just before intervention
Definition of an up-to-date strategy (possible revision)
DECISION: definition of the operating strategy
Guiding system Control
Guiding the intervention :
- passively: to superimpose real / simulation
=> video display, localization through a pointing device
semi-actively: indications to follow (guiding constraints)
=> optical or simple mechanical devices
actively: sub-tasks robotization
=> redundant sensors controlled robots
=> advanced teleoperation
ACTION: performing a medico-surgical intervention

3. The instance of Stereotactic Neurosurgery

Stereotactic Neurosurgery consists in introducing a tool into the brain with the help of pre- and intra-operative images, such as CT, NMR and conventional X-ray pictures [Talairach 52]. This technique has many clinical applications. According to the clinical application, various methods have been proposed to help the surgeon to plan and to perform a stereotactic operation. One key issue is the use of pre-operative images during the intervention. This raises the problem of the matching between these 3D images (NMR, CT, ...) and operative devices.

One solution is based on the "Talairach method" [Talairach 52 & 57] and it enhances the use of intra-operative X-ray images. In the operating theatre, two X-ray systems allow to acquire several pairs of orthogonal angiograms, ventriculograms and "simple" images, while the head of the patient is maintained in a stereotactic frame (submillimiter accuracy is achievable only at this cost). Then, some geometrical considerations allow the surgeon to match these X-ray projections with CT and NMR images, interactively. The advantage of such methods is that they can take into account pre-operative images which are acquired with no stereotactic frame fixed on the head of the patient. But some inaccuracies can result from interactive processes and from particular geometrical assumptions. Work is in progress in our team to make these matching procedures based on anatomical landmarks more accurate and easier : the purpose is to develop automatic (or semi-automatic) procedures that match segmented 3D anatomical structures observed on X-ray projections and on 3D images (CT or NMR).

Another method uses a stereotactic frame during both the operation and the preoperative exams. So, new stereotactic frames have been adapted first to CT-scan, [Jacques 80], [Gibaud 83],... and more recently to NMR devices [Kelly 86]. These methods have proven reliable and accurate, but they are quite cumbersome for they need to instal a specific stereotactic frame on the patient during all pre-operative examinations, and their extension to other clinical applications are not possible. So they still remain limited to particular sites.

Some authors haved proposed to perform the surgical intervention directly in the CT-scan [Rhodes 82], [Iseki 85],... These methods provide intraoperative control and they allow to perform fast operations but they require to allocate a specific CT-scan to the stereotactic interventions.

Moreover, NMR and angiograms are not taken into account. [Kwoh 88], [Young 87] have completed such systems with a robot to guide the introduction of the probe into the brain. This work must be considered as pioneer of the introduction of robots in the medical field.

Recently, some approaches, that do not use the stereotactic frame, have been proposed to guide the neurosurgeon to localize anatomical structures with respect to the 3D pre-operative images [Kosugi 88]. The principle is to use a "neuronavigator", that looks like a passive robot, for it consists of a probe mounted on 6 encoded but non motorized articulations. First, the surgeon matches pre-operative images with the pointing device by an adjustment of the tip of the probe on at least 3 anatomical landmarks taken on the skin surface, then the surgeon is able to visualize the extremity of the probe on pre-operative CT or NMR images. The accuracy of this method is quite poor (about 5 mm), due to the matching process. It also suffers from a lack of intra-operative sensors (that would provide information on vessels or on displacements of structures during the intervention). But it is very easy to use and it provides a useful help when a good accuracy is not the first requirement.

Intraoperative Ultrasound images can also be used [Rubin 80]. Work is in progress to correlate such images with a preoperative model obtained with CT or NMR.

In the case of Functional Neurosurgery, the use of the anatomical knowledge which is contained in an Atlas of the brain raises the problem of the matching betwen two different brains that have different 3D representations... Anatomical reference points such as the anterior and posterior commissures of the brain ventricles can be used [Talairach 57]. But local elastic matching procedures look very interesting [Bajcsy 89].

In conclusion, in order to deal with all the clinical applications of stereotactic neurosurgery, the computerized systems must take multi-modality images into account, integrate flexible matching procedures, and benefit from the use of robots. Such systems are more or less developed in the world, according to the chosen method. A preliminary system that links a robot with intraoperative X-ray images has been developed [Lavallée 89 a, b, c]. It has been successfully used (in the department of Neurosurgery of Grenoble's University Hospital, Pr. Benabid) for more than 100 interventions : 78 biopsies of brain tumors, 12 implantations of stimulating electrodes into a precise point of the thalamus (in order to stop Parkinson tremor), 3 implantations of a serial of recording electrodes (in order to reconstruct an epileptic focus) and 4 interstitial brachytherapies (I125 and Ir192). Figures 2 and 3 illustrate these results.

4. Conclusion

Many kinds of medical or surgical interventions can be "computer assisted", and some instances are already in common use (see [MURIM 90] for a review of existing applications and of work under development). The information on which CAMI are based are multi-modal, which stresses the importance of the development of Image Information Systems in hospitals. Such systems begin to exist, and will be more easily accepted if they contribute to a qualitative improvement of the use of medical images, which will be the case if they are used for CAMI.

REFERENCES

AYACHE N., BOISSONNAT J.D., BRUNET E., COHEN L., CHIEZE J.P., GEIGER B., MONGA O., ROCCHISANI J.M., SANDER P. Building Highly Structured Volume Representations in 3D Medical Images. Proceedings CAR'89, Berlin, juin 89, Lemke H.U. ed., Springer-Verlag, 765-772, 1989

BAJCSY .R. KOVACIC.S. Multiresolution Elastic Matching. Computer vision, graphics, and image processing **46**, 1-21, 1989

CINQUIN P., LAVALLEE S, GORET C, DEMONGEOT J., CROUZET G, PELTIE P. Computer Assisted Intervertebral Puncture, Proceedings 7 th International Congress on Medical Informatics in Europe, Springer-Verlag, A. Serio ed., Rome (Italie), 847-852, 1987.

DALTON B.L., DU BOULAY G. Medical image Matching. Proceedings SPIE vol **914**, Medical Imaging II, 456-464, 1988

GAMBOA-ALDECO A., FELLINGHAM L., CHEN G. Correlation of 3D surfaces from multiple modalities in medical imaging. Proceedings SPIE Vol. **626** Medecine XIV / PACS IV, 467-473, 1986

GIBAUD B. Etude et réalisation d'un systèmes d'analyse d'informations radiologiques tridimensionnelles tomodensitométriques stéréotaxiques et stéréoscopiques. Thèse de l'Université de Rennes I, 1983.

HOHNE K.H. Pictorial Information Systems in Medecine, Series F : Computer and Sytem Science. Vol 19, NATO, Advanced Science Institute Series, Springer, 1984.

ISEKI H., AMANO K., KAWAMURA H. et al., A New Apparatus for CT-Guided Stereotactic Surgery, Proc. 9th Meeting World Soc. Stereotactic and Functional Neurosurgery, Toronto 85, Appl. Neurophysiol. **48**, 50-60, 1985

JACQUES S., SCHELDEN H., MCCANN G.D., FRESHWATER D.B., RAND R. Computerized three-dimensional stereotaxic removal of small central nervous system lesions in patients, J. Neurosurg **53**, 816-820, 1980

KELLY P.J., Technical approaches to identification and stereotactic reduction of tumor burden, Walker, MD & Thomas, DGT(eds), Biology of brain tumour, 237-343, 1986

KOSUGI Y., WATANABE E., GOTO J. et al., An Articulated Neurosurgical Navigation System Using MRI and CT Images, IEEE Transactions on Biomedical Engineering **35**, 147-152, 1988

KWOH Y.S., HOU J., JONCKHEERE E.A., HAYATI S., A Robot with Improved Absolute Positioning Accuracy for CT Guided Stereotactic Brain Surgery, IEEE Transactions on Biomedical Engineering **35**, 2, 153-160, 1988

LAVALLEE S., CINQUIN P., DEMONGEOT J., BENABID A.L., MARQUE I., DJAID M., Computer Assisted Driving of a Needle into the Brain, Proceedings CAR'89, Berlin, juin 89, Lemke H.U. ed., Springer-Verlag, 416-420, 1989 a

LAVALLEE S., A New System for Computer Assisted Neurosurgery, Proceedings of IEEE Engineering Medicine Biology Society conf.,Yongmin Kim ed., Seattle (USA), nov. 89 , part 3, 926-927, 1989 b

LAVALLEE S., Gestes médico-chirurgicaux assistés par ordinateur : application à la neurochirurgie stéréotaxique, Thèse Génie Biologique et Médical, Grenoble, France, décembre 1989 c

MARQUE I., CINQUIN P. Segmentation d'Images 3D par Construction d'un Système Différentiel. Proceedings 7ème congrès AFECT- Reconnaissance des Formes et Intelligence Artificielle, Ed. Dunod, Paris, 907-922, 1989.

MOSGES R., SCHLONDORFF G., KLIMEK L., MEYER-EBRECHT D., KRYBUS W., ADAMS L., CAS - Computer Assisted Surgery. An Innovative Surgical Technique in Clinical Routine, Proceedings CAR'89, Berlin, juin 89, Lemke H.U. ed., Springer-Verlag, 413-415, 1989

MURIM Multidimensional Reconstruction and Imaging in Medicine : State of the Art report, EEC DG XIII, Advanced Informatics in Medicine, January 1990.

PELIZZARI C.A,CHEN G.T.Y, SPELBRING D.R, WEICHSELBAUM R.R, CHEN C.T, Accurate Three Dimensional Registration of CT, PET and / or MR Images of the Brain, Journal of Computer Assisted Tomography **13**, 1, 20-26, 1989

RHODES M.L., GLENN W.V., AZZAWI Y., HOWLAND R.S., An Improved Stereotactic System for CT Aided Neurosurgery, First IEEE Int. Symp. on Medical Imaging and Image Interpretation, Berlin 1982

RUBIN J.M., MIRFAKHRAEE M., DUDA E.E. et al., Intraoperative Ultrasound Examination of the Brain , Radiology **137**, 831-832, 1980

SCHIERS.C, TIEDE.U, HOHNE.K.H, Interactive 3D Registration Of Image Volumes From Different Sources, Proceedings CAR'89, Berlin, juin 89, Lemke H.U. ed., Springer-Verlag, 765-772, 667-669, 1989

TALAIRACH J., AJURIAGUERRA J. (de), DAVID M., Etudes Stéréotaxiques des Structures Encéphaliques Profondes chez l'Homme. Technique. Intérêt Physiologique et Thérapeutique. Presse Méd., 28, 605-609, 1952

TALAIRACH J., DAVID M., TOURNOUX P., CORREDOR H., KVASINA T., Atlas d'Anatomie Stéréotaxique des Noyaux Gris Centraux, Masson et Cie Edit., Paris, 1957.

TAYLOR R.H., PAUL H.A., MITTELSTADT B.D., GLASSMAN E., MUSITS B.L., BARGAR W.L., A Robotic System for Cementless Total Hip Replacement Surgery in Dogs, Proceedings of the 2nd Workshop on Medical & Healthcare Robotics, d.t.i. ed., Newcastle, UK, sept. 1989, 79-89, 1989

TOENNIES K.D., HERMAN G.T., UDUPA J.K., Surface Registration for the Segmentation of Implanted Bone Grafts, Proceedings CAR'89, Berlin, juin 89, Lemke H.U. ed., Springer-Verlag, 381-385, 1989

YOUNG R.F., Applications of Robotics to Stereotactic Neurosurgery, Neurological Research **9**, 123-128, june 1987

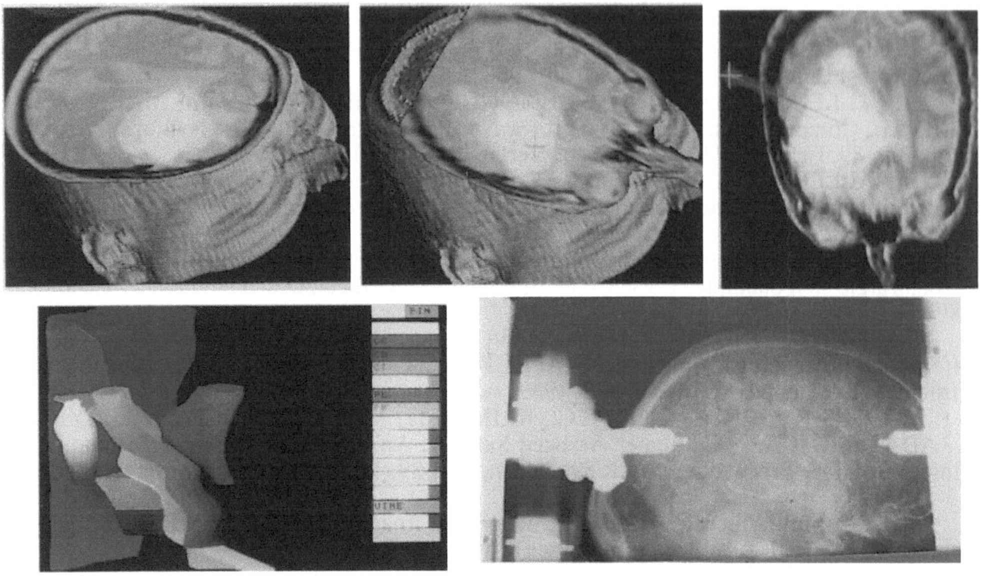

Figure 2 : Definition of the strategy on multimodality information. A target point and a linear trajectory are selected on reformatted slices of 3D NMR images (up), on an anatomical Atlas which has been digitized and segmented into surfaces of structures (down left) and on orthogonal X-ray images such as ventriculograms and angiograms (down right).

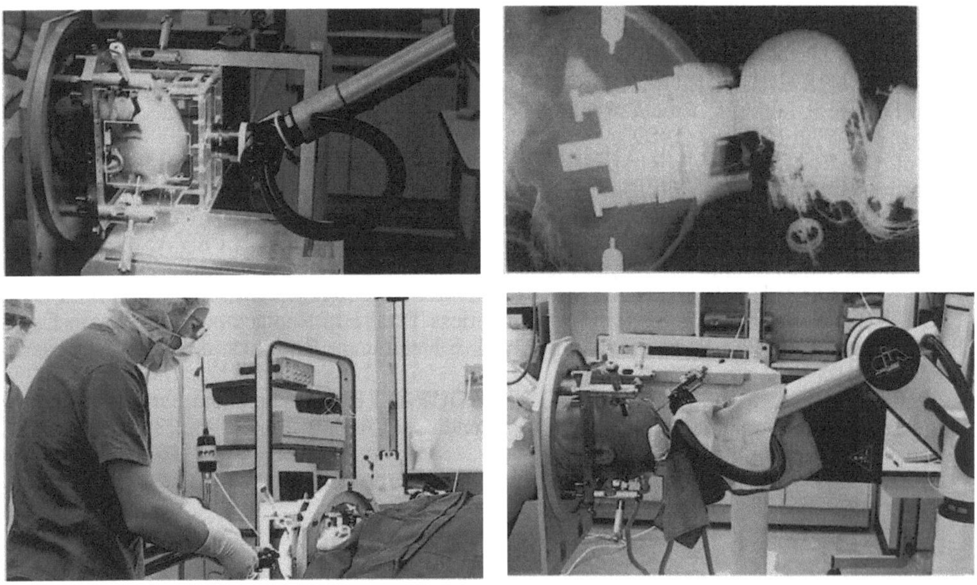

Figure 3 : positioning of a linear guide through a 6-axis robot. At the begining of each intervention, the robot carries a calibration cage on the head of the patient and reference X-ray images are acquired in order to calibrate the X-ray device and to link the robot coordinates with the image coordinates (up left). Then, the robot positions a linear guide in the location defined on multimodality images. Feedback control is achieved by checking the final location on X-ray images (up right). The power supply and the servo-control of the 6 axes is cut off, which is possible since reduction ratios have been increased.The surgeon finally introduces the tools (drills and probes) into the guide. Figures show the performance of a tumour biopsy (down

Systems

Systems for Display of Three-Dimensional Medical Image Data

Henry Fuchs

Department of Computer Science
University of North Carolina
Chapel Hill, NC 27599-3175 USA

Abstract

This paper briefly describes several research projects at UNC Chapel Hill involved with the display of 3D medical image data. A number of display methods is reviewed, and several new approaches outlined. Despite dramatic progress in display capability in recent years, even the most powerful current systems are inadequate for many daily medical imaging tasks. More powerful display systems in the near future may dramatically increase comprehensibility of complex 3D medical image data by enabling smooth interaction by direct manipulation, display without noticeable lag, and presentation of more powerful 3D depth cues such as head-motion parallax and stereopsis. These new systems will consist not only of a more powerful display engine, but also of an improved user interface with which the user can more easily interact than with current systems. One promising such interface for the future is based on a head-mounted display, one that allows the user to observe synthetic 3D structures superimposed on the world around him, allowing him to walk about these structures and modify them by direct manipulation using hand-held (simulated) tools. The poor quality of current miniature video displays prevents this approach from becoming a useful display method for 3D medical image data.

Keywords: 3D display, medical display, volume rendering, head-mounted display, parallel computation, VLSI

NATO ASI Series, Vol. F 60
3D Imaging in Medicine
Edited by K. H. Höhne et al.

1. Introduction and Previous Work

The explosive development of 3D medical image acquisition modalities in the past two decades such as CT, MRI, and ultrasound has stimulated the development of display technologies for these new kinds of data. This image-based data, with a large number of slice images in each single patient study, poses a difficult problem for the designers of 3D display machines. For decades these designers have been able to build machines with the assumption that, almost always the final image on the screen was represented and manipulated in the machine in some compact, geometric form. Until the 1980's and still today the vast majority of display devices for 3D data are based on the display of discrete geometric primitives such as points, lines, polygons and spheres [Molnar and Fuchs 1990] [Silicon Graphics 1990].

Increasingly 3D medical image data violates this assumption. The size, in terms of the number of bits, of the data set to be displayed from a single patient study regularly exceeds the size of the final image. This and other characteristics of the new 3D medical imaging modalities are forcing a re-examination of the basic organization of display systems. The re-examination is still far from complete. For example, although a number of systems specially structured for the display of volume image data have been developed (see, for example [Goldwasser 1985 and 1986] and [Kaufman 1988]), there has been no consensus reached on even the fundamental organization for this kind of a system. Compare this uncertainty to the rather standard organizational pipeline of most display systems using geometric primitives: transformation, clipping, scan conversion, visibility determination using a depth buffer, and pixel painting. Most researchers in the area of 3D medical image display have concentrated on algorithmic improvements for either object definition (for example, [Herman and Liu 1979] and [Lorensen and Cline 1989]) or for display quality (for example, [Frieder, Gordon and Reynolds 1985], [Hoehne et al. 1989], and [Levoy 1988 1989 1990] and [Tiede, Hoehne, et al 1990]). Several groups with access to graphics accelerators have developed improved algorithms that might work on a variety of platforms, but are optimized for the platform of choice: [Drebin, Carpenter and Hanrahan 1988] and [Ney et al, 1990] for the Pixar Image Computer, and [Johnson and Mosher 1989] for the SUN/TAAC accelerator board. In general, the current state of development is such that either a high-quality image can be generated with some wait, or a poorer quality image (with noticeable sampling artifacts) can be generated in real time on some specialized hardware.

mounted display (see Fig. 4b), a descendant of Ivan Sutherland's pioneering ideas on virtual reality [Sutherland 65] and on head-mounted displays [Sutherland 68].

2. Related Work at UNC-Chapel Hill

Here at UNC-Chapel Hill, a number of research groups have been working on various aspects of interactive 3D display, for medical and other applications. We will review in this paper only those related to medical display applications. (For information on the 3D object definition work here, see [Pizer, Cullip and Fredericksen 1990] in these proceedings.) These projects are concerned with image generation (hardware and software improvements), display devices, and interaction (devices and techniques). For a discussion of rendering and display hardware issues, see [Fuchs, Levoy and Pizer].

2.1 Image generation algorithms

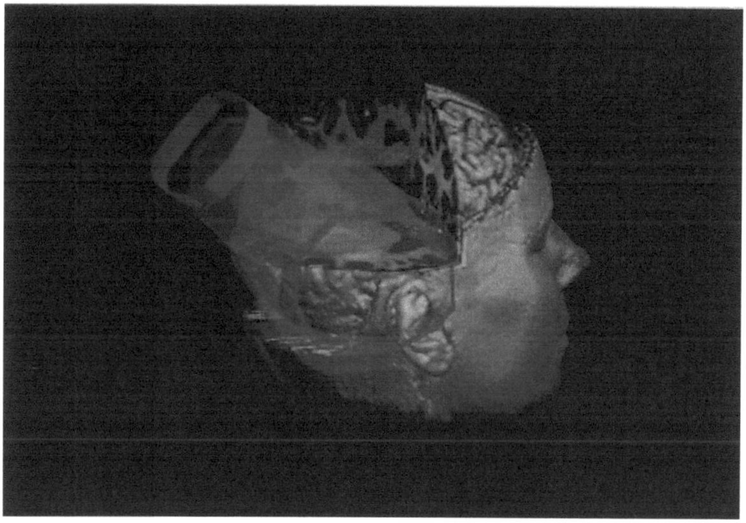

Fig. 1: Volume rendering of MR data with polygon-delineated tumor region in purple and beam boundaries in red. Image by Marc Levoy and Victoria Interrante [Levoy 1990]. MR data courtesy of Siemens AG, edited by Julian Rosenman. Treatment plan courtesy of NC Memorial Hospital.

Figure 1 illustrates the state of the local art in volume rendering that superimposes polygon and volume data sets. The next development should be the

2.2 Image Generation Architecture

Image generation architecture work has focused in recent years on designing and developing architectures that take advantage of the massive parallelism available with custom-designed VLSI circuits. The latest of these machines, Pixel-Planes 5, is currently under construction [Fuchs, Poulton, et al 1989]. Its predecessor, the 1986 Pixel-Planes 4, consists of a general-purpose, front-end processor that traverses a display list, performs transformations and lighting calculations, and generates instructions to a 512 x 512-pixel frame buffer with a one-bit processor at each pixel. The 262,144 processors are controlled in a SIMD fashion. Pixel-Planes 4 has been a mainstay in our Graphics and Image Lab, being used daily for nearly four years.

Pixel-Planes 5, a considerably more complicated machine, features at the outermost level a large-grain MIMD parallel organization that allows concurrent execution of independent programs at each of several dozen board-level processing units (see Fig. 2). Each of the 128x128-element SIMD processing arrays can be independently processing a different stream of commands. Although most often these processing arrays are organized by software to map onto the final image one processor per pixel, the processing arrays can be organized in other ways. For instance two different volume rendering methods have been proposed, one optimized for fast interaction [Levoy 1989], the other optimized for progressive refinement of the volume rendering as new (ultrasound) data keeps arriving [Ohbuchi and Fuchs 1990].

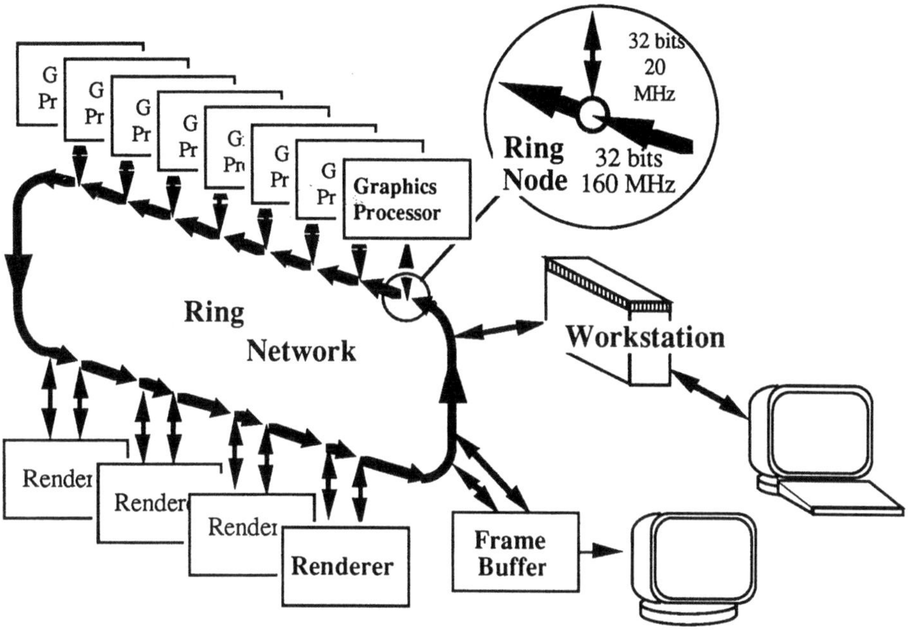

Fig. 2: Pixel-Planes 5 overview, showing each of the board-sized processing units. Graphics Processors each have an Intel i860 processor, each Renderer has a 128 x 128 processor SIMD array.

We expect that for many if not most applications, performance of Pixel-Planes 5 will be limited by the communication speed between board-level units. Next generation architectures may explore organizations that have even higher bandwidth between certain selected "partners" within a MIMD organization (Fig. 3). For example, it may be more efficient for many applications to have a fixed association between a general purpose board and a SIMD processor array board. Consider the "sort last" version [Molnar 88]; in this, the data set is randomly distributed among the general purpose processors, G. Each G and R pair process and completely render their assigned fraction of the data set. A Sorting network following all the R's calculates the final image by selecting for each final pixel x,y the pixel from all the partial-data R's at this x,y location that has the smallest z (nearest surface). A more sophisticated version of this sorting network would do proper compositing by considering *all* the partial-data pixels and their associated z values at the current x,y location Notice that the Rasterization units need not each have a complete frame buffer; indeed, it may be advantageous to have them generate the image of their part of the data set in scan-line order, in which case a buffer of only a few scan lines might be adequate.

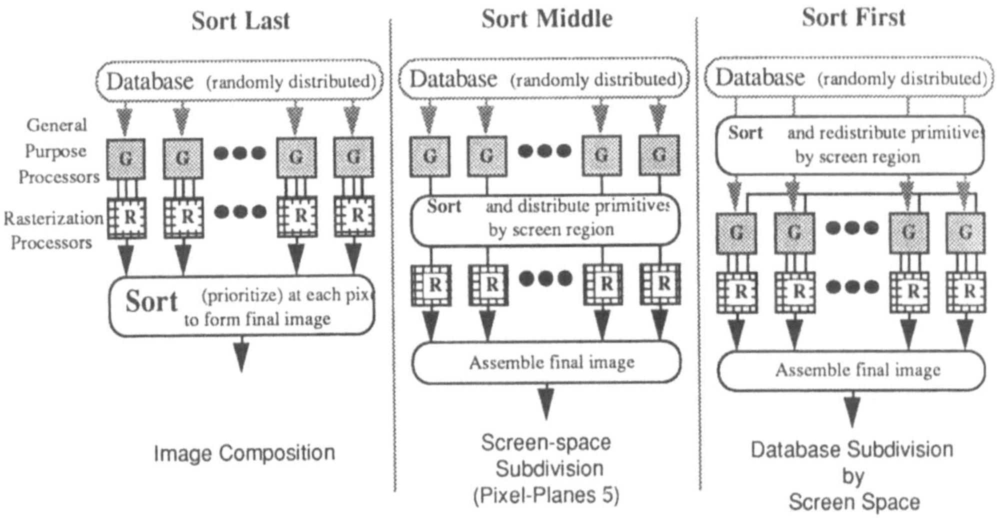

Fig. 3: Variety of multicomputer organizations that may offer improvements in performance over Pixel-Planes 5.

The "sort first" machine, suggested by John Poulton, would distribute G-R pairs about the final image, so each pair covered a small fraction of the screen. Each primitive in the data set would be sent to the G-R pair that covered its current screen location. As each frame is generated, primitives that no longer fall into their G-R's screen region are sent to the G-R pair that does cover their current location. The working assumption is that between any consecutive frames, only a small fraction of the primitives would move to a different screen region and thus inter-processor communication needs may be modest.

2.3 Display devices

A variety of display devices can ease the comprehension of complex 3D images -- devices that present more 3D cues than a conventional video monitor. Among the most popular are various devices that deliver a stereo image pair, a separate image to each eye. Over the years, we have experimented with a variety of stereo display devices, settling currently on the full-screen polarizing plates produced by Tektronix. These switch polarization in response to vertical synchronization signal. The plate is mounted immediately in front of a video monitor. The image generator is instructed to display the left and the right image on alternate fields, so that any user wearing (passive) glasses with appropriately different polarization in each eye will see, at any instant, a dark monitor image in one eye and the proper image in the other eye. Multiple viewers can observe a stereo image simultaneously.

Unfortunately, the first reaction of many viewers upon trying such lenses is to move their head from side to side. Of course, the image does not change, but appears to distort as the user's head moves. We have found, however, that if the user's head is tracked and the proper new image is displayed instantly (that is, the image of the 3D object as it would look from the user's proper point of view), the user perceives a much stronger 3D percept than when only a stereo still image is displayed. Surprisingly, if the head tracking is on and stereo is off, the image movement is disturbing—the image appears to rotate away from the user in a rather unnatural fashion.

A "true 3D" display device which gives a very strong 3D percept without the need for any head-worn aids, is the vibrating, so called "varifocal" mirror [Mills, Fuchs and Pizer, 1984]. With this display, the user looks into a vibrating reflecting drum-head sized diaphragm to observe a changing image. (This vibrating surface is the "varifocal mirror.") The image is changed cyclically in synchrony with the mirror vibrations, which are usually set to about 30Hz. Since a particular image is always displayed when the mirror is in the same position, the image appears to float in the same position in space. Having several dozen or more images flashed in 1/30 of a second gives the appearance of a stack of transparent images floating in space. Having the images be computer generated allows them to be changed interactively.

Perhaps the most exciting display for the future is one that is small enough to be mounted on a head-gear (Figs. 4a and 4b). If the user's head is tracked in real time and the 3D image on the miniature displays can be changed in a small fraction of a second, then images can be generated which account for the user's head position, allowing the user to "walk around" in a simulated 3D environment [Chung et al 1989].

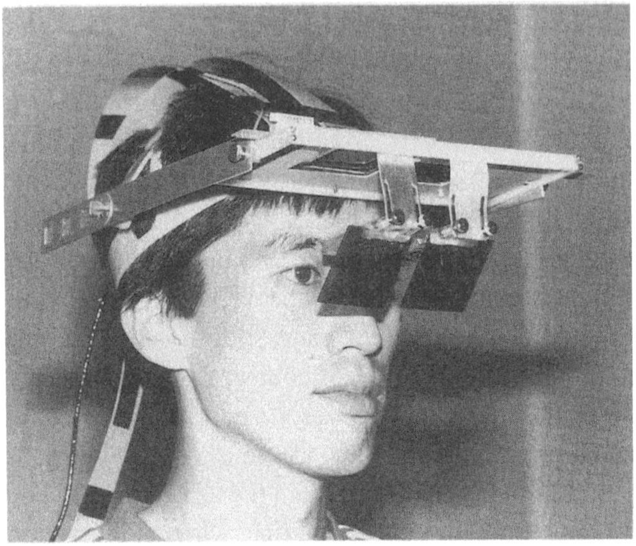

Fig. 4a: Head-mounted display at UNC (1986 vintage). The left and right eye displays are mounted in the visor, pointed down, so that the user observes the computer-generated scene superimposed on his surroundings.

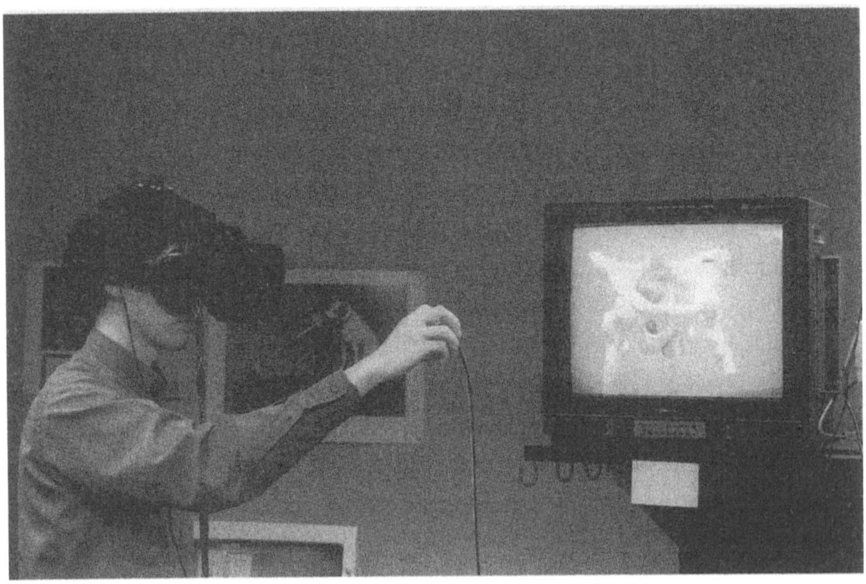

Fig. 4b: Head-mounted display device (by VPL, Inc.) at UNC. For this photo, the image seen by the user is also displayed on the nearby monitor. The image of the hand-held cursor is a faintly-seen ball on the top right portion of the computer-generated image.

2.4 Input devices

It sometimes appears that the state of the art in input devices for interactive 3D graphics has been declining rather than moving forward. In the 1970's and 1980's when interactive 3D displays were quite expensive, they were often fitted with an array of input knobs, buttons, joysticks, and trackballs. Now that interactive 3D displays are relatively inexpensive and rather widespread, seldom does one have more input devices than a conventional 2D mouse. Our experience is that a variety of carefully designed translation and rotation *positional* inputs are often very useful for interactive work.

Recently we have been concentrating our efforts on developing a tracking device for the head-mounted display that would allow the user to move easily about the work room. Our experience with the most common 3D tracking device, by Polhemus, indicates that tracking speed and accuracy deteriorates quickly at distances greater than one meter from the tracking source. We would like to have a working volume of a large room, to allow the user to move about a patient, to see computer generated data superimposed within the patient (see Fig. 7). Toward this end, we have been working on an optical system using high speed miniature cameras that are worn on the head and are pointed at the ceiling. In the ceiling is mounted an array of infra-red LEDs. At any image gathering cycle, of about 100 microseconds, exactly one LED is turned on [Wang, Chi, and Fuchs 1990]. The head-mount location is determined from the LED's known location in space and its 2D location in the camera's image (see Fig. 5a and 5b). In theory only a single camera and 3 observed LEDs are needed, but to allow a wide range of user rotations while some camera is still pointing at a nearby patch of the ceiling, we are currently using three cameras [Wang et al 1990].

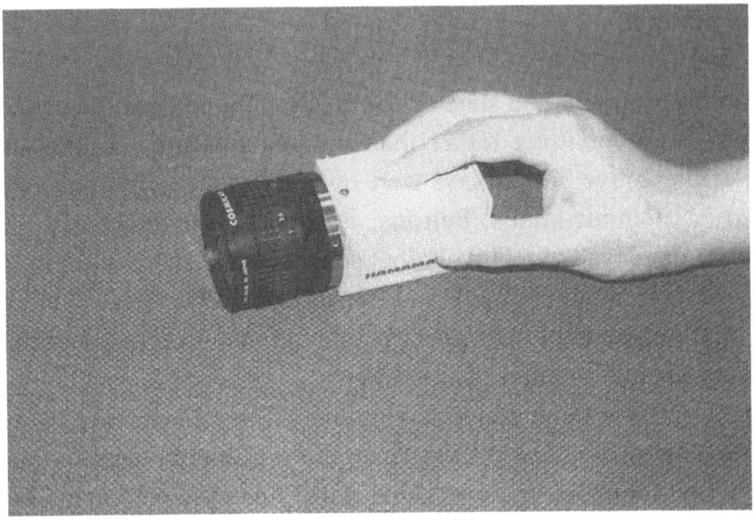

Fig. 5a: Small camera used for experimental head tracking in a room-sized environment. Camera can capture the x,y location of an infra-red LED in approximately 100 microseconds.

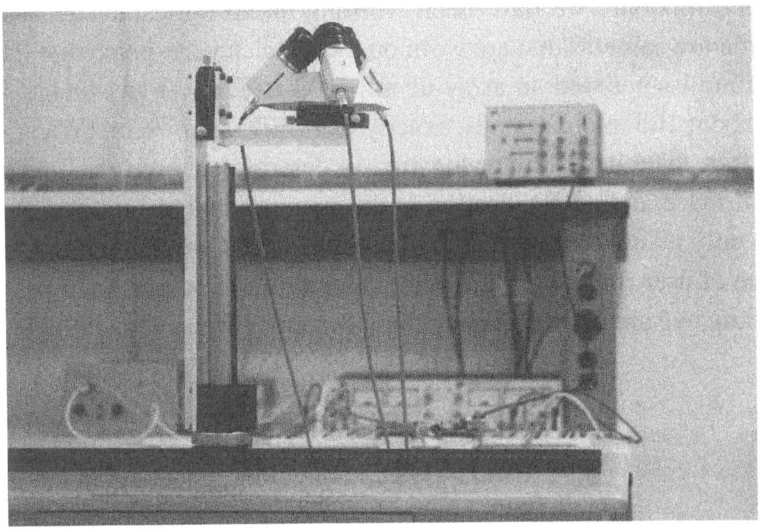

Fig. 5b: Prototype for head-mounted tracking system. The three-camera cluster will be helmet-mounted. With improved (holographic) optics, we expect to reduce the number of cameras needed in the system to two or possibly even one.

3. Applications and Systems

The above-described technological developments should be able to support a variety of medical imaging systems and applications. Two such applications are 1) interactive display of ultrasound data and 2) interactive radiation therapy planning. They are both characterized by the need comprehend very complicated 3D structures and to interact with them for diagnosis or planning.

3.1 3D Display of Ultrasound Data

The results of our initial experiments in the display of ultrasound data are shown in Figures 6a through 6d [Ohbuchi and Fuchs 1990]. It is important to remember that these images were not generated in real time, but that the algorithm and system developments are in that direction.

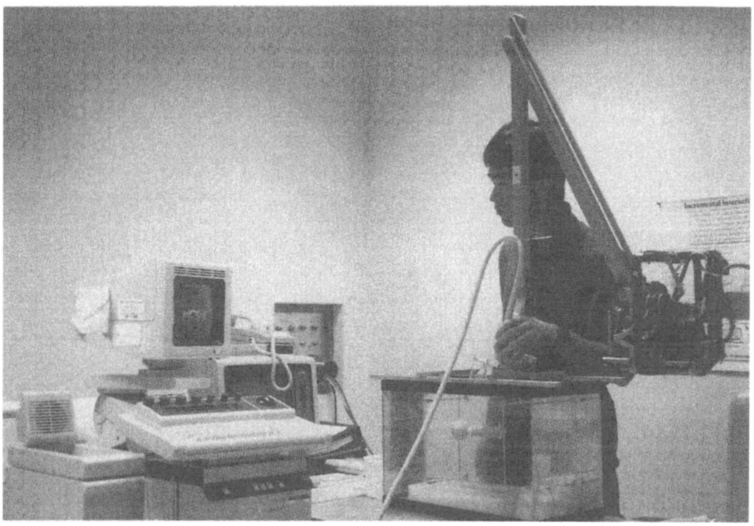

Fig. 6a: Experimental setup for 3D reconstruction of ultrasound data. Computer captures position and orientation of the mechanical arm and the video signal from the ultrasound imager.

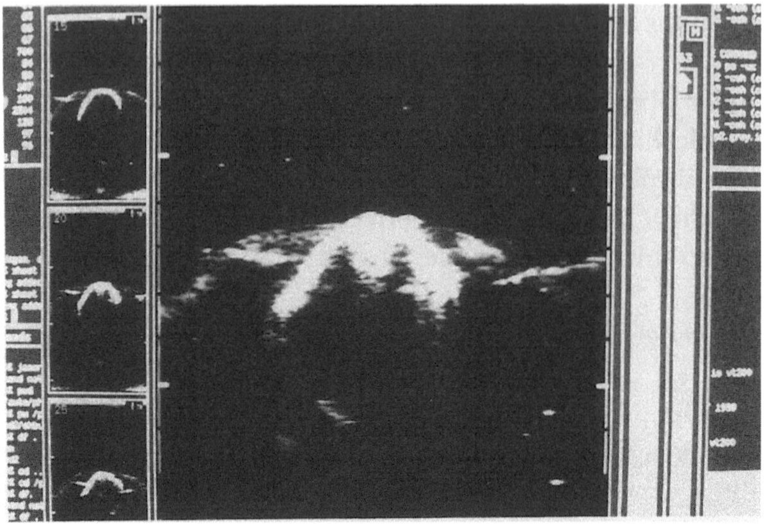

Fig. 6b: Sample ultrasound image from the doll's head region.

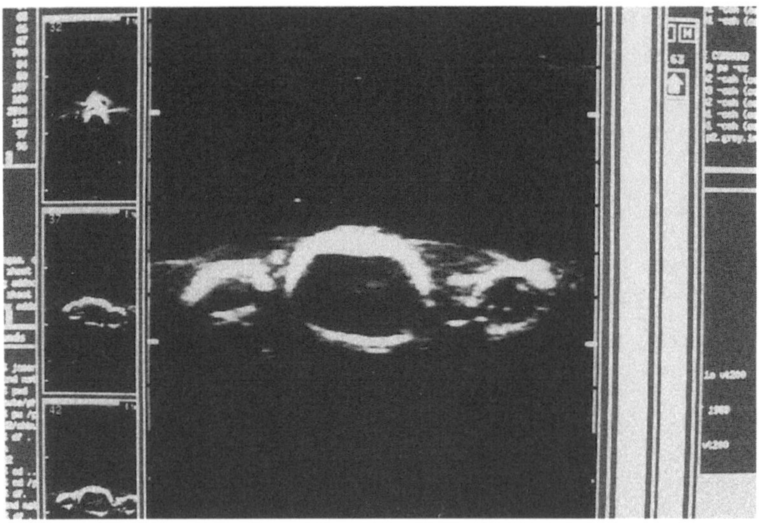

Fig. 6c: Sample ultrasound image of doll's torso, showing cross sections of body and arms.

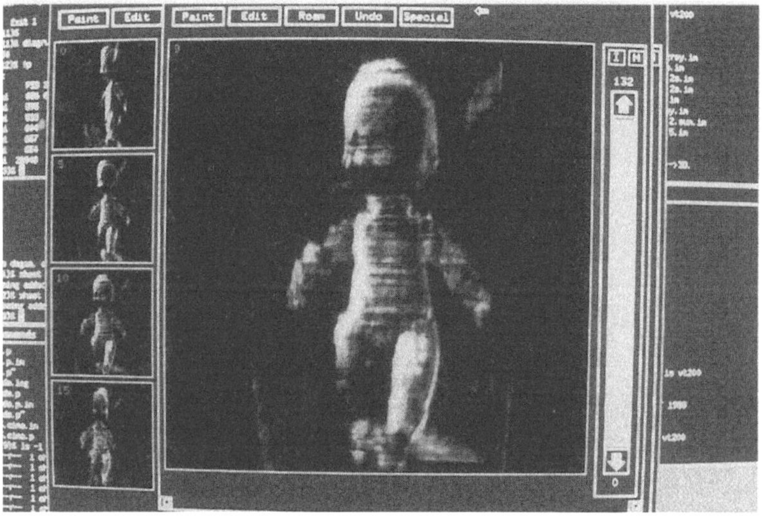

Fig. 6d: Three-dimensional reconstructed images of a doll from a set of ultrasound scans. Notice various views of the 3D data on left side of photo.

Figure 7 illustrates our conception of an ultrasound display system we plan to complete in several years. The system consists of a head-mounted display, a real-time tracker for the ultrasound transducer, and an image generation system that reconstructs in real time a 3D image from a succession of arbitrarily-oriented 2D ultrasound slice images. We hope that the ability to observe the data registered within the patient will make the ultrasound data and exam more helpful to the physician in correlating the ultrasound data with palpation and using the ultrasound data to help with amniocentesis and surgical tasks.

3.2 Interactive Radiation Therapy Planning

One of our most recent projects aims to test the efficacy of radiation therapy planning when done interactively rather than by an off-line batch process [Rosenman et al 1989].

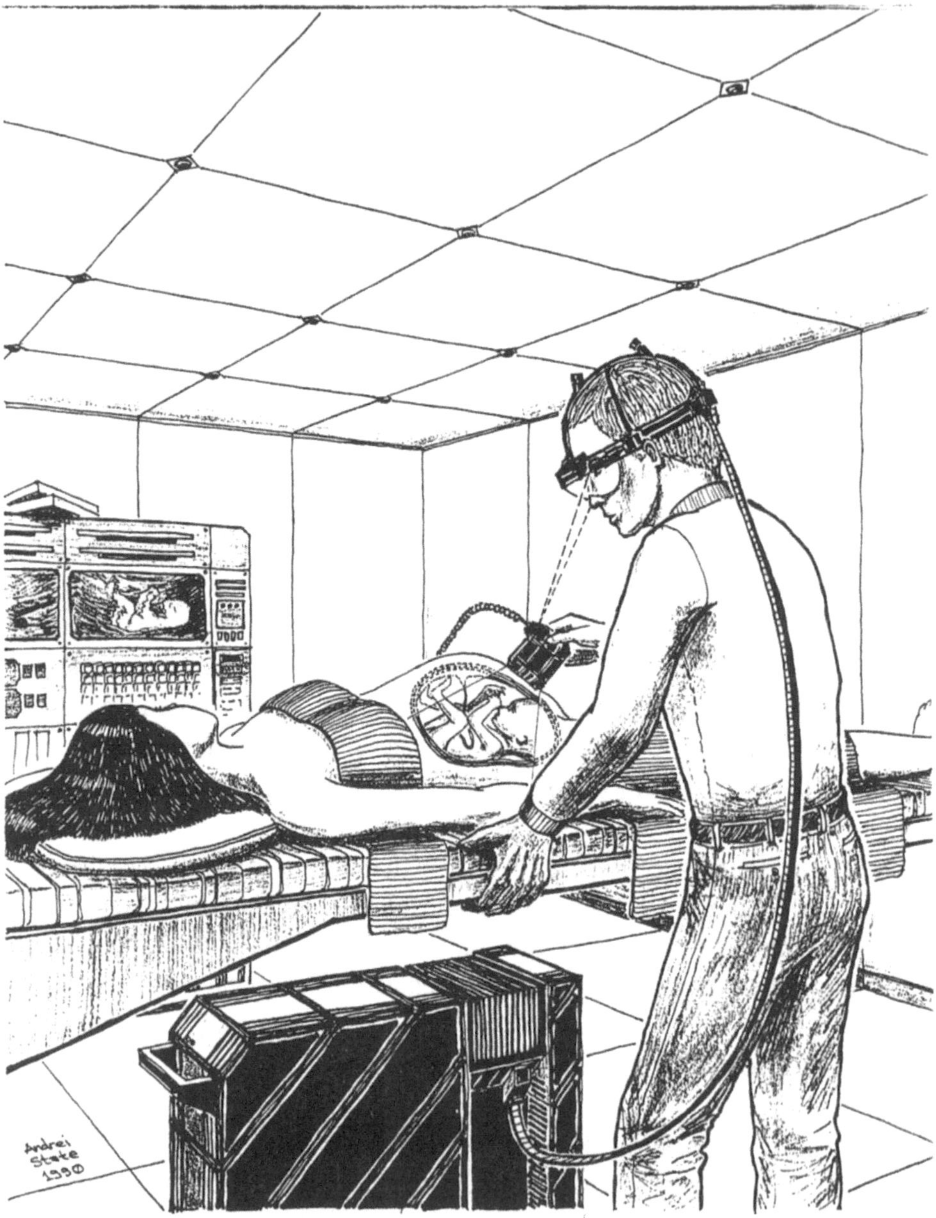

Fig. 7: Conceptual sketch of a future real-time 3D ultrasound display system with a head-mounted display. Physician observes the ultrasound data registered within the patient.

We are working to have a Cray Y-MP supercomputer at the North Carolina Supercomputer Center connected via a high-speed network to the interactive graphics system at UNC-Chapel Hill. As the therapy planner changes parameters of the plan, such as the orientation and strength of a radiation beam, the new dose distribution will be instantly recalculated and displayed. We expect that such a capability will allow the therapy planner to more rapidly explore the multi-dimensional design space than current systems, which take 30 minutes or more to evaluate a single potential plan.

4. Acknowledgements

The work described in this paper involves dozens of individuals on perhaps a dozen distinct projects in the Graphics and Image Lab and the Microelectronics Systems Lab, both in the Department of Computer Science at UNC Chapel Hill. Support for the projects comes partially from NIH Grant P01 CA 47982 (Stephen M. Pizer, Project Director), NIH RR 02170-05 (Frederick P. Brooks, Jr., PI), NSF Grant DCI-8601152 and DARPA ISTO No. 6090 (both with Fuchs and John Poulton, PI's), ONR Contract N0014-86-K-0680, NSF Grant CDR-86-22201 (Theo Pilkington of Duke University, Project Director).

5. References

Chung, J.C., M.R. Harris, F.P. Brooks, H. Fuchs, M.T. Kelley, J. Hughes, M. Ouh-young, C. Cheung, R.L. Holloway, M. Pique, "Exploring Virtual Worlds With Head-Mounted Displays," *Proceedings of SPIE: Non-Holographic True Three-Dimensional Displays*, Vol. 1083, 1989.

Drebin, R.A., L. Carpenter, and P. Hanrahan, "Volume Rendering," *Computer Graphics*, Vol. 22, No. 4 (Proceedings of Siggraph '88) pp. 65-74, August 1988.

Frieder, G., D. Gordon, and R. A. Reynolds, "Back-to-Front Display of Voxel-Based Objects," *IEEE Computer Graphics and Applications*, Vol. 5, No. 1, pp. 52-59, January 1985.

Fuchs, Henry, John Poulton, John Eyles, Trey Greer, Jack Goldfeather, David Ellsworth, Steve Molnar, Greg Turk, Brice Tebbs, Laura Israel, "A Heterogeneous Multiprocessor Graphics System Using Processor-Enhanced Memories," *Computer Graphics*, Vol. 23, No. 4 (Proceedings of SIGGRAPH '89) pp. 79-88, August 1989.

Fuchs, Henry, Marc Levoy, and Stephen M. Pizer, "Interactive Visualization of 3D Medical Data," *IEEE Computer*, Vol. 22, No. 8, pp. 46-51, August 1989.

Goldwasser, Samuel, et al., "Physicians Workstation with Real-Time Performance," *IEEE Computer Graphics and Applications*, Vol. 5, No. 9, pp. 44-57, September 1985.

Goldwasser, Samuel., "Rapid Techniques for the Display and Manipulation of 3D Biomedical Data," Tutorial presented at 7th Annual Conference of the NCGA, Anaheim, CA, May 1986.

Herman, G.T. and H.K. Liu, "Three-Dimensional Display of Human Organs from Computer Tomograms," *Computer Graphics and Image Processing*, Vol. 9, No. 1, pp. 1-21, January 1979.

Hoehne, K.H., M. Bomans, A. Pommert, M. Riemer, C. Schiers, U. Tiede, and G. Wiebecke, "3-D Visualization of Tomographic Volume Data Using the Generalized Voxel-Model," *Proceedings of Chapel Hill Workshop on Volume Visualization*, Chapel Hill, NC, pp. 51-57, May 1989.

Johnson, E.R. and C.E. Mosher, Jr., "Integration of Volume Rendering and Geometric Graphics," *State of the Art in Data Visualization*, Siggraph'89 tutorial series,

Kaufman, A. and R. Bakalash, "Memory and Processing Architecture for 3D Voxel-Based Imagery," *IEEE Computer Graphics and Applications*, Vol. 8, No. 6, pp. 10-23, November 1988.

Levoy, Marc, "Display of Surfaces from Volume Data," *IEEE Computer Graphics & Applications*, pp. 29-37, May 1988.

Levoy, Marc, "Design for a Real-Time High-Quality Volume Rendering Workstation," *Proceedings of Chapel Hill Workshop on Volume Visualization*, Chapel Hill, NC, pp. 85-92, May 1989

Levoy, Marc, "A Hybrid Ray Tracer for Rendering Polygon and Volume Data," *IEEE Computer Graphics and Applications*, Vol. 10, No. 2, pp. 33-40, March 1990.

Lorensen, W.E., and H.E. Cline, "Marching Cubes: A High-Resolution 3D Surface Construction Algorithm," *Computer Graphics*, Vol. 23, No. 3 (Proceedings of SIGGRAPH '87) pp. 163-169, July 1989.

Mills, Peter H., Henry Fuchs and Stephen M. Pizer, "High-speed Interaction on a Vibrating Mirror 3D Display", *Processing and Display of Three-Dimensional Data II*, Proceedings of SPIE, Vol. 507 (August 1984) pp. 93-101.

Molnar, Steven, "Combining Z-buffer Engines for Higher-Speed Rendering," *Advances in Computer Graphics Hardware III*, Eurographics Seminars, 1988.

Molnar, Steven, Henry Fuchs, "Advanced Raster Graphics Architecture," Chapter 18 in *Computer Graphics: Principles and Practice*, Second Edition, by J. Foley, A. van Dam, S. Feiner and J. Hughes, Addison-Wesley Publishers, Reading, MA, 1990.

Ney, Derek R., Elliot K. Fishman, Donna Magid, and Robert A. Drebin, "Volume Rendering of Computed Tomography Data: Principles and Techniques," *IEEE Computer Graphics and Applications*, Vol. 10, No. 2, pp. 24-32, March 1990.

Ohbuchi, Ryutarou and Henry Fuchs, "Incremental 3D Ultrasound Imaging from a 2D Scanner", *Proceedings on the First Conference on Visualization in Biomedical Computing*, May.22-25, 1990, Atlanta, GA (also available as UNC-CH Dept. of Computer Science Technical Report, TR90-013).

Pizer, S.M., T.J. Cullip and R.E. Fredericksen, "Toward Interactive Object Definition in 3D Scalar Fields," in these proceedings, 1990.

Rosenman, Julian, George Sherouse, Henry Fuchs, Stephen M. Pizer, Andrew L. Skinner, Charles Mosher, Kevin Novins, Joel Tepper, "Three-Dimensional Display Techniques in Radiation Therapy Treatment Planning," *International Journal of Radiation Oncology Biology Physics*, Vol. 16, No. 1, 1989.

Silicon Graphics Computer Systems, "POWERpath (tm) Processor Architecture," Technical Report, Silicon Graphics Computer Systems, Mountain View, CA, February 1990.

Sutherland, Ivan E., "The Ultimate Display," *Proceedings of the IFIP Congress '65*, pp. 506-508, 1968.

Sutherland, Ivan E., "A Head-Mounted Three Dimensional Display," *Proceedings of the AFIPS Fall Joint Computer Conference*, 1968.

Tiede, Ulf, Karl Heinz Hoehne, Michael Bomans, Andreas Pommert, Martin Riemer, and Gunnar Wiebecke, "Investigation of Medical 3D-Rendering Algorithms," *IEEE Computer Graphics and Applications*, Vol. 10, No. 2, pp. 41-53, March 1990.

Wang Jih-fang, Vernon Chi, Henry Fuchs, "A Real-time Optical 3D Tracker for Head-mounted Display Systems," *Computer Graphics*, Vol. 24, No. 2 (Proceedings of 1990 Symposium on Interactive 3D Graphics), pp. 205-216, March 1990.

Wang, Jih-fang, Ron Azuma, Gary Bishop, Vernon Chi, John Eyles, Henry Fuchs, "Tracking a Head-mounted Display in a Room-sized environment with Head-mounted Cameras," to appear in *Proceedings: SPIE '90 Technical Symposium on Optical Engineering & Photonics in Aerospace Sensing*, Orlando, Florida, April 1990.

A SOFTWARE SYSTEM FOR INTERACTIVE AND QUANTITATIVE ANALYSIS OF BIOMEDICAL IMAGES

Richard A. Robb
Mayo Foundation, Rochester, MN 55905

ABSTRACT

A comprehensive software system called ANALYZE has been developed which permits detailed investigation and structured evaluation of 3-D and 4-D biomedical images. This software system can be used with any 3-D imaging modality, including x-ray computed tomography, radionuclide emission tomography, ultrasound tomography, magnetic resonance imaging, and both light and electron microscopy. The system is a synergistic integration of fully interactive modules for direct display, manipulation and measurement of multidimensional image data. Several original algorithms have been developed to optimize the tradeoffs between image display efficiency and quality. One of these is a versatile, interactive volume rendering algorithm. The inclusion of a variety of semi-automatic segmentation, quantitative mensuration, and process design tools (macros) significantly extends the usefulness of the software. It can be used as a "visualization workshop" to prototype custom applications. ANALYZE runs on standard UNIX computers without special-purpose hardware, which has facilitated its implementation on a variety of popular workstations, in both standalone and distributed network configurations.

1. INTRODUCTION

Human vision provides an extraordinarily powerful and effective means for acquiring information. Much of what we know about ourselves and our environment has been derived from images – images produced by various instruments, ranging from microscopes to telescopes which extend the range of human vision into realms beyond that which is naturally accessible. However, the full scientific, educational, and/or biomedical value of these images, although profoundly significant, remains largely unexploited. This is due primarily to the lack of objective, quantitative methods to fully analyze the intrinsic information contained in the images. We have largely relied on subject interpretations, and only recently have begun to recognize and unearth the rich treasures of image recordings.

The traditional disciplines of biological and medical science are significantly grounded in the observation of living structures and in the measurement of various properties of

NATO ASI Series, Vol. F 60
3D Imaging in Medicine
Edited by K. H. Höhne et al.

these structures (e.g., their functions). These observations and measurements are often recorded as images. Ever since the invention of the microscope and discovery of x-rays, physicians, surgeons, and life scientists have been using images to diagnose and treat disease and to better understand basic physiology and biology. The value of biomedical images depends upon the context in which they are obtained, and the scientific or medical interests and goals which motivate their production and use. The imaging modalities used in biology and medicine are based on a variety of energy sources, including light, electrons, lasers, x-rays, radionuclides, ultrasound and nuclear magnetic resonance. The objects imaged span orders of magnitude in scale, ranging from molecules and cells to organ systems and the full body. The advantages and limitations of each modality are primarily governed by the basic physical and biological principles which influence the way each energy form interacts with tissues, and by the specific engineering implementation for a particular medical or biological application. The variety of disease processes and abnormalities affecting all regions of the human body are so numerous and different that each imaging modality possesses attributes that make it uniquely helpful in providing the desired understanding and/or discrimination of the disease or abnormality, and therefore no single method has prevailed to the complete exclusion of others. In general, the methodologies are complementary, together providing a powerful armamentarium of clinical diagnostic, therapeutic and biomedical research capabilities which has potential to significantly advance the practice of medicine and the frontiers of biological understanding.

The process of forming an image involves the *mapping* of an object, and/or some property of an object, into or onto what is called image space. This space is used to visualize the object and its properties, and may be used to quantitatively characterize its structure and/or its function. Imaging science may be defined as the study of these mappings and development of ways to better understand them, to improve them and to productively use them. There are four postulates fundamentally important in imaging science which derive from this definition. First, image mappings, varied as they are, provide a direct measurement of form and/or function; second, an image, especially a digital or computer image, is truly synergistic – that is, the gestalt or whole of an image is greater than the sum of its parts; third, the contextual "glue" that holds discrete images together contains important information – therein is contained subtle details fundamental to understanding the mapping of the real object and its properties into image space; and fourth, there are many ways, mostly arbitrary, to manipulate and use images – a rational basis for their selection is important to the biomedical scientist and clinician alike.

The challenge of imaging science is to provide advanced capabilities for acquisition, processing, visualization and quantitative analysis of biomedical images in order to significantly increase the faithful extraction of both scientific and clinical information which they contain. This is a formidable task, one which consistently suggests that existing approaches and constructs are not adequate to address it effectively. The need for resolution of this problem will become increasingly important and pressing as advances in imaging and computer technology enable more complex objects and processes to be imaged and simulated. The system described in this paper represents a software paradigm sufficiently robust and flexible to meet this need.

2. BACKGROUND

The revolutionary capabilities provided by new 3-D and 4-D imaging modalities (CT, MRI, PET, SPECT, Ultrasound, Confocal Microscopy, etc.) now provide data for direct visualization and study of structure and function of internal organs and organelles in man. However, the ability to extract objective and quantitatively accurate information form 3-D biomedical images has not kept pace with the ability to produce the images themselves. Modern advances in the microelectronics industry have provided remarkable hardware capabilities to help close this gap (1). We now can put the supercomputers of the recent past on our desktops for a fraction of the cost. Powerful graphics and image display hardware, recently the exclusive purview of high-end CAD/CAM or image processing workstations, are now available on personal desktop computers. What is needed additionally is comprehensive, integrated software for interactive and quanitative display and analysis of multi-modality, multidimensional biomedical images.

Techniques for processing and displaying 3-D information include integrated projections of the volume onto the display screen, sometimes with prior dissolution of structures from the volume (2), stereo displays generated from projection images (3-4), shaded surface display (5-7), and volume rendering (8-12). One of the most popular techniques for presenting 3-D structures is to use depth-shading of object surfaces. An important precursor to shaded surface display is the extraction of a 3-D surface description from a raw 3-D volume image. This segmentation step is often required not only for display but for measurement as well. However, the segmentation step is an area which is in need of improved algorithms (7,13-16), particularly for the extraction of soft tissue structures, and for images such as those produced by MRI and ultrasound. Shaded surface display algorithms for medical applications have been based on conventional polygonal models (17), a cuberille model for the surface (18), a contour model for the surface (19), and an octtree model of the 3-D object (18). Another 3-D display technique, often called ray casting or volume rendering, has been implemented in several applications (5,8-12,20-21). This technique features the advantage of using the entire 3-D data set ("volume image") in the process. The challenge heretofore has been to exploit this powerful technique in an interactive implementation. This paper describes and illustrates a solution to this problem.

The Biotechnology Computer Resource and the Biodynamics Research Unit (BRU) at the Mayo Clinic has been involved in the development of biomedical image display and analysis algorithms for 3-D and 4-D images since the early 1970's. The successful development of the Dynamic Spatial Reconstructor (22-23) has provided a ready source of 3-D and 4-D imagery from a variety of biomedical investigative and clinical studies. During this time, several unique and useful analysis techniques were developed. These include projection displays (3), stereo displays (2-3), interactive oblique sectioning algorithms (24), shaded surface display algorithms (19), and space-filling virtual 3-D display using a vibrating mirror system (25-26). Many of these capabilities have been developed on different computer systems as the need for them arose. The programs were developed, to a large extent, independent of one another by programmers who have individual styles of programming. Most programs were initially run on mainframe-based computer systems in a serial fashion, i.e., a single program was executed for each display and analysis task needed to address the imaging application, with intermediate image data written to disk or tape for storage. This was an inefficient means of analyzing and displaying multidimensional image data and led to the search for a new approach to the implementation of these techniques. Therefore, these established tools and several new important capabilities have been generalized and made applicable to a wide variety of multi-dimensional biomedical images (e.g., CT, MRI, PET, ultrasound, microscopy, etc.) and have been integrated into a single comprehensive software package which runs on dedicated, powerful image analysis workstations and networks (1,12,27).

3. SOFTWARE DESIGN CONSIDERATIONS

3.1 Architecture and Implementation

The implementation of a software package to integrate multidimensional, multi-modality image display and analysis capabilities requires the design of an efficient shell to synergistically relate such programs. The primary design goal for our software has been to provide the user with an integrated, interactive means of accessing all of the display and analysis tools required to completely evaluate multidimensional image data sets. The key features of such a package include an intuitive user interface, access to all of the programs necessary to complete any particular application task (with intermediate steps stored in memory rather than other storage media), and efficient implementation of each of the program tasks to provide truly interactive display and analysis. Furthermore, from a software development perspective, certain standards are required for the representation of image data and program formats to facilitate and accommodate development by multiple programmers. To allow for expandability, the architecture of any such package should be designed in a modular fashion.

The ANALYZE software system is a manifestation of this design criteria. Each ANALYZE program presents the user with a common base of functions and a common user interface and display format. These functions are the basic building blocks of all the ANALYZE programs and are contained in libraries that can be linked to each of the individual programs when they are compiled. This organization provides a common implementation style for the ANALYZE modules, and facilitates parallel program development simultaneously by different programmers. ANALYZE development by multiple programmers is facilitated by transparent access and seamless organization of the ANALYZE software system. The programming environment available in the Biotechnology Computer Resource provides a powerful and flexible user interface to the programming tools on the systems, and in conjunction with the ANALYZE software organization, provides a well-organized and integrated software development environment. All the programs are built around common functions contained in linkable libraries for all programmers to access. Each programmer is responsible for the contents of a single library, and each is required to use the library functions in applications programs to provide a common base for these programs. If changes in library routines are necessary, the programmer responsible for that library is requested to make the change and build the library. These libraries provide common utilities for the user interface, image manipulations, graphics, and peripheral I/O. These functions are well documented with manual pages describing calling parameters and use of each function. Each programmer is responsible for a set of ANALYZE programs. These individual programs are built from the library functions to provide a common software base, but are written for a specific analysis or display task. As previously described, each ANALYZE program has a specific directory in the ANALYZE main directory from which a programmer can access the modules associated with that program, alter the modules using an editor, compile the new program, and have the executable code automatically placed on the appropriate target system. Multiple windows for an editor, compiler, and debugger can be opened on the same screen to facilitate parallel development of tasks.

The ability to share disk files across a network facilitates simultaneous development of ANALYZE software for several different machines. A single set of source files provides the development team with a simple means of managing changes across this collection of varied computer architectures. Program compilation is performed on the target machines from this common source code, with the resulting binary copies of the programs placed in suitable working directories on either the server or the target machine. In a similar manner, users who log into one type of system or another are provided with the correct working code automatically. ANALYZE can utilize the "rsh" application program, one of the standard TCP/IP programs, to permit users to access remote printers and tape drives from within an ANALYZE program running on networked workstations. Machines with these resources are specified by the user in the ANALYZE configuration module. By using "rsh", requests for printing or tape operations are handled with no specific details required.

3.2 Program Organization and User Interface

The ANALYZE software package is written entirely in the "C" programming language and runs on many standard UNIX workstation systems. There are over 60 programs in the package, made up of over 250,000 lines of source code. The subprogram structure is modular and parallel, with each display and analysis function incorporated as a separate program. This architecture permits systematic enhancements and modifications which has fostered development of a readily extensible and portable package.

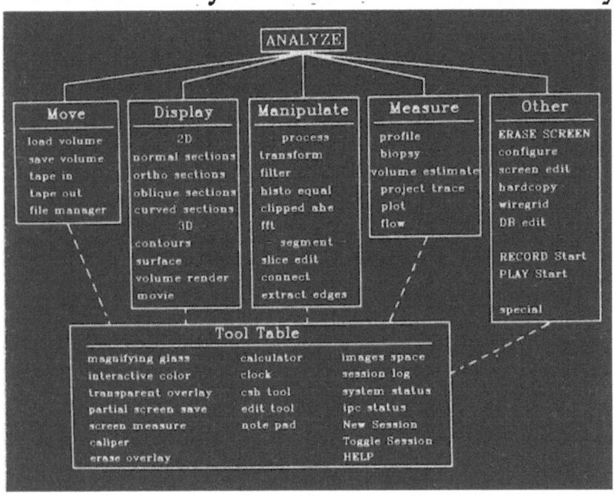

Fig. 1 Diagram of the hierarchial structure of programs in ANALYZE.

The infrastructure of the ANALYZE architecture is based on a hierarchical arrangement of display, manipulation, and measurement tools through which the user invokes single processes from the main ANALYZE menu, as depicted in Figure 1. Each process or program has access to one or more image data sets loaded by the user. Major emphasis has been placed on optimizing the tradeoffs between speed (interactivity) and accuracy (image fidelity) in all ANALYZE procedures.

Figure 2 is a montage of examples illustrating several facets of the ANALYZE user interface. Pop-down menus for selection of individual processes is provided. The interface tools include numeric field entries (using sliders, if desired), scroll bars for selection of items from a list, and confirmation messages. Other user interface tools include the use of "hot keys" for invoking selected options, such as interactive windows for the display, review, and editing of text files. ANALYZE uses color extensively and intelligently to facilitate clarity and ease of use.

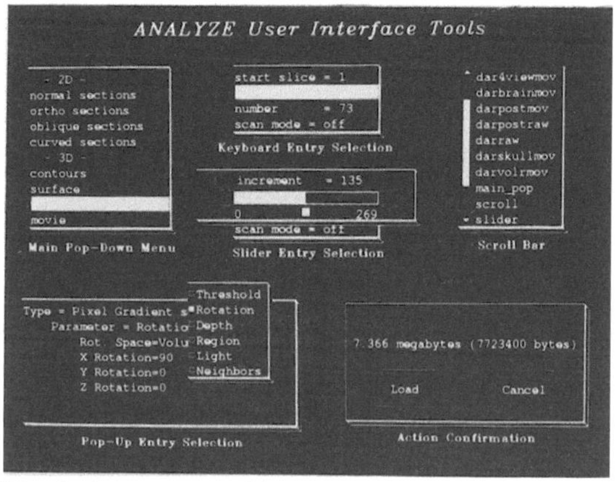

Fig. 2 A montage of interactive tools employed in the ANALYZE user interface.

4. SYSTEM FEATURES

4.1 Image Data Management

The ANALYZE system consists of modules which share image memory segments and which communicate with each other to pass related image information. The integration of these subprocesses allows for multiple analyses whereby the output of one process may be in the input to another. Image data is input and output in the system using the programs in ***MOVE***. Several image data formats archivally stored on 1/2" tape can be loaded onto an ANALYZE image data base file using the ***tape in*** program. Similarly, image data base files can be written out to 1/2" tape using the ***tape out*** program. The ***load volume*** program allows the interactive selection of data base files, review of selected image data by displaying the images on the screen prior to loading, and control of the word size and dimensions of the image data when loading into memory. The ***save volume*** program stores image data in memory to a new or existing image data base file on disk, with control over the spatial region of the data in memory that is saved.

The ANALYZE image data base supports the following image data types: 1 bit binary (packed 8 bits/byte); 8 bit unsigned integer; 16 bit signed integer; 32 bit signed integer; 32 bit single precision floating point; 64 bit double precision floating point; and 64 bit complex (32 bit real and 32 bit imaginary). These image data types are further supported by representations in the shared image memory, i.e., ANALYZE programs know and can work with images composed of any of these data types. The ANALYZE image database currently supports up to four dimensions, e.g., three spatial dimensions plus time or three spatial dimensions plus modality, etc. When loading or storing image data, the bounds of all dimensions can be controlled along with specified increments within each dimension. A "bounding box" can be used to automatically reduce the data array to the smallest subregion containing non-zero data. A robust ***file manager*** program provides image data base file management utilities, including backup and restore, delete, copy, rename, compress, and uncompress.

4.2 Ancillary Utilities

A rich set of useful ancillary utilities are available in ANALYZE. Some of these are loaded under the ***other*** option on the root menu. A ***configure*** program allows the user to design the ANALYZE working environment to specific preferences, including font size and type, colors for text highlights and background, single/double click for mouse selections, specification of networked resources to be used, and the numeric format of any computed data for compatibility with popular statistical analysis packages. The ***screen edit*** program provides extensive capabilities for screen preparation, including generation of text and graphics, image scaling, and coloring, cut-and-paste, and file merging, all of which can be formatted and displayed in a variety of ways. Textual, numeric, graphic and image information can be printed on a variety of supported printers using the ***hardcopy*** program. The ***macros*** tools can be used to generate repetitive routines, customized processing, tutorials, and demonstrations. This tool provides capabilities for interactive input from the operator upon playback of the ***macros*** (e.g., pauses, entry of file names, numbers, etc.). Programs and utilities that exist outside of the ANALYZE shell can be run by using the ***special*** options. Upon selection of this tool, another menu is generated, giving the user selections for these external programs. This menu is generated from a text file that can be easily edited to add new programs.

A helpful ***TOOLTABLE*** can be invoked from any screen in ANALYZE to provide even more tools. A ***magnifying glass*** can be invoked for magnification of any region of the screen as illustrated in Figure 3 (see appended color figure). The mouse buttons control the size of the area being magnified and the magnification factor. The magnifying mode can be toggled to display histograms or numeric pixel values in the region selected. The ***interactive color*** tool can be used to interactively and flexibly specify the mapping of colors to images as shown in Figure 3. The ***transparent overlay*** tool allows the user to select a region of the screen and transparently move it around to another area on the screen. The ***screen measure*** tool makes distance measurements in pixels along an interactively specified straight line or curvilinear trace. A line profile for the line or trace can also be interactively displayed. Another screen measuring tool, called ***caliper***, can be used to measure linear segment distances and angles between specified lines as shown in Figure 3.

The ***C-shell*** tool can be used to enter operating system (UNIX) commands, such as deleting files, moving files, checking process status, etc. Several different ***status*** tools are

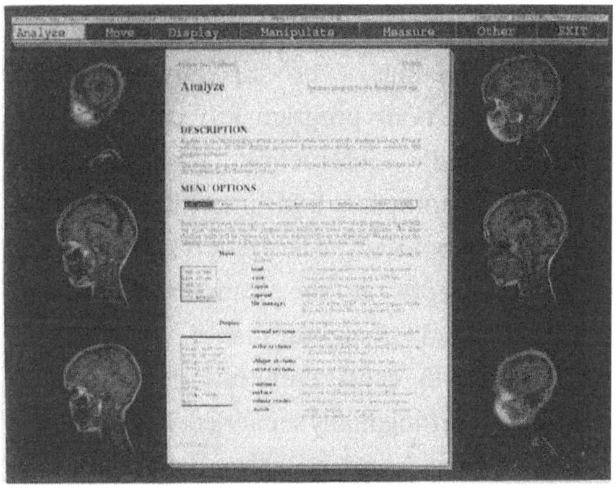

Fig. 4 Example of page from rasterized ANALYZE manual using context sensitive *help* tool from *TOOLTABLE*.

available to display run-time information. The ***new session*** and ***toggle session*** options can be used to start multiple sessions of the ANALYZE package and allow the user to rapidly switch between these sessions. This allows multiple tools to be run on the same or different image data set(s) loaded into memory. A complete, on-screen ***text editor*** has been developed for entry and review of text notes typed into a specific notepad file. Extensive ***help*** is available throughout ANALYZE. A comprehensive User's Manual has been developed for the ANALYZE package, and the pages of this manual have been rasterized into ANALYZE image files and placed on-line, as shown in Figure 4. The electronic manual is context sensitive - that is the pages describing the program currently running are automatically displayed, and the pages for that section can be turned using the mouse buttons. Keyboard options exist to allow the user to change to any portion of the ANALYZE manual, including the Introduction and Appendices.

4.3 Image Display

The ***DISPLAY*** routines in ANALYZE provide capabilities for a variety of graphic presentations of multi-formatted 2-D sections and of computed 3-D objects extracted from the volume image data sets. Included in each display module are tools for interactive manipulation of the presentation style and format.

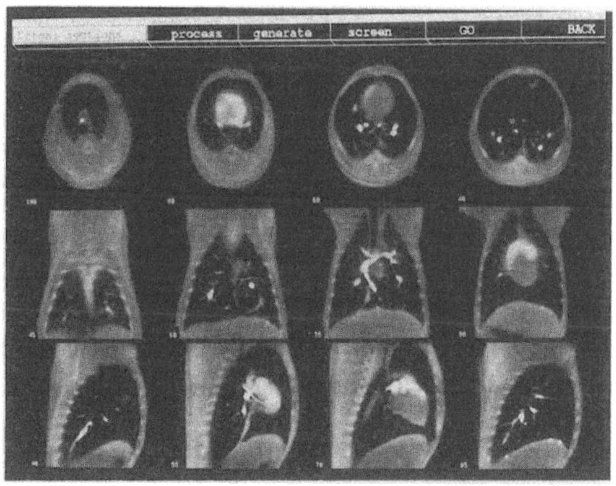

Fig. 5 Multiplanar, orthogonal images of chest displayed with *normal sections* program.

The ***normal sections*** program permits 2-D images to be arbitrarily sized, flexibly formatted and rapidly displayed on the screen. Multiple sections through a 3-D volume image can be displayed by selection of a starting position and automatic increment. Figure 5 shows a display produced in ***normal*** sections. Figure 6 shows a

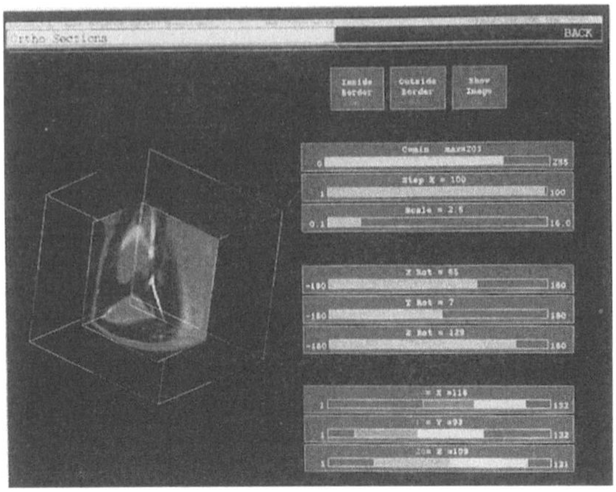

Fig. 6 Interactive orthogonal sectioning and 3-D visualization using the *ortho sections* program.

different way to display orthogonal sections - provided by the ***ortho sections*** program - a cubic volume with interactive scaling (both geometric and intensity), rotation about any axis, and dissection at any depth in the transverse, coronal, and/or sagittal planes. In Figure 6, 3-D chest data has been rotated, orthogonally sectioned, and intensity-windowed so as to better visualize the contrast-filled heart and pulmonary arteries. The ***oblique sections*** program provides an interactive method for arbitrary viewpoint selection by providing the user with pictorial feedback cues to indicate the current orientation of the volume image data. Newly oriented 3-D images can be stored in the data base for subsequent analysis. Figure 7 shows a display from the ***curved sections*** module in which a curved line can be drawn on one of the standard orthogonal planes (transverse, coronal, or sagittal) and sections perpendicular to this line interactively computed and displayed adjacent one to another to yield an image of any curvilinear section through the volume. Modes that can be used to generate the output image include free-hand trace, rubber band, and rays.

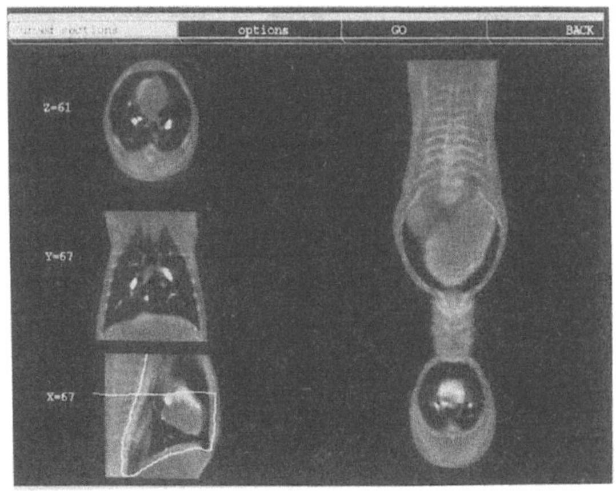

Fig. 7 An example of interactive *curved sections* showing a trace along an arbitrary path on one section (bottom left) with corresponding section displayed (right).

Frequently the output of segmentation programs (described subsequently) are contours defined by 3-D coordinates on the surface of the segmented structures. ANALYZE supports a data structure of this type, called a "surface description file", wherein the 3-D coordinates are stored along with a normal to the surface at each point. The ***connect*** and ***extract edges*** programs automatically format such 3-D coordinate files into ***contour*** files. The ***contours*** program provides interactive display capability for these data files.

The *surface* program takes as input any number of surface description files for multiple objects in an image data set and produces a shaded surface display of this data. The surface description algorithm (19) uses flat patches of a user-defined size which can be oriented in many directions to generate faces for each point in the surface description file. Sequences of shaded surface displays can be generated and stored in an image database file for later display in the *movie* program in order to visualize sequences of parameter changes, i.e., rotation. Figure 8 (see appended color figure) shows shaded surface images of segmented objects in the abdomen of a monkey scanned in the Mayo DSR (23). The *surface* program has been used extensively for visualization of objects to convey shape and relative dimensions, and for verifying that a given object has been properly identified.

One of the most versatile and powerful image display and manipulation tools in ANALYZE is the *volume render* program. Volume rendering techniques (8-12) are able to display surfaces with shading and other parts of the volume simultaneously. An important advantage is to display data directly from the grayscale volume. The selection of the data which will appear on the screen is done during the projection of the voxels (sometimes called ray-casting). A function of different attributes of the voxels, such as their density or gradient values and/or spatial coordinates can be invoked during the projection process to produce "on the fly" segmented surfaces, cutting planes anywhere in the volume, and/or selected degrees of transparency/ opacity within the volume. Volume set operations (union, intersection, difference of volume) can also be invoked during the projection process. Angle of view and illumination direction can be interactively adjusted. The implementation of volume rendering in ANALYZE has several unique features and advantages in comparison to shaded surface display and other volume rendering methods. The algorithm is optimized to be fast without compromising fidelity (12). The goal of the ray-tracing implementation is to interactively render an anatomical volume with different degrees of segmentation, and with image quality sufficient to convey the information in the data set, no more or no less.

Figure 9 shows six volume images rendered using pixel gradient shading. A full head CT scan of a head trauma patient was used to produce these images. By interactively changing the voxel density threshold for surface definition within the volume, successive surfaces within the volume were rendered, including an image of the bandaged face (top left), the face alone (top right), and finally the skull (bottom center). A transparent combination of the soft tissue and skull from the lateral view-point has been generated and is shown in the lower right image. These images demonstrate the capability to use interactive thresholding for tissue dissolution, rotation for 3-D visualization of structural

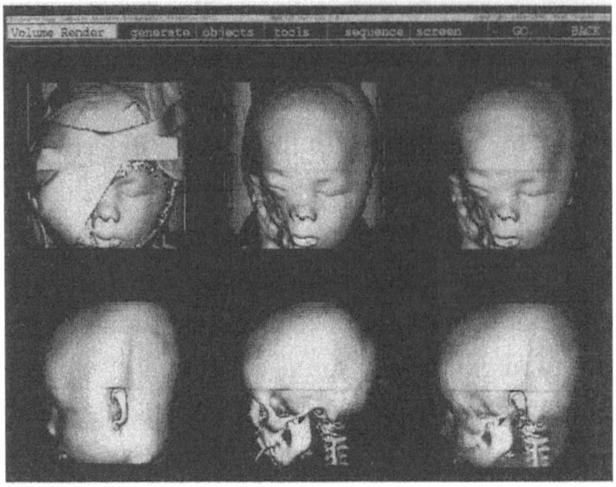

Fig. 9 Three-dimensional images of trauma patient produced by *volume render* showing effect of thresholding, rotation, and transparency.

dimensions and geometric relationships, and transparency for simultaneous viewing of exterior and interior tissues.

The ***volume render*** program includes tools for individual object discrimination and manipulation. Figure 10 (see appended color figure) illustrates this capability. The images in the top row have been rendered from the raw x-ray CT data using an object map to render the full image (top left), remove the bandage and head holder (top second), remove the skin (top third), and finally remove the bone to reveal the brain (top right). The bottom left image in Figure 10 illustrates the capability to translate a section of the skin away from the rest of the image, exposing the underlying skull. The bottom right image illustrates a combination of translation and rotation of multiple objects to visualize interior structures. These images depict the capability of the ***volume render*** module to interactively explore 3-D images, much as a surgeon or pathologist would an actual organ or body part. Each image in Figures 9 and 10 required less than two seconds to compute from a volume image of dimensions 160x160x160. Table I shows a comparison of times required to render different size images on different workstations.

TABLE I

Timing* of ANALYZE Volume Rendering Algorithm For Different Image Sizes and Processors

Size / *Processor*	128^3	200^3	256^3
SUN 3/110	4.08	12.79	39.83
SUN 4/260	1.06	3.33	6.79
SUN SPARC 1	0.83	2.76	5.98
SGI Personal IRIS	1.15	3.53	13.02
SGI 4D/70	1.03	3.37	9.90

*** Seconds in CPU Time.**

Figure 11 (see appended color figure) illustrates two utilities in the ***volume render*** module for interactively positioning mutually orthogonal or arbitrary oblique planes which represent the raw x-ray CT data and which intersect familiar can be surfaces in the 3-D volume. Figure 12 (see appended color figure) shows clipping planes in the front and back of the current viewing angle, interactively adjusted to cut through the 3-D surface image (top center and right). Masked regions defined on the surface image by interactive tracing (bottom row), and any selected rendering parameters can be applied exclusively to this region to visualize "inlaid" structures surrounded by the current image.

Both the ***surface*** program and the ***volume render*** programs can automatically generate sequences of images stored in an image file where consecutive images differ by selected combinations of rendering parameters applied in specified increments, such as thresholding, dissecting, rotating, etc. The ***movie*** program provides the capability to rapidly display these sequences of images in a cine movie loop. This program is also useful for redisplay of images acquired through time in order to visualize changes in the images.

4.4 Image Manipulation

Image processing functions for enhancement of images or as a precursor to application of segmentation tools are important components of an image display and analysis package. The programs in the ***MANIPULATE*** group of modules in ANALYZE allow the user to apply several types of image processing and segmentation algorithms to the image data.

The ***transform*** program allows the user to enter any desired formula, where the operands are variables assigned to image files or constants, and the operators are the standard arithmetic or logical operators (a limited set of transcendental functions are also included) to form a linear combination of image sets, i.e., to do "image algebra". Several common image transforms can also be applied in the ***transform*** program. These include trilinear interpolation, image rotation and translation, intensity windowing and scaling, and image volume subregioning and flipping.

A variety of three-dimensional convolution and frequency space filters can be applied to the image data using the ***filter*** program. Several edge enhancement filters (e.g., Sobel) and histogram equalization can be performed on 2-D or 3-D image data sets. Clipped adaptive histogram equalization can also be applied to the image data for contrast enhancement. The package provides both forward and inverse Fourier transforms on image data and can represent frequency space as a complex image (i.e., magnitude and phase, real and imaginary) in the shared image memory. Filters can be applied non-destructively (without altering the image memory) and inversely applied to display the spatial domain. An interactive, versatile ***digital filter design*** tool is also available for customized processing of images. Figure 13 shows an example of this capability. The interactive graphic tool specifying the custom-designed frequency domain filter is shown at the bottom of Figure 13 and the original and filtered images of an MRI head scan are shown at the top of Figure 13.

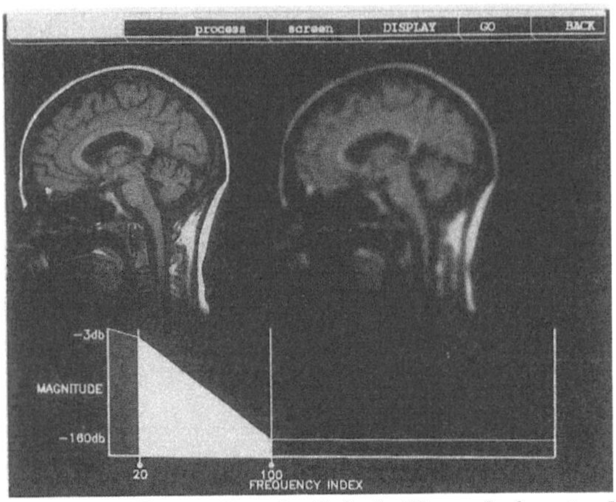

Fig. 13 Example of *digital filter design* tool (bottom) used to interactively specify a frequency domain filter. Top left = original, top right = filtered.

The *slice edit* program provides both efficient manual segmentation tools as well as capabilities for semi-automated object definition and feature extraction. The top row of images in Figure 14 illustrate two of the techniques available, including "painting" on the image data with a selected cursor shape and size and tracing area(s) to be deleted. A depth can be set in both of these modes to cause the edit to penetrate into the volume image. Interactive thresholding can also be used as a precursor to editing. A semi-automatic mode for finding structure boundaries is also available. The user defines a seed point from which region growing is to originate, and then moves a slider to alter the threshold range within which pixels will be included in the growth region. The center row of images in Figure 14 depicts this technique applied to multiple MRI sections of the head leaving only the segmented structure of interest, as shown in the bottom row.

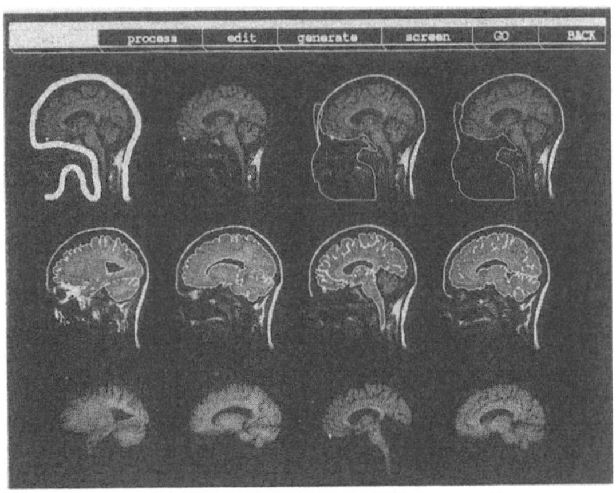

Fig. 14 Examples of image editing and segmentation using *slice edit*. Both manual (top) and automatic (center) methods are available.

4.5 Image Mensuration

The programs in ***MEASURE*** are all interactive, and provide quantitative sampling capabilities, making no assumptions about the significance of these measurements, which must be determined by the application. Each program can produce output that is stored in an ANALYZE data file structure called a "stats" file. Stats files can be generated in several different formats which match the formats accepted as input to some of the common numerical analysis packages, such as SAS, RS/1, S, and DIF files for PC programs.

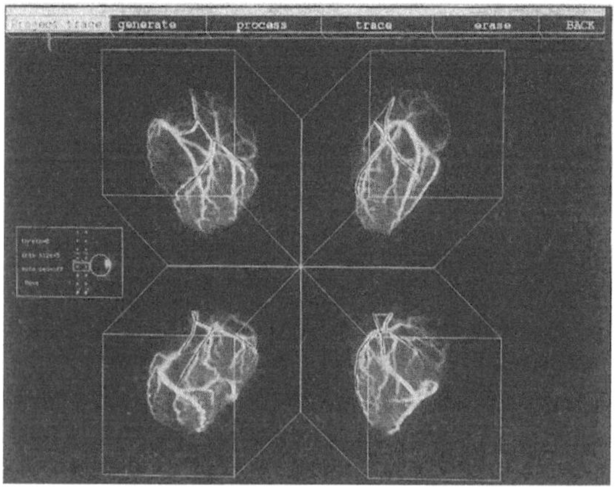

Fig. 15 Four brightest voxel projection images computed from different angles are used in the *project trace* program to track structures in 3-D.

The *project trace* program provides a method of unambiguously tracing 3-D tree structures, such as dye-filled coronary arteries or pulmonary arteries. Interactive traces on any one projection are simulutaneously displayed on all projections to generate the trace. Such traces can be used to compute segment lengths, branching angles, and oblique images. Figure 15 shows an example of this program used to trace the coronary arterial tree.

Single dimension profiles can be interactively sampled from image data and plotted using the *profile* program. The line profiles can be applied to multiple images as shown in Figure 16. In automatic sampling mode, multiple lines through multiple images can be sampled and stored in a stats file. The *volume estimate* program measures and estimates the volume of an arbitrarily shaped three-dimensional object using a probabilistic model (28). A user-specified number of voxels are randomly marked in the volume image as shown in Figure 17, and a cursor of varying

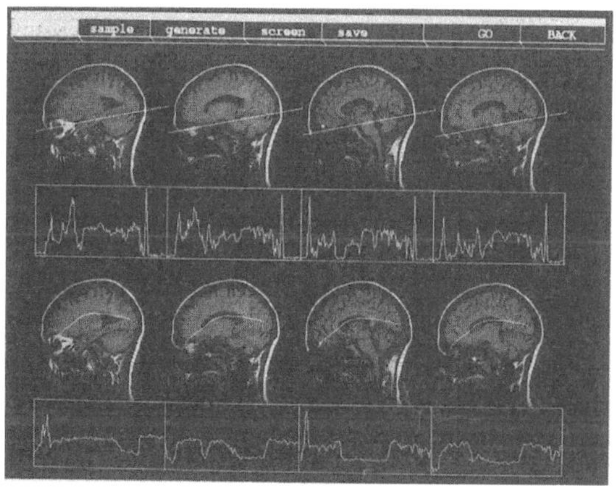

Fig. 16 Line profiles can be interactively sampled and displayed on multiple images using the *profile* program .

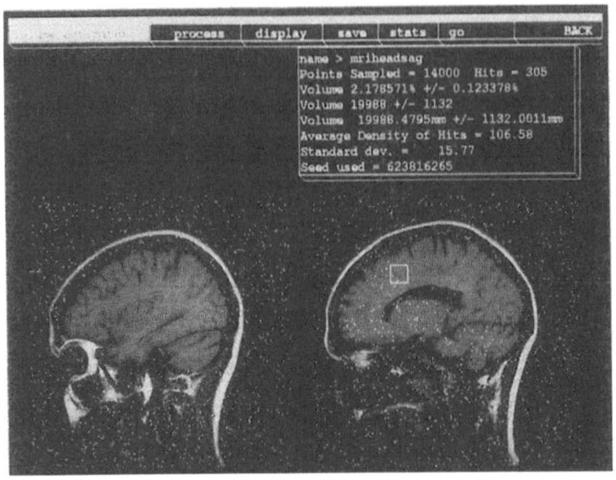

Fig. 17 The *volume estimator* program generates randomly marked voxels in the 3-D volume which are "vacuumed" up in the subvolume of interest.

size is then used to interactively select the marked voxels overlying the structure for which the volume is to be calculated. The ratio of the selected voxels in each section to the total number of marked voxels is used to estimate the volume of the desired object. A significant advantage in speed of volume estimation is realized by comparison to manual tracing of the structure of interest, with similar accuracy (greater than 95% (28)).

The ***biopsy*** program can be used to interactively define and sample regions-of-interest (ROIs) on the image data. Regions can be defined using analytic shapes (for example, rectangles and ellipses), by tracing a free-form area on the image, or by semi-automated boundary definition using region-growing. Information provided by the area sampling includes the maximum and minimum values, mean and standard deviation of the values, the calibrated area within the region, and the integrated brightness-area product, as shown in the stats panel in the upper right of Figure 18. Areas can be automatically sampled through a series sections to provide a volumetric sampling function if the third dimension is spatial, or a time sampling function if the third dimension is time. Individual ROIs can be summed together to provide an arbitrary shaped volumetric sample through the multidimensional image data.

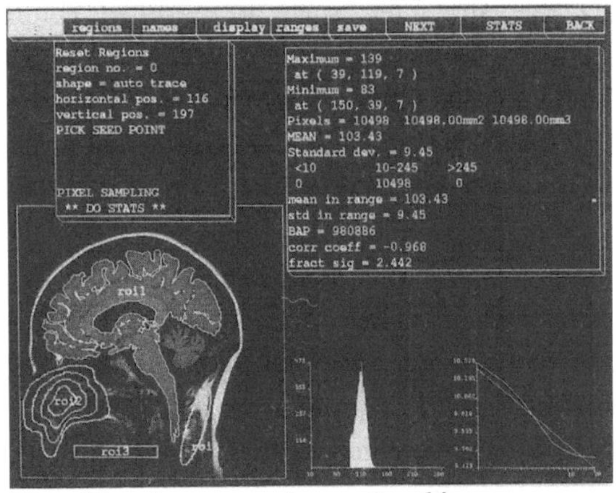

Fig. 18 A screen from the *biopsy* program showing different types of regions of interest and quantitative sampling of regions, including fractal signatures/graphs.

Another measurement provided by the ***biopsy*** program is the fractal signature for ROIs. The fractal dimension (29) is calculated for the specified regions by placing an envelope both above and below the grayscale surface, and then calculating the volume that is contained between them. This value is a scalar multiple of the enclosing surface area, which is logarithmically proportional to the fractal dimension. The distance between the "blankets" of the envelope and the surface can be varied to generate values at different scales. When these calculations are completed over a desired range of scales, a log-log plot of surface area vs. scale is generated and a least squares estimate of the linear regression line for this plot is calculated, as shown in the lower right of Figure 18. The linear correlation coefficient of this fit is given in the stats output, and the slope of the fit is reported as the fractal signature for the region. These data can be used as input to the ***plot*** program for data plotting and numerical analysis. The ***plot*** program is very flexible and provides the user with many options for data manipulation and graphing.

5. MULTI-MODALITY APPLICATIONS

The ANALYZE software is being used in a wide variety of applications in a number of institutions. These applications include both basic science and clinical investigations, ranging from exploring the world of the living cell to simulating surgery (i.e., planning an actual operation with a computer). ANALYZE has been used extensively for analysis of many organ systems and their function - the heart, the lungs, the kidneys, the brain, etc. There are several potential applications of ANALYZE that are non-medical - for example, in oil exploration, meteorology, and industrial inspection. The following sections indicate only some of the ongoing biomedical uses of ANALYZE.

5.1 3-D Microscopy

New technologies in digital microscopy can generate tomographic images of serial sections of specimens. Therefore, the study of microscopic structures such as cells and tissue sections can now be significantly facilitated with 3-D visualization tools. Figure 19 shows the application of the ANALYZE volume rendering program to several microscopic structures obtained from different microscopy modalities. The upper row of 3-D images were computed from sections of a single corneal cell acquired using a digital scanning fluorescence microscope. The 3-D images were generated at four different angles of view. The center row of images computed from

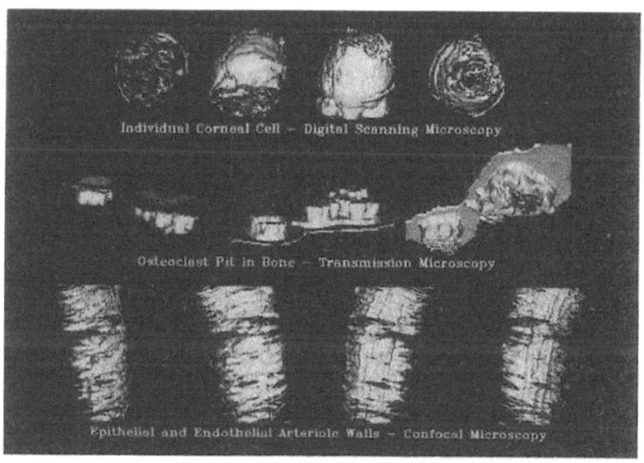

Fig. 19 Examples of applications of ANALYZE in 3-D microscopy and 3-D display of cells and tissue specimens.

standard light transmission microscopic data are 3-D displays of bone surfaces and pits within the surfaces caused by osteoclast cells. The top section has been left intact to give a reference surface for the pits descending into the bone. Confocal microscopy has provided the capability to create serial sections of cells and tissues without the blurring resulting from out-of-focus structures. The focal plane of the confocal system can be accurately specified and finely controlled to create the multiple thin sections through a structure necessary for 3-D rendering. The bottom row of images in Figure 19 are of a cerebral arteriole acquired using confocal microscopy. The longitudinal and lateral oriented epithelial and endothelial cells of the arteriole can be visualized as the images are rotated.

5.2 X-Ray Computed Tomography

Since the advent of X-ray computed tomography in the early 1970's, many years of research into display and analysis techniques for CT data have been accomplished. As scanning resolutions in all dimensions became higher, the possibility of accurately rendering 3-D structure from x-ray CT became real. Figure 20 shows ANALYZE volume rendered images of a wrist from multiple angles of view. The raw x-ray CT image data was created at high resolution in all dimensions, including an axial slice thickness of 1 mm, and was generated to clinically detect and measure joint erosions in rheumatoid arthritis. These images are part of a movie file that can be displayed to show the bones in the wrist smoothly rotating around varying axes. The images in Figure 21 (see appended color figure) show ANALYZE shaded surface displays of x-ray CT image data acquired for tumor visualization, specifically an acoustic neuroma located behind the right ear. Multiple objects have been segmented and shaded in different colors: the soft tissue, skull, brain (cut half way through the axial extent of the scan to reveal the tumor), and tumor (in red). Each one of these objects can be independently manipulated and turned on or off to provide interactive tumor visualization. Linear and curvilinear distance measurements and volume calculations can be performed directly on these 3-D images.

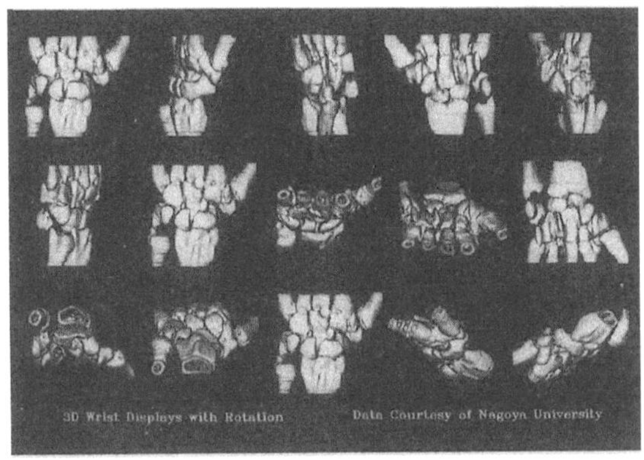

Fig. 20 Three-dimensional display of wrist bones rendered from high-resolution x-ray CT image data.

Figure 22 shows the application of the ANALYZE volume rendering program to x-ray CT data acquired for craniofacial surgery planning and simulation. The images show data from a pre-operative scan of a patient with a facial feature. Through the use of the editing, measurement, and display tools available in the volume rendering program, a surgical procedure for this patient was planned and simulated using ANALYZE, including completion of a prosthesis. ANALYZE has been used to plan many such craniofacial surgery cases both within Mayo Foundation and at other institutions.

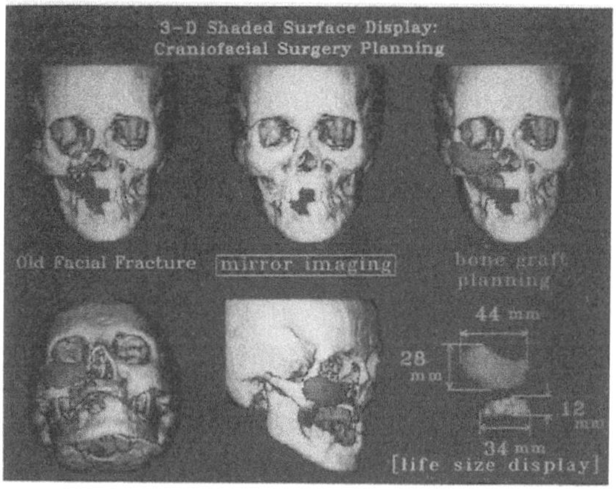

Fig. 22 Example of object manipulation in volume render for prosthesis design in craniofacial surgery planning.

5.3 Magnetic Resonance Imaging

Magnetic Resonance Imaging (MRI) has proven useful for detailed visualization of soft tissue structures often unrevealed by x-ray CT imaging. For instance, MRI can be used to image white vs. gray matter in the brain, or to image soft tissue structures in joints, including the muscles and ligaments.

Figure 23 (see appended color figure) shows the application of ANALYZE to an MRI scan of the knee. The muscles, bone, and cartilagineous structures in the knee have been volume rendered, shown at the left, and the oblique plane tool used to generate radial sections through the knee joint. The 3-D rendered display is used as a reference for accurate sectioning. The images on the right show 3-D displays from multiple angles of view of the menisci and cruciate ligaments segmented from the MRI volume image of the knee.

New scanning modes in MRI have allowed for the acquisition of isotropic 3-D image data sets at high spatial resolution. Figures 24 and 25 show volume rendered images from a 3-D acquired MRI scan of a head for visualization of brain structures. In Figure 24 (see appended color figure), the editing tools in ANALYZE have been used to segment the brain from the other tissues, and rendered to visualize the brain in position with the facial soft tissue as reference. The objects can be rotated together, or can be separately manipulated as shown by the image depicting the brain lifted out of the soft

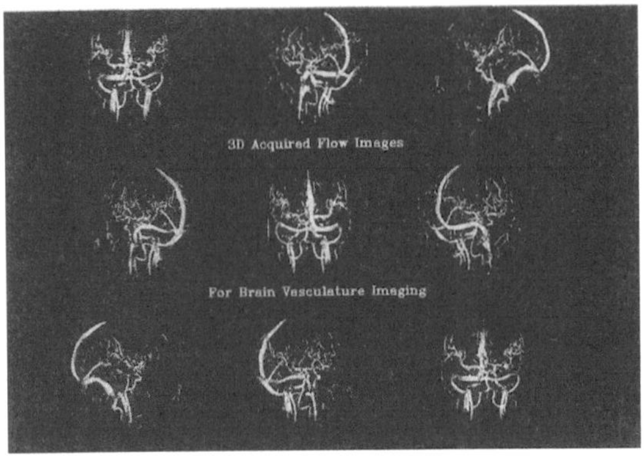

Fig. 25 Three-dimensional images of brain vasculature rendered by ANALYZE from 3-D acquired MRI angiographic images.

tissue and rotated opposite of the facial structures. The striking detail in the convolutions of the brain has been the impetus for use of this kind of data acquisition and rendering in the planning of neurosurgery cases at Mayo Foundation, including mapping of epileptic foci in the gyra for therapeutic resection. Another new scanning mode being developed in MRI imaging can be used to image vasculature structures through which blood is flowing. This "MR Angio" acquisition method was used to produce the data used for the 3-D rendered images of brain vasculature in Figure 25. The vessels in the brain have been rendered from multiple angles of view to generate a rotational movie for visualizing the 3-D anatomical relationships of these vessels.

5.4 Nuclear Medicine Imaging

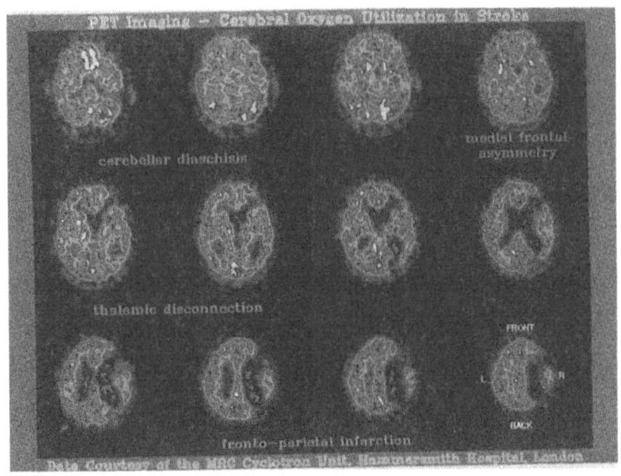

Fig. 26 Positron Emission Tomography (PET) images of cerebral oxygen uptake.

Physiological and metabolic imaging of the body has been accomplished using radioactively labelled chemicals that are preferentially absorbed by certain tissues. The radioactive emissions are measured with detectors specific to the radiation type and tomographic images produced. Such imaging devices form pictures of physiologic and metabolic function of organs and organ systems of the body, but do not generally form highly-detailed images of the structures of these organs. Figure 26 shows images acquired using Positron Emission Tomography (PET) scanning to study cerebral oxygen utilization in stroke victims. The images are multiple serial sections of a brain where the images are generated using radiolabelled oxygen, and the different shades correspond to levels of activity in the brain. The interactive display tools in the ANALYZE package have been used to generate these functional images.

ANALYZE has been used in nuclear medicine imaging to visualize areas of metabolic activity, and, in conjunction with the other imaging modalities, can be used to integrate the functional images with detailed anatomical images obtained from other scan modalities, as shown for PET and MRI in Figure 27 (see appended color figure).

5.5 Ultrasound Tomography

Another imaging modality in which ANALYZE has found application is ultrasound tomography. The images in Figure 28 (see appended color figure) depict 3-D displays of a in-vitro heart scanned with an ultrasound imaging system developed in the Biodynamics Research Unit at Mayo. The upper row of images show different angles of view of the myocardium. Half of the myocardium has been edited away and rendered with the chambers in a different color, as shown in the middle row of images. The chambers can be turned off and the images rendered again to show the interior walls of the chamber, as depicted in the bottom row of images.

5.6 Dynamic Spatial Reconstructor

The Dynamic Spatial Reconstructor is a unique x-ray scanning device developed in the Biodynamics Research Unit at Mayo to generate time sequences of synchronous 3-D volume images in studies of structural to functional relationships in moving organs. The DSR is capable of generating an isotropic 3-D volume image 60 times per second, truly a four-dimensional imaging system.

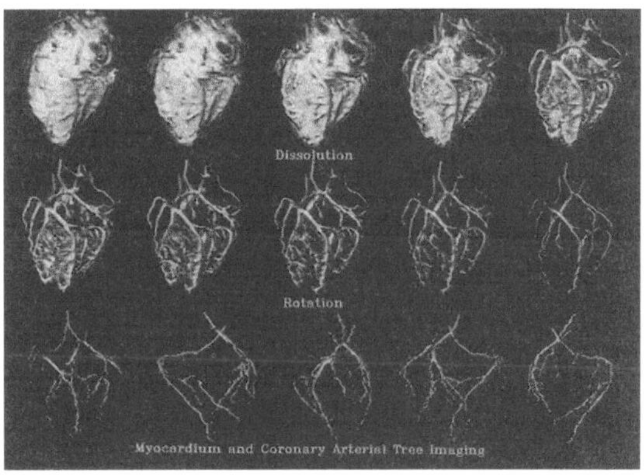

Fig. 29 Volume rendered images of the myocardium and coronary arterial tree from DSR images, showing dissolution of myocardium.

The images in Figure 29 have been generated from a DSR scan of a heart with a dye injection to visualize the coronary arterial tree. These images were rendered using the ANALYZE volume rendering tools and demonstrate the capability to use interactive thresholding for tissue dissolution and rotation for 3-D visualization of the structural dimensions and geometric relation-ships in the coronary arterial tree. Figure 30 (see appended color figure) is a montage of images demonstrating the 4-D scanning capabilities of the DSR. Each image is a 3-D shaded surface display of the major chambers of the heart, red being the left ventricular chamber and blue the right ventricular chamber (essentially the surface of the blood pool in the chambers). Since multiple volume images can be acquired through time, these shaded surface displays can be generated for multiple time points throughout the cardiac cycle. Here 15 time points through a single heart beat are depicted starting at end diastole, going through end systole (approximately image 11), and back to near end diastole. The "hole" in the left ventricular chamber is caused by invagination of the papillary muscle from the surrounding myocardium. These images are part of a movie sequence that shows the change in 3-D structure of the heart through time from several different viewing angles. Such a movie is more than just a picture of the heart, it is a picture of the heart beat, a study of structure and function simultaneously.

6. WORKSTATION IMPLEMENTATIONS

ANALYZE has been successfully ported to a number of workstation platforms. Generally these implementations are straightforward if the platform has a standard UNIX operating system and C compiler. It is only required to modify the ANALYZE graphics libraries to read and write screen pixels. (Some minor nuances in compiled code may need to be reconciled in different vendor products.) Typically ANALYZE, which is composed of over 250,000 lines of C source code, can be ported to a new workstation platform within 2-3 weeks. If special system constraints exist, as presented by some special graphics options (e.g., a windows manager), the port may take longer. Table II provides a listing of some of the workstation platforms to which ANALYZE has been successfully ported, along with comparative performance measures for several of the modules in ANALYZE. These ANALYZE-specific benchmarks test rates of 1) disk transfer, 2) screen display, 3) integer computation, and 4) floating point computation in various ANALYZE processes. Table II also indicates an "overall performance" factor obtained by applying certain weights to different performance functions and normalizing to unity (the higher the number, the better the performance rating). Changing the values of weights would change the overall outcome, but the values selected for Table II are those considered to be appropriate in prioritizing functions used in typical ANALYZE applications (e.g., speed of display is critical for interactive usage).

TABLE II

Performance of Some ANALYZE Program Functions on Different Systems

SYSTEM	Disk Xfer[1] (Mbytes/sec)	Pixel Display[2] (Mpixels/sec)	Integer[3] (Images/Sec)	Float Pt.[4] (Images/Sec)	Overall[5] Average
DECStation 3100	0.27	0.83	0.18	2.08	0.19
HP 360	0.27	2.30	0.20	0.63	0.23
IBM RT 135	0.48	0.75	0.18	0.47	0.14
MIPS RS2030	0.43	3.76	0.62	2.94	0.54
SGI 4D-25 Pers. Iris	0.84	1.17	0.69	2.66	0.42
SGI 4D-70/GT	0.42	0.56	0.46	1.73	0.25
SGI 4D-120/GTX	Not Tested	0.90	0.55	2.20	0.31
Solbourne 4/620	1.03	1.54	0.33	0.99	0.28
SUN 3-160	0.39	1.48	0.11	0.05	0.16
SUN 3-280	1.66	2.13	0.22	0.08	0.31
SUN 3-470	2.98	3.33	0.36	0.17	0.51
SUN 386i-250	0.14	2.41	0.17	0.38	0.22
SUN SparcStation1	0.82	4.03	0.57	1.37	0.53
SUN 4-260	0.97	1.36	0.46	1.02	0.32
SUN 4-330	2.87	4.57	0.73	2.05	0.75
SUN 4-490	2.90	6.67	1.00	3.08	0.99

1. ***Load*** program timing in megabytes/sec to transfer 4 Mbyte volume image
2. ***Movie*** program timing in megabytes/sec to display 5-cycle sequence of 160x160 size images
3. ***Volume render*** program timing in images/second for 160x160x160 size images
4. ***Adaptive Histogram Equalization*** program timing in images/second for 256x256 images
5. Averaged performance with weighting on disk transfer = 20%, pixel display = 40%, integer computation = 30%, and floating point = 10%.

7. NEW DEVELOPMENTS

The ANALYZE software package has been developed and continues to be refined and extended to provide comprehensive, detailed quantitative investigations and evaluation of 3-D and 4-D biomedical images. Under development are two customized applications, one to facilitate use of ANALYZE specifically for simulation of surgical procedures employed in many aspects of craniofacial and orthopedic surgery planning, and the other for interactive manipulation and display of "beam's eye" views of 3-D treatment plans in radiation therapy. Both extended and new user interfaces (based on X windows) with comprehensive data-based management capabilities, including a structured query language, are in development. Arguably the two most important problems inhibiting major progress in biomedical imaging are automated segmentation of features of interest and accurate fusion of information from multiple modality images. New algorithms for these "differentiation and integration" problems in imaging science are being developed and evaluated. All of these capabilities will combine, in an evolutionary way, to provide increased sensitivity and specificity in quantitative analyses of cells, tissues, organs and complete biological systems.

8. SUMMARY

Major segments of the biological sciences and the practice of medicine are based on the study and knowledge of the relationships of anatomic structure to biological function. Traditionally, this knowledge has been gained either indirectly or by inference and, in the final analysis, by direct surgical vivisection or by postmortem examinations. Direct visualization and study of anatomic structure and function of internal organ systems in man have, up to the present, been the preserve of the surgeon and pathologist. The revolutionary capabilities provided by new 3-D and 4-D imaging modalities (30) for obtaining similar information non-invasively, non-destructively and painlessly can now provide this data to the internist, surgeon, and researcher for direct reproducible examinations of individual patients or experimental subjects without disturbing the physiology of the organ system under study or altering its normal integration into the physiology of the body as a whole. The ANALYZE software system facilitates comprehensive exploitation of these capabilities.

9. ACKNOWLEGEMENTS

The author is grateful to his colleagues in the Mayo Biotechnology Computer Resource and Biodynamics Research Unit without whom this work would not be possible. Special recognition is given to the programmers who develop and maintain the ANALYZE software system, namely Dennis Hanson, Ron Karwoski, Al Larson, Margret Ryan and Ellis Workman. Workstation and network support is provided by Bruce Cameron, Mark Korinek, Russ Moritz, Ann Nielsen and Mahlon Stacy. Appreciation is expressed to Marge Fynbo and Christine Welch for preparation of the manuscript. This work is supported in part by NIH grants RR-02540 and HL-04664.

10. REFERENCES

1. R. A. Robb, "A workstation for interactive display and analysis of multi-dimensional biomedical images," in Computer Assisted Radiology, H. U. Lemke, M. L. Rhodes, C. C. Jaffee, and R. Felix, eds., Proceedings of the International Symposium CAR '87, 642-656 (1987).
2. L. D. Harris, R. A. Robb, T. S. Yuen, and E. L. Ritman, "Non-invasive numerical dissection and display of anatomic structure using computerized x-ray tomography," Proc. SPIE, 152, 10-18 (1978).
3. L. D. Harris, R. A. Robb, S. A. Johnson, and A. S. Khalafalla, "Stereo display of computed tomographic data," in Challenges and Prospects for Advanced Medical Systems, H. A. Emlet, Jr., ed., Symposia Specialists, Inc., 127-135 (1978).
4. L. F. Hodges and D. F. McAllister, "Stereo and alternating-pair techniques for display of computer-generated images," IEEE CG&A 5, 38-45 (1985).
5. K. H. Hohne, M. Riemer, U. Tiede, and M. Bomans, "Volume rendering of 3D-tomographic imagery," Proc. Xth IPMI Intl. Conf., Utretch, The Netherlands, 403-412 (1987).
6. G. T. Herman, "Computer produced stereoscopic display in radiology," Proc. NCGA '86, 3, 71-79 (1986).
7. P. B. Heffernan and R. A. Robb, "Display and analysis of 4-D medical images," Proc. Intl. Symp. CAR'85, 583-592 (1985).
8. R. Drebin, L. Carpenter, and P. Harrahan, "Volume rendering," SIGGRAPH '88 65-74 (1988).
9. M. Levoy, "Display of surfaces from volume data," CG&A 8(3), 29-37 (1988).
10. R. A. Robb and C. Barillot, "Interactive 3-D image display and analysis," Proc. in Proceedings of SPIE, Hybrid Image and Signal Processing, 939, 173-202 (1988).
11. K. H. Hohne and R. Bernstein, "Shading 3-D images from CT using grey-level gradients," IEEE Trans. Med. Imag., MI-15, 45-47 (1986).
12. R. A. Robb and C. Barillot, "Interactive display and analysis of 3-D medical images," IEEE Trans. Med. Imag., MI-8, 217-226 (1989).
13. S. M. Pizer, W. R. Oliver, and S. H. Bloomberg, "Hierarchial shape description via the multi-resolution symmetric axis transform," Tech. Report, Dept. Comp. Science, Univ. North Carolina, (1986).

14. S. M. Pizer, J. M. Gauch, and L. M. Lifshitz, "Interactive 2D and 3D object definition in medical images based on multiresolution image descriptions," The Univ. of North Carolina Technical Report 88-005 (1988).
15. S. Peleg, J. Naor, R. Hartley, and D. Avnir, "Multiple resolution texture analysis and classification," IEEE PAMI 6(4), 518-532 (1984).
16. S. Grossberg and E. Mingolla, "Neural dynamics of perceptual groupings: textures, boundaries, and emergent segmentation," Percept. Psychophys. 38, 141-171 (1985).
17. L. L. Fellingham, J. H. Vogel, C. Lau, and P. Dev, "Interactive graphics and 3-D modeling for surgical planning and prosthesis and implant design," Proc. NCGA '86, 3, 132-142 (1986).
18. G. T. Herman and H. K. Liu, "Display of three-dimensional information in computed tomography," JCAT 1, 155-160 (1977).
19. P. B. Heffernan and R. A. Robb, "A new method for shaded surfaced display of biological and medical images," IEEE Trans. Med. Imag., 4, 26-38 (1985).
20. D. A. Talton, S. M. Goldwasser, R. A. Reynolds, and E. S. Walsh, "Volume rendering algorithms for the presentation of 3D medical data," Proc. NCGA '87, III, 119-128 (1987).
21. E. J. Farrell, T. J. Watson, R. A. Zappulla, and M. Spigelman, "Imaging tools for interpreting two and three dimensional medical data," Proc. NCGA '87, III, 60-68 (1987).
22. E. L. Ritman, J. H. Kinsey, R. A. Robb, B. K. Gilbert, L. D. Harris, and E. H. Wood, "Three-dimensional imaging of heart, lungs, and circulation," Science, 210, 273-280 (1980).
23. R. A. Robb, "High-speed three-dimensional x-ray computed tomography: the Dynamic Spatial Reconstructor," Proc. IEEE 71, 308-319 (1983).
24. L. D. Harris, "Identification of the optimal orientation of oblique sections through multiple parallel CT images," JCAT 5, 881-887 (1981).
25. L. D. Harris, J. J. Camp, E. L. Ritman, and R. A. Robb, "Three-dimensional display and analysis of tomographic volume images utilizing a varifocal mirror," IEEE Trans. Med. Imag. 5, 67-72 (1986).
26. J. J. Camp, M. C. Stacy, and R. A. Robb, "A system for interactive volume analysis (SIVA) of 4-D biomedical images," J. Med. Sys. 2, 287-310 (1987).
27. R. A. Robb, D. P. Hanson, R. A. Karwoski, A. G. Larson, E. L. Workman, and M. C. Stacy, "ANALYZE: A comprehensive, operator-interactive software package for multidimensional medical image display and analysis," Comput. Med. Imag. Graph. 13, 433-454 (1989).
28. M. D. Bentley and R. A. Karwoski, "Estimation of tissue volume from serial tomographic sections: A statistical random marking method," Investigative Radiology 23(10), 742-747 (1988).

29. B. B. Mandelbrot, Fractals: Form, Chance, and Dimension, (1977).
30. R. A. Robb, Three-Dimensional Biomedical Imaging, CRC Press, Inc., Boca Raton, FL (1985).

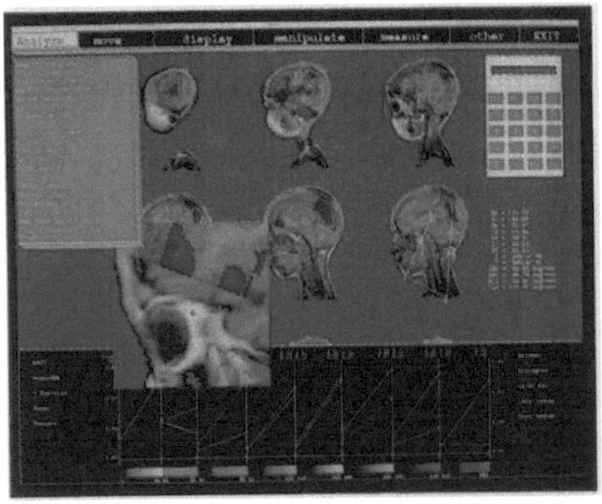

Figure 3. A montage of *TOOLTABLE* utilities including *magnifying glass, interactive color, caliper, and calculator.*

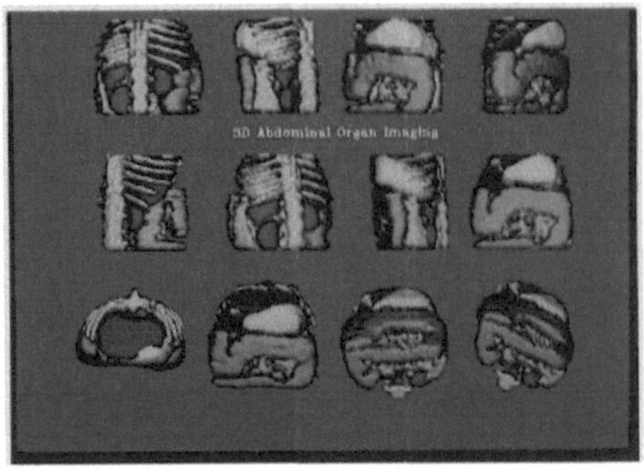

Fig. 8 Three-dimensional abdominal organ segmentation and shaded surface display using 3-D scan data from Dynamic Spatial Reconstructor.

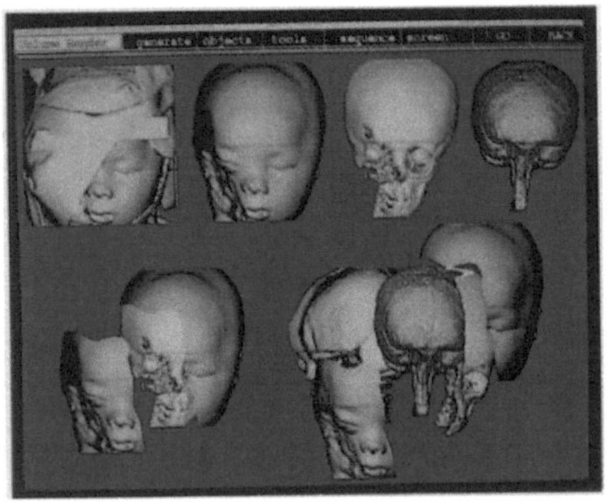

Fig. 10 Montage of images demonstrating the capability to segment and color objects in *volume render* and individually manipulate them.

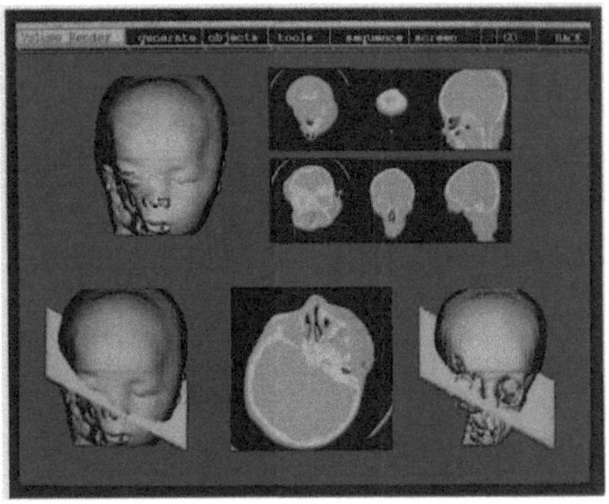

Fig. 11 Interactive tools for orthogonally (top) and obliquely (bottom) sectioning the 3-D image data in *volume render* using selected 3-D surfaces.

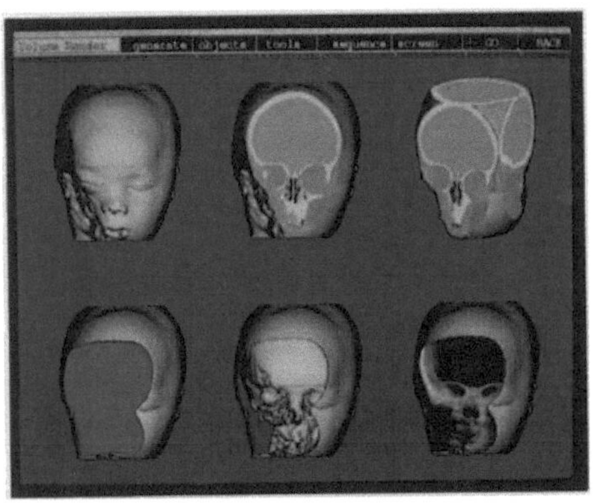

Fig. 12 Arbritrary clipping planes and windows can be interactively placed in volume rendered images to visualize internal values.

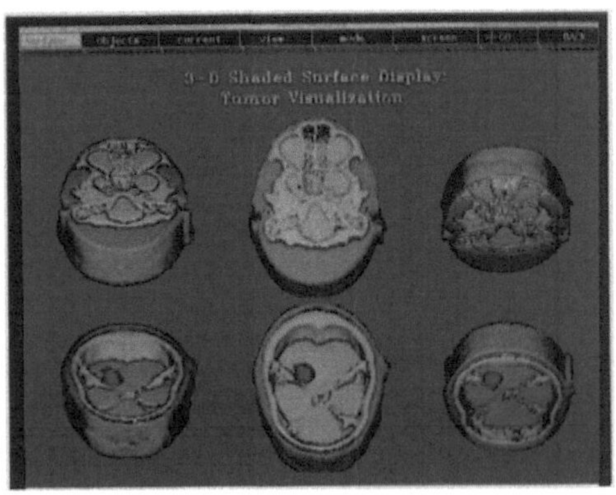

Fig. 21 Shaded surface displays of x-ray CT image data for acoustic neuroma tumor visualization.

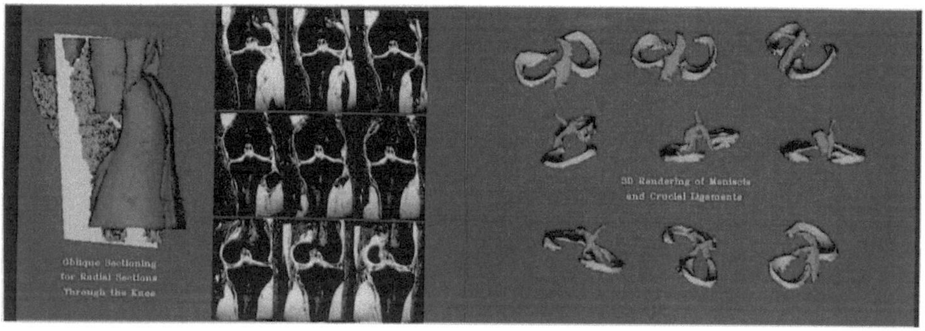

Fig. 23 ANALYZE application to magnetic resonance images of the knee. Radial sections can be generated (left) and substructures segmented and displayed (right).

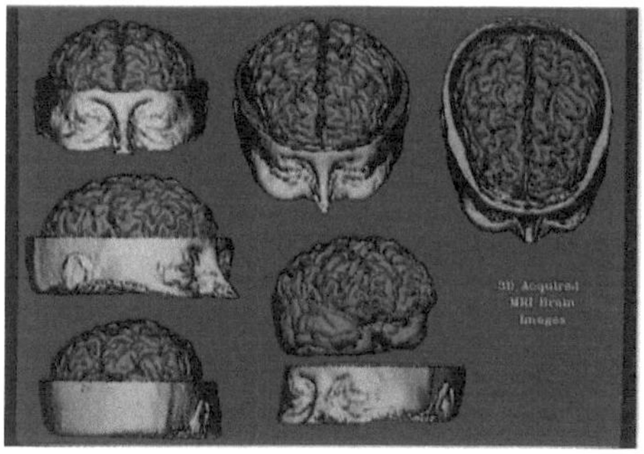

Fig. 24 Volume rendered images from 3-D MRI head scan showing segmentation, rotation, and object manipulation.

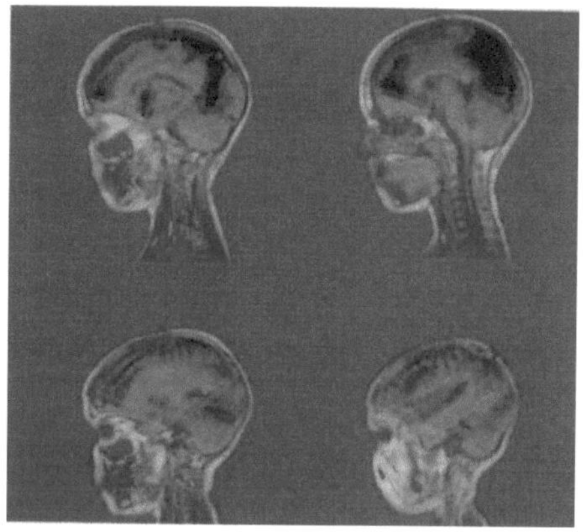

Fig. 27 Integration of PET (color) and MRI (grayscale) images for simultaneous analysis of cerebral structure and function.

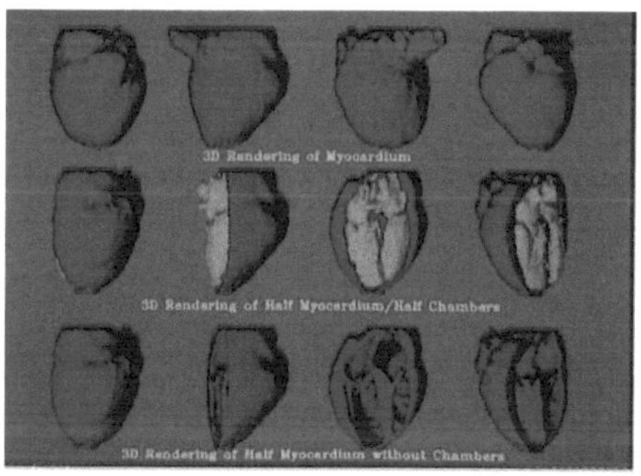

Fig. 28 Shaded surface displays of heart computed and displayed from ultrasound tomographic images.

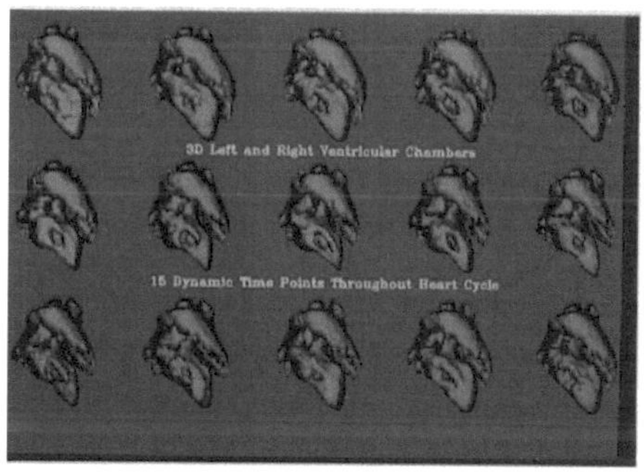

Fig. 30 Time sequence of 3-D shaded surface displays of heart chambers throughout one complete heart cycle.

CARVUPP: Computer Assisted Radiological Visualisation Using Parallel Processing

Farzad E. Yazdy, Jon Tyrrell, Mark Riley
and Norman Winterbottom

IBM United Kingdom Scientific Centre, Winchester
Hampshire SO23 9DR, United Kingdom

Abstract

A parallelised volume rendering algorithm has been constructed and implemented on a network of INMOS' transputers. The technique offers great flexibility in terms of expansion to any number of processors available to the user and achieves a good speed-up factor. It is well suited to objects represented by a number of 2-dimensional slices and our main use for it has been Computerised Tomography (CT) data of human anatomical structures. At present using 16 T8 transputers we can, on average, render and display a volume of 6 MBytes of data in about 12 seconds with full shading. The original voxel densities are retained throughout the process so that different features within the volume of data can be identified, rotated in 3-d and displayed.

In this paper we describe the basic principles and performance of CARVUPP and its potential as a clinically usable tool. The system architecture and the parallelised algorithm we have developed will be outlined in detail.

Keywords: Volume rendering, parallel processing, transputer

1. Introduction

Imaging and visualisation of scientific data has recently become a very important area of research as it often provides and furthers the understanding of the behaviour of certain variables with respect to others. Recent advances

NATO ASI Series, Vol. F 60
3D Imaging in Medicine
Edited by K. H. Höhne et al.

in computer graphics have made a substantial contribution in this area. A large class of scientific data to be visualised is the case where the data is a sampled function of three variables. When these three variables are spatial coordinates the sampled data is often referred to as volume data and the regular grid of three dimensions has been called by some authors the cuberille environment [Herman 1979].

During the last two decades similar advances have been made in the area of medical imaging. Images of extremely high quality are now obtained using a variety of imaging modalities. Amongst these modalities are devices such as CT (Computerised Tomography), MRI (Magnetic Resonance Imaging), SPECT (Single Photon Emission Computed Tomography), PET (Positron Emission Tomography) and Diagnostic Ultrasound. All of the above modalities produce tomographic images in the form of slices of the human body.

As a result of the excellent quality of these images and the fact that a 3-d volume of data can be generated, the quest for 3-d data visualisation in medicine has become that much greater. Even though one may argue that the information contained in the cross-sectional slices themselves is sufficient for many practical purposes, these have not necessarily been presented to the clinician as a choice. The interpretation of these slices requires very specialised training and is typically performed by a diagnostic radiologist. Other clinicians and surgeons may have difficulty interpreting the complex cross-sectional anatomy represented by these two dimensional slices. The representation of medical data in three dimensions has become a useful tool in presenting the data in another form that is much more acceptable to a wide variety of physicians.

On the other hand, because of the very large amount of data that is usually required, it is not feasible to transfer the data to a personal computer (PC) that other clinicians may have in their offices and reconstruct the images in this way for normal clinical use. Several PC implementations of 3-d visualisation have been reported in the literature [Russell 1987]. However, these systems, some of which produce very high quality images, are of no practical everyday use to clinicians mainly due to the long time that they take to produce each picture (typically tens of minutes).

With the amount of data involved for the reconstruction of three dimensional images, fast creation of these images only seemed a dream some time ago. However, with the rapid progress made in parallel processing in very recent years, faster and faster reconstruction has become possible. We have recently been investigating parallelisation of volume rendering, primarily for use as a clinical tool.

The processor we have chosen is INMOS' transputer. In the rest of this paper we briefly describe some sequential techniques that have been used by various researchers for 3-d reconstruction. We also outline possible ways of parallelising these techniques for use on a multi-processor system. The parallel technique that we have developed and its implementation on our multi-transputer machines will be discussed in detail.

2. Sequential Volume Rendering Techniques

As the data to be rendered in many applications is collected in the form of a series of slices, it only makes sense to represent the entire data in terms of voxels. In this case the total number of slices put together represents a total digital scene as a three-dimensional array of values. Objects within this scene can then be defined by selecting the appropriate values.

With the data represented in terms of voxels, the main tasks that need to be accomplished in order to generate a 3-d rendered picture are as follows.

1. Rotation of the volume of data in order to view it at a desired orientation.

2. Determination of the object of interest within this volume.

3. Determination of the visible part of this object (Hidden Surface Removal).

4. Shading and display of the visible surface.

Rotation of the volume of data is achieved by transforming the coordinates of every point in the three-dimensional array of data to a new point within another three-dimensional array whose bounds are large enough to contain the entire rotated volume of original data. We will briefly describe this transformation of data in a later section but a comprehensive description of these transformations is available in the literature [Frieder 1985, Reynolds 1987].

The object of interest within the total volume is normally determined by choosing a set of voxels whose values fall within a given threshold range [Herman 1979]. Levoy, on the other hand, defines a classification operator that detects isovalue contour surfaces [Levoy 1988]. Once the object of interest has been classified there are mainly two most common ways of determining its visible part. The first method called Back-to-Front or Painter's Algorithm is where the three dimensional cube containing the rotated object of interest is scanned from the plane furthest from the viewing position. Once a full voxel is encountered, its (x, y) positions are recorded and the (x, y) position of the screen is given the shaded value of the encountered voxel. This scanning procedure continues until the front of the cube is reached with the screen values constantly being updated by new ones. Frieder gives a detailed explanation of this technique [Frieder 1985].

The second method is Front-to-Back, where the cube containing the data is scanned from the plane nearest to the viewing position and is traversed in the z-direction (into the screen). In this case once a filled voxel is encountered again its (x, y) position is recorded and the corresponding (x, y) position of the screen is given the shaded value of this voxel. Unlike the previous method once an opaque voxel is encountered no further examination of the voxels behind it is necessary as they are covered by this first opaque voxel. The two techniques mentioned differ in that the former is much slower but allows the hidden surfaces to be seen as the final picture builds up from the back of the viewing scene. Reynolds gives a good review of these and other techniques [Reynolds 1987].

With the visible surface having been determined, shading is applied to this surface to create the impression of a true three dimensional picture. This is accomplished by calculating a surface normal for each visible point and using Lambert's cosine law to calculate the amount of reflected light [Foley 1984]. In volume rendering The surface normal can be calculated in two ways. The first method works out the normal by looking at the voxels immediately surrounding the visible one in each direction. There are several good reviews of this technique available in the literature [Robb 1988, Hoehne 1986, Barillot 1985, Chen 1985]. Another technique first suggested by Gordon and Reynolds uses the buffer containing the depth of each visible surface (z-buffer) to calculate a sham normal [Gordon 1985].

3. Parallelisation

A distribution of the tasks mentioned in the previous section among a number of processing units is usually realised in two ways. The array containing the final picture that is to appear on the screen can be broken up into a number of regions. This is effectively the same as allocating a section of the screen to each processor. In this way each processor can work out what part of the final picture appears on its section of the screen (Fig. 1). This technique is efficient and relatively simple to implement but only works best in cases where each processor has enough random access memory to hold all the object data. If this is not the case the technique becomes I/O bound. This situation arises because in going from one view (set of rotation angles) to another, different data is needed for the calculation and fetching this data may require some time.

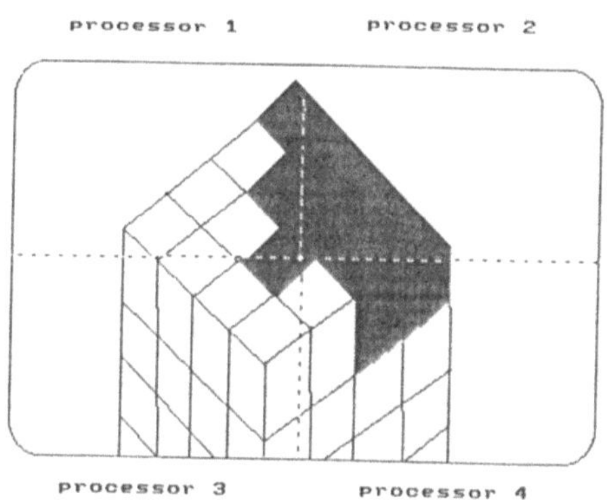

Figure 1: Partitioning of the screen for 4 processors. Each processor works out a fraction of the total picture that would appear in its part of the screen.

A second method of parallelisation can be achieved by splitting the original data cube into a number of smaller sub-cubes and assigning each processor the task of rendering one of these (Fig. 2). This is usually a less efficient technique compared to the first (if there are no severe I/O restrictions associated with the first method). The reason for this is that depending on the angles of rotation of the object, one small sub-cube may partially or in extreme cases even completely cover another sub-cube by appearing in front of it. Of course each processor working on its sub-cube is not aware of this and creates a sub-picture which it thinks is visible. As a result, when at a later stage all these sub-pictures are merged to create a final visible picture, part or in some cases all of each sub-picture that has been that part of the work done by each processor may be unnecessary. However, as each processor usually does not possess enough memory to hold the entire data base and the amount of data can vary considerably in different applications, this technique is still preferred. Goldwasser and Reynolds have produced one implementation of it using a number of processors [Goldwasser 1987].

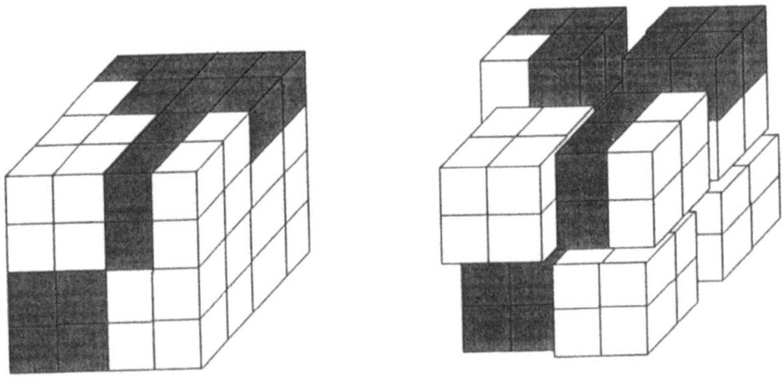

Figure 2: Partitioning of the data. The original cube of data is broken down to smaller sections and each one is passed to a separate processor for rendering.

4. Principles of CARVUPP

CARVUPP makes the necessary choices between the foregoing techniques and addresses the special implications of parallel processing systems. It can logically be divided into three separate parts.

- Message routing within the network of transputers.
- Rendering.
- Combining the results and display.

4.1 Message Passing

Message passing around the network, whether it is an instruction to a processor asking for a task to be done or the transfer of data itself, is done by Generalised NetWorking (GNW). This is a software package of routines that we have developed which simplifies the programming of a network of transputers by providing user friendly functions [Winterbottom 1990]. By including a library of these functions at each transputer, all transputers become capable of sending or receiving requests for work or data from anywhere within the network. Tyrrell describes in more detail some of the capabilities of GNW [Tyrrell 1990].

4.2 Rendering

We have parallelised the rendering by splitting the total cuberille data into smaller sub-cubes which we call voxel space data. Our rendering algorithm expects the following parameters:

- Number of cuberille voxels in the x, y and z directions, vox.x, vox.y and vox.z respectively.
- The lowest indices (x.min, y.min and z.min) in cuberille coordinates of the sub-cube of data that has been allocated to the algorithm.
- The data in 1 byte per pixel for the sub-cube.

We use the standard matrix of rotation about the three axes (x,y,z). This matrix is written as:

$$T = \begin{bmatrix} \cos\alpha\cos\gamma + \sin\alpha\sin\beta\sin\gamma & -\sin\alpha\cos\gamma + \cos\alpha\sin\beta\sin\gamma & \cos\beta\sin\gamma & 0 \\ \sin\alpha\cos\beta & \cos\alpha\cos\beta & -\sin\beta & 0 \\ -\cos\alpha\sin\gamma + \sin\alpha\sin\beta\cos\gamma & \sin\alpha\sin\gamma + \cos\alpha\sin\beta\cos\gamma & \cos\beta\cos\gamma & 0 \\ 0 & 0 & 0 & 1 \end{bmatrix}$$

As the rotations are applied about the centre point of the object (our chosen origin), the object is first translated by values $-(nvx/2)$ about the x-axis, $-(nvy/2)$ about the y-axis and $-(nvz/2)$ about the z-axis, where nvx, nvy and *nvz* are the number of voxels in the x, y and z directions respectively. This is achieved by the transformation matrix:

$$-ve\ Shift = \begin{bmatrix} 1 & 0 & 0 & -(nvx/2) \\ 0 & 1 & 0 & -(nvy/2) \\ 0 & 0 & 1 & -(nvz/2) \\ 0 & 0 & 0 & 1 \end{bmatrix}$$

Having been rotated, each data point is then translated forward to fit into the new transformed array. This is accomplished by multiplying each data point by the matrix:

$$+ve\ Shift = \begin{bmatrix} 1 & 0 & 0 & \sqrt{3}\,Nv/2 \\ 0 & 1 & 0 & \sqrt{3}\,Nv/2 \\ 0 & 0 & 1 & \sqrt{3}\,Nv/2 \\ 0 & 0 & 0 & 1 \end{bmatrix}$$

where Nv is the largest of the values nvx, nvy and nvz previously defined and $\sqrt{3}\,Nv$ is the length of one side of the cube that contains the transformed data. The $\sqrt{3}$ factor ensures that the bounds of the cube are large enough to cover the rotation of the original sub-cube of data.
Hence the total transformation matrix is given as:

$$M = \begin{bmatrix} +ve\ Shift \end{bmatrix} \begin{bmatrix} T \end{bmatrix} \begin{bmatrix} -ve\ Shift \end{bmatrix}$$

Multiplying each old coordinate (x, y, z) of the sub-cube of data by the above matrix M will result in its new rotated coordinates (x', y', z').

We use the inverse of this matrix which yields the transformation from the rotated cube back to the original sub-cube. In this way we are able to start at the position:

$$(x', y', z') = (0, 0, 0)$$

and proceed along the z' direction encountering voxels of the original sub-cube. If the voxel encountered falls within the threshold range of interest, the process of incrementing in the z' direction stops.The depth of this voxel is recorded in a z-buffer and its surface normal is calculated from the object space as mentioned previously. Using the surface normal the shaded value of the encountered voxel is held in a separate buffer. The process of scanning is again resumed from the beginning of the array ($z' = 0$) and this time x' is incremented by 1 and scanning along the z' direction resumes.

This procedure is effectively the famous Front-to-Back technique. However, unlike the sequential method, there is now the need for a z-buffer to store the depth of each voxel encountered. The reason for this additional array is that as the original cuberille data has been split up into subcubes, a further comparison of the visible surfaces determined for each sub-cube is necessary in order to determine the final picture. Storing the real depth is necessary for a further comparison of all visible surfaces.

4.3 Creating the Screen Picture

There is an additional process that combines all visible pictures and decide which part, if any, of each visible picture is covered by another. From the information supplied to the render process (section 3.2), it can be interpreted which region of the final (screen) picture the visible picture produced by each processor falls onto.

It is also possible to determine the screen depth (Global z' value) of the visible part for each picture. As a result, if any two points of two visible pictures produced by separate processors fall onto the same point of the screen, the merger process chooses the one that has the lower depth, as this hides the higher depth value behind it.

5. Implementation

The above algorithms have all been implemented in INMOS' OCCAM. Figure 3 shows the basic components of our processing environment. We have used MEIKO's M10 computing surface which provides the processing and graphics display power, connected to an IBM PS/2 computer acting as the host and providing file system services. The host communicates with the first transputer (called the host node) by running a program called the server. The network includes a Mass Store (MS) which is a transputer with 8MBytes of memory and a Graphics Node which is another transputer with 2MBytes of Video RAM connected to an IBM 5081 display. The rest of the transputers, called worker nodes, are arranged in a ring as shown in Figure 3.

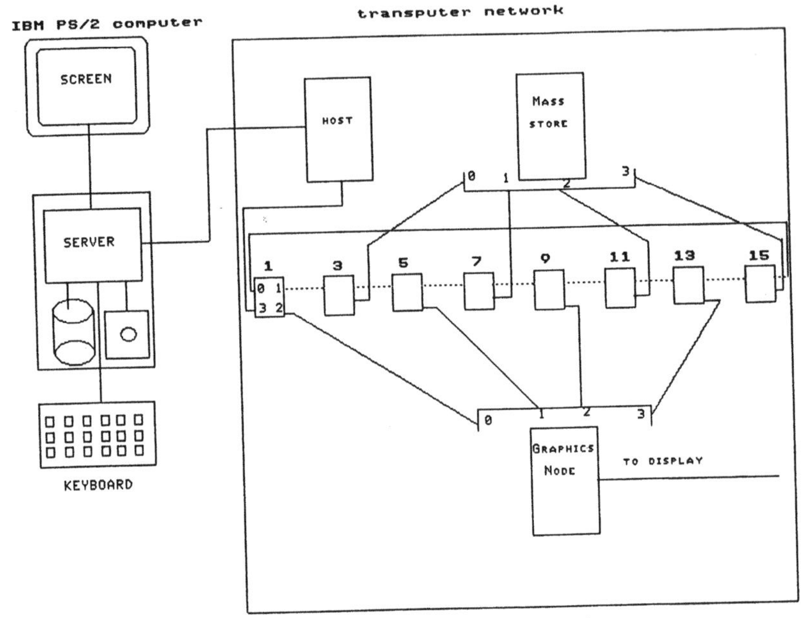

Figure 3: Processing environment, the transputer network is Meiko's computing surface. The small boxes within the network represent the worker nodes connected in a ring. The dotted lines represent links to workers not shown in the diagram. The small numbers 0,1,2 and 3 represent asynchronous transputer links.

We run a Driver program on the MS which acts as the central controller for CARVUPP. This program runs the user interface (asking the user for file names, angles of rotation etc), fetches the data from the PC disk and keeps it on the MS. It then splits up this data into smaller sub-cubes and transmits one sub-cube together with the extra information needed (see section 3.2) to each worker node. All worker nodes contain copies of the program that renders a sub-cube and after local rendering has been completed each worker node returns a z-buffer and a shaded image with a number of additional parameters, to the MS.

The procedure that combines all individual shaded images into a final picture also resides on the MS. The Driver passes all these incoming images with their additional parameters to this process where they are combined into a final picture. The Driver then sends this picture to the Graphics node, which in turn displays it on the screen. All the above message passing and data transmission is done using GNW.

6. Experimental results

We have experimented with a set of 44 CT scans, each of resolution 256×256 pixels. As there was originally a physical gap in between two consecutive scans, these were linearly interpolated to create a total data base of roughly 6MBytes. The original distribution of this data from the MS to the worker nodes takes roughly 12 seconds. Once this is done, the time taken for subsequent rotations of the object, or redefining the object of interest is about 12 seconds on average.

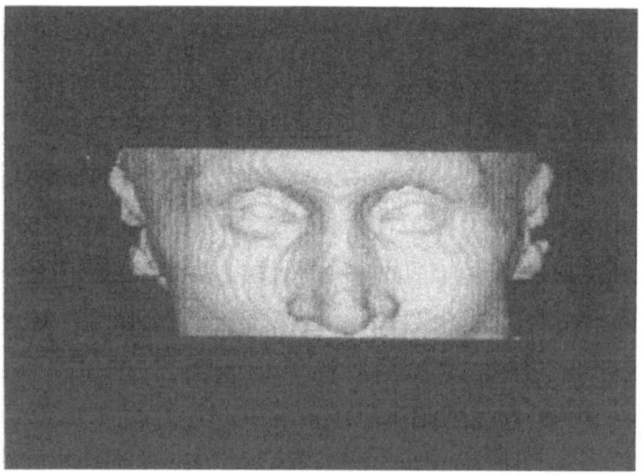

Figure 4: 3-d rendered picture of soft tissue reconstructed from 44 CT scans.

Figures 4 and 5 show rendered images of soft tissue and bone reconstructed from the CT data mentioned above. We have also experimented with rendering the same data using 8 worker nodes and have established that doubling the number of workers to 16 offers on average a 1.5 fold increase in speed.

7. Conclusions

Implementing a parallel volume rendering algorithm on INMOS' transputer has been a successful exercise. Although our system in its present

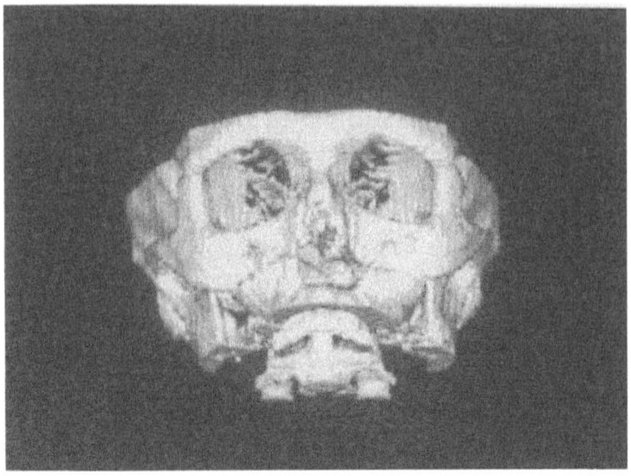

Figure 5: Rendered picture of boney structures reconstructed from the same data set as Figure 4.

form lacks the speed and functionality of some of the more specialised systems built specifically for medical data visualisation [Robb 1989, Goldwasser 1987], there are good reasons for us furthering our investigations into the use of the processors we have used for this task. The transputer can be used with host computers that most clinics may well possess already so there is no need for major initial investment into a specialised system.

Because of its nature and the fact that any number of transputers can be coupled together to enhance processing power, our system is easily upgradable in steps. Thus its value can be initially investigated by the interested clinician with the minimum of additional equipment to speed up performance. Furthermore, transputer boards can easily be obtained as there are numerous hardware suppliers producing them. Also the fact that the transputer is a general purpose processor, once acquired, it can be used for other applications that may be required in the clinic.

The idea of this system being accessible to clinicians has already generated profound interest amongst them and we do feel that we have contributed to making the use of 3-d visualisation more popular in the medical community. We are enhancing the functionality of the system by working in areas such as user interface and investigating further speeding up the algorithms. These should in future make the system much more interactive and hence desirable.

References

Barillot, C., Gibaud, B., Luo, L. M., and Scarabin, J. M. 3-D Representation of Anatomic Structures From CT Examinations. SPIE Biostereometrics 85, Vol 602:307-314, 1985.

Chen, L., Herman, G. T., Reynolds, R. A., and Udupa, J. K. Surface Shading in the Cuberille Environment. IEEE Transactions on Computer Graphics and Applications, pages 33-43, December 1985.

Foley, J. D. and Van Dam, A. Fundamentals of Interactive Computer Graphics, chapter 16. Addison Wesley, 1984.

Frieder, G., Gordon, D., and Reynolds, R. A. Back-to-Front Display of Voxel-Based Objects. IEEE Transactions on Computer Graphics and Applications, pages 52-60, January 1985.

Goldwasser, S. M. and Reynolds, R. A. Real-Time Display and Manipulation of 3-D Medical Objects: The Voxel Processor Architecture. Computer Vision, Graphics and Image Processing, 39:1-27, 1987.

Gordon, D. and Reynolds, R. A. Image Space Shading of 3-dimensional Objects. Computer Vision, Graphics and Image Processing, (29):361-376, 1985.

Herman, G. T. and Liu, H. K. Three-Dimensional Display of Human Organs from Computed Tomograms. Computer Graphics and Image Processing, 9:1-21, 1979.

Hoehne, K. H. and Bernstein, R. Shading 3D-Images from CT Using Gray-Level Gradients. IEEE Transactions on Medical Imaging, MI-5(1):45-47, March 1986.

Levoy, M. Display of Surfaces from Volume Data. IEEE Transactions on Computer Graphics and Applications, pages 29-37, May 1988.

Reynolds, R. A., Gordon, D., and Chen, L. A Dynamic Screen Technique for Shaded Graphics Display of Slice-Represented Objects. Computer Vision, Graphics and Image Processing, 38.:275-298, 1987.

Robb, R. A. and Barillot, C. Interactive 3-D image display analysis. SPIE Hybrid Image and Signal Processing, Vol 939:173-202, 1988.

Robb, R. A. and Barillot, C. Interactive Display and Analysis of 3-D Medical Images. IEEE Transactions on Medical Imaging, 8(3):217-226, September 1989.

Russell, G. and Miles, R. B. Display and perception of 3-D space-filling data. Applied Optics, 26(6):973-982, March 1987.

Tyrrell, J. IBM UKSC Internal Report, work in progress, 1990.

Winterbottom, N. Generalised NetWork (GNW) IBM UKSC Internal Report, work in progress, 1990.

Applications

A Volume-Rendering Technique for Integrated Three-Dimensional display of MR and PET Data

Xiaoping Hu[1], *Kim K. Tan*[1], *David N. Levin*[1], *Charles A. Pelizzari*[2], *George T.Y. Chen*[2]

Departments of Radiology[1] and Radiation Oncology[2]
University of Chicago Hospitals Chicago, Illinois

Abstract

Methods have recently been developed for using MR data to create 3-D computer models of the gyral anatomy of the brain. This paper describes image processing and computer graphic techniques for using PET images to produce 3-D models of brain surface metabolism. Then, an existing algorithm for retrospective image registration is used to fuse the two constructs into a unified 3-D model of brain structure and function. Neurosurgical planning can benefit from this display technique which localizes PET-detected metablolic abnormalities with respect to MR-defined gyral anatomy.

1. Introduction

As medical technology advances, there is a tendency for each patient to undergo more imaging procedures. This has created a growing need for integrated methods of displaying the many resulting images. For example, neurology patients are often diagnosed and monitored with magnetic resonance (MR), x-ray computed tomography (CT), and/or positron emission tomography (PET) scanning, which produce cross-sectional images of a variety of brain properties. Brain anatomy is best depicted by MR, which creates images of the proton density and relaxation times (T1, T2) of the brain before and after administration of contrast agents. Furthermore, experimental MR techniques are being developed to image vascular anatomy, perfusion, and diffusion. Skeletal anatomy and calcifications are best delineated by CT, which is often done before and after infusion of an intravenous contrast agent. Maps of brain function (e.g. metabolism, blood flow, etc.) are provided by PET. After several of these procedures

NATO ASI Series, Vol. F 60
3D Imaging in Medicine
Edited by K. H. Höhne et al.

are done, physicians are faced with many (more than 100) cross-sectional images, depicting various aspects of anatomy and function in different slices of the brain. They must mentally integrate this mass of fragmented information into a coherent 3-D picture. Since the data is digital, it is natural to ask if a computer can synthesize this information into a multimodality, multivariable 3-D model. Then a physician could interact with the model in a natural fashion and try to understand its medical significance without having to consider the exact geometry of data acquisition and other technical details. Fortunately, it is now possible to create this type of multimodality 3-D display by using high speed computer workstations and new algorithms for 3-D rendering and image registration.

Surface-rendering and volume rendering are the archetypal algorithms for 3-D reconstruction although the distinction between them is blurred in actual practice. Surface-rendering (1) is performed by modeling the object as a well-defined, infinitessimally-thin surface. After a description of this surface is extracted from the gray scale data, its local properties (e.g. emitted and reflected "light") are projected onto the viewing screen to create 3-D views. Volume-rendering (2,3) is based on a "fuzzy" model of the imaged object, which is assumed to be bordered by an indistinct transition zone of voxels containing mixtures of materials. 3-D views are created by integrating along rays which cross this transition zone and by projecting those integrals onto the viewing screen. Proponents of surface-rendering point out that it is a more practical algorithm to implement in many hardware environments since it requires less memory and processing speed. The computer's memory need only hold a list of surface voxels, defined by pre-processing the gray scale data with a discrete tissue characterization algorithm; then, 3-D views are created by applying projection operations to this 2-D manifold. Volume-rendering enthusiasts reply that relatively inexpensive parallel processors will soon be capable of performing ray tracing rapidly enough to produce volume renditions in "real time" (4,5). Volume-rendering also benefits from its greater generality. Gray scale data is first subjected to a continuous tissue characterization algorithm which uses a ramp-like or sigmoidal look-up table to assign fractional tissue contents to each voxel; if the operator changes this look-up table to resemble a step function, discrete tissue characterization and surface-rendering can be recovered as a special case. Thus, the user of a volume-rendering scheme has the freedom to "tune" the algorithm to produce surface-rendered views as needed. Volume-rendering also requires less stringent pre-processing than surface-rendering. It is only necessary to identify the approximate configuration of the "fuzzy" interface to be rendered; voxels at the margins of this transition zone make small, fractional contributions to the integrals along rays and can be included or excluded with little effect. In contrast, the quality of a surface rendition is critically dependent on the identification of the surface by discrete tissue characterization since every surface point makes a full contribution to 3-D views. Notice that the two methods handle partial volume averaging in different ways. Volume-rendering is a completely continuous mapping of image data onto the 3-D viewing plane; voxels, which are affected by partial volume averaging, make fractional contributions to the 3-D image depending on their tissue content. Thus, partial volume averaging naturally leads to a "fuzziness" or indistinct appearance of the 3-D model. On the other hand, surface-rendering methods must use some type of over-sampling method (6,7) in order to estimate the position of the actual surface within a voxel which straddles it. If this is not done carefully, crevices may be "filled in" or jagged protrusions may result (8).

Both methods have long been used to create 3-D views of skeletal anatomy from CT data (1). Since bones have the highest CT numbers, they can be identified automatically by look-up tables for either discrete or continuous tissue characterization. Then, surface- or volume-rendering leads to 3-D views, which are useful for planning surgical treatment of craniofacial malformations (9), complex hip fractures (10), or other skeletal disorders. Volume-rendering advocates (8) maintain that small surface features like hair line fractures are best preserved by a continuous rendering technique.

The higher soft tissue contrast of MR data makes it the preferred starting point for attempts to define and render the surfaces of soft tissues such as the brain (6,7, 11-15). It is not straightforward to identify brain in MR data since it does not have a unique intensity unlike CT-depicted bone. Therefore, it is necessary to use spatial information as well as gray scale values for this purpose. Investigators have tried a variety of algorithms which utilize edge detectors (12), connectivity or region-growing (6), and contour generation by manual methods (13) or threshold-tracking (11,14,15). All of these techniques require some interactive image editing by the user who must correct occasional errors made by the computer. The isolated brain is then subjected to surface- or volume-rendering to create 3-D views. The resulting images depict important landmarks on the brain surface, which are associated with movement, sensation, speech, and hearing (14); these areas cannot be identified with certainty on the cross-sectional views. Clinical trials (15) suggest that neurosurgeons can use this new information to avoid disruption of these structures during resections of nearby lesions.

It is often necessary to correlate information from a pair of 3-D data sets which were acquired at different times and, possibly, with different modalities. In order to do this quantitatively, it is necessary to find the coordinate transformation which maps each voxel in one volume of data onto the corresponding voxel of the other. In soft parts of the body a complex non-linear transformation will be required to account for the variable deformations of body shape resulting from differences in patient positioning. On the other hand, since the skull is rigid, a linear transformation (translation, rotation, and rescaling) should suffice to relate brain images acquired at different times. This linear mapping can be derived by matching extrinsic or intrinsic features which are identified in the two data sets. The use of extrinsic markers (e.g. fiducial marks) is not feasible in a routine clinical setting since it requires prospective manipulations of the patient; i.e. a trained individual must be present during both imaging procedures to make sure that the markers are positioned in exactly the same way. It is more practical to match intrinsic landmarks since this can be done retrospectively after routine scans. Most investigators have attempted to match specific anatomical points which can be identified on the two scans. This approach is problematical for several reasons. First, it is often difficult to identify the same point-like feature in images which represent different attributes (e.g. anatomy vs. function). In fact, an expert in imaging anatomy may be required to find these points. Second, serious errors can result since only a few points are used, and these may be imaged with different degrees of partial volume averaging. Recently, two of us (CAP and GTYC) have developed a retrospective image registration technique (16,17) based on matching intrinsic surface features in two data sets. This method is robust and less susceptible to partial volume errors since hundreds of points are used to describe the surfaces to be matched. Furthermore, it can be executed automatically or with limited help from a technician who need only recognize large surfaces in the data.

An earlier report (18) discussed the clinical value of using this new image registration technique to create integrated cross-sectional images from MR, PET, and CT scans. In the following sections we describe tissue characterization and volume-rendering algorithms for using PET data to create a 3-D model of brain surface metabolism; retrospective image registration was then used to fuse this construct with an MR-derived model of brain surface anatomy. The result was an integrated 3-D model of brain anatomy and function (19,20) which served as a framework for the unified display of MR and PET data. This program will be illustrated with one clinical example; a more extensive discussion of clinical applications was reported elsewhere (21).

Methods

As indicated in Figure 1, we have written a semi-automatic threshold-tracking program for isolating the brain in MR images. Then, continuous tissue characterization and gradient-shaded volume-rendering are used to produce 3-D views of the brain surface. Since these methods have been discussed in detail elsewhere (14,15,19), only a brief description is presented here.

MR images of the head were acquired during a single 10-minute scan on our 1.5 T system (Siemens Medical Systems, Iselin, NJ). This gradient echo, volumetric pulse sequence (TR/TE = 40/13 ms; 40° flip angle) produced T1-weighted images of 63 thin (3 mm) contiguous parasagittal sections (Fig 2a). Each image was a 256 x 256 matrix of pixels with 1.2 mm sides. Since this pulse sequence only took 10 minutes, problems with patient motion were uncommon, and it was easy to perform during a routine brain imaging protocol. Better 3-D renditions would result if 127 thinner (1.5 mm) sections were acquired. However, this can be accomplished in ten minutes only if TR (and TE) can be reduced or if 128 x 256 matrices are used; these options are available on machines with the latest software and hardware.

The brain and a fringe of voxels containing cerebrospinal fluid (CSF) were isolated in each cross-sectional image by means of an image-editing program of our own design. The program had a mouse-driven "point and click" interface so that it could be run by technologists with no

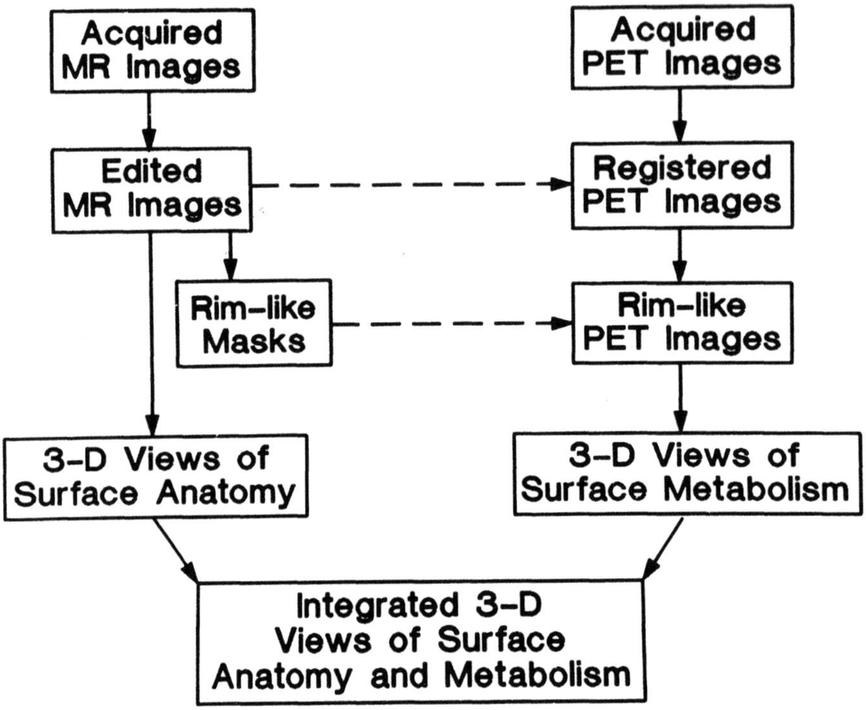

Figure 1. Diagram of the algorithm for creating integrated 3-D models of brain anatomy and metabolism from MR and PET data (reprinted from Radiology; see reference 21).

knowledge of UNIX, C, or X-windows. The user selected a gray scale threshold value which was between the typical signals of CSF (dark) and brain (bright). Then, he placed a "seed" in the CSF which was adjacent to the surface of the brain. The program generated a contour around the brain by tracking the chosen threshold in counter-clockwise fashion, starting with the seed pixel (Fig 2b). An edited image was then created by omitting all pixels outside the contour (Fig 2c). The exact shape of the contour was unimportant since marginal pixels with little brain content made only small fractional contributions to the volume-rendered 3-D images. Specifically, the contour did not have to follow the complex convolutions of the brain surface. The contour was only forbidden to cut across brain tissue or to include high signal extra-axial tissues. Approximately 15 percent of the computer-drawn contours did not satisfy these requirements. Typical problems included deflection of the contour by high signal pixels of dural membrane, optic nerve tissue, or bone marrow at the skull base. In these cases, the operator used

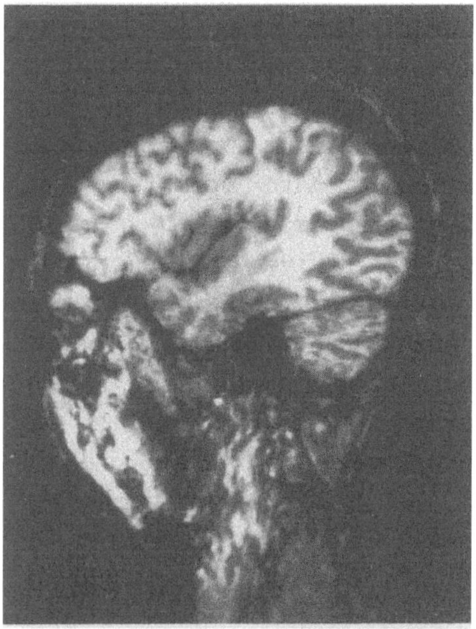

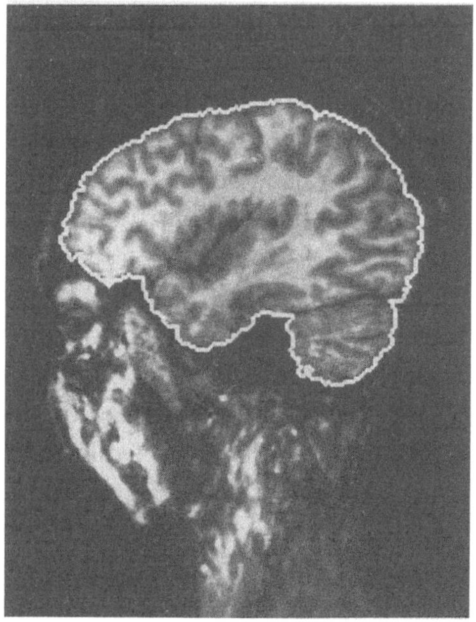

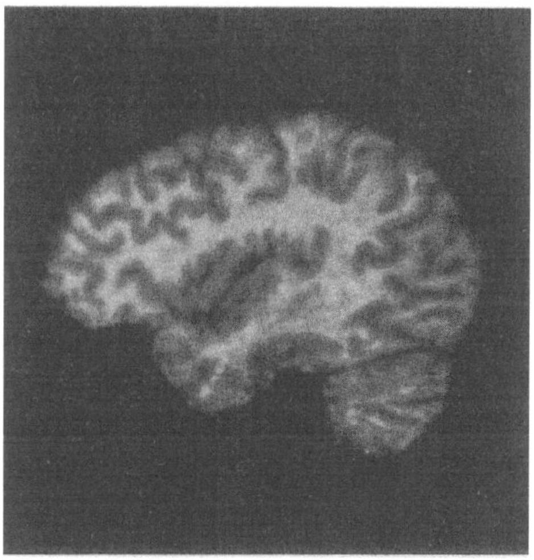

Figure 2. MR images of a 16-year-old male who had suffered from medically- intractable epilepsy since infancy. Electroencephalographic recordings from scalp electrodes showed a poorly-defined left temporal abnormality.
a) (top left) MR image of the left temporal lobe from a T1-weighted volumetric gradient-echo scan. MR imaging with multiple pulse techniques revealed no anatomic abnormalities.
b) (top right) Computer-drawn contour enclosing brain and adjacent CSF in a.
c) (bottom left) Edited MR image obtained by removing pixels outside the contour in b.

"cutting and painting" tools to lower the signal of these pixels before redrawing the contour. It took about 45 minutes for interactive editing of all 63 images on a Sun 3/280 workstation (Sun Microsystems, Sunnyvale, CA). However, we are now developing a three-dimensional version of this segmentation algorithm which should require much less user interaction.

Tissue characterization was done by using a look-up table to assign fractional brain content to each pixel in the processed images. The exact shape of the look-up function was unimportant as long as it was a continuous monotonic mapping which was nearly equal to 0 and 1 at typical signals of CSF and brain, respectively. Measurements of CSF and brain signals were used to choose a sigmoidal function with appropriate "center" and "width". Then, a few 3-D views were made by subjecting the brain fraction images to volume- rendering on a Pixar Image Computer (Pixar, San Raphael, CA). The user could adjust the look-up table to "burn away or expand" the brain by raising or lowering the "center"; similarly, the interface zone between brain and CSF could be made more or less "fuzzy" by increasing or decreasing the "width". Thus, the user could adjust the look-up table to create 3-D views which emphasized different aspects of the original data; this was analogous to "windowing" operations commonly used to process cross-sectional images before interpretation by a radiologist. Of course, in both situations the exact nature of the look-up table must be considered before using the images to make geometric measurements. It took approximately 20 minutes to adjust the tissue characterization function in this interactive fashion. We look forward to volume-rendering at frame rates (5) so that the lookup table can be "tuned" in "real time".

After these adjustments were finalized, volume-rendering was used to create a digital "movie loop", consisting of 32 or 64 3-D views separated by equal angles (Fig 3). These could be "played back" at frame rates to simulate the rotation or tumbling of the brain. We utilized Pixar's ChapVolume software, which was commercially distributed in 1986 and later described in detail by Drebin (2). The user selected intensity, opacity, and "density" attributes for brain (bright white, 100% opaque, 100% dense) and CSF (colorless, transparent, 0% dense). The fractional content of brain and CSF in each voxel was then used to calculate the linear combination of these attributes which was assigned to that voxel. The intensity and opacity distributions were multiplied by the magnitude of the local three-dimensional gradient of the density distribution; this had the effect of suppressing the attributes of homogeneous tissues and enhancing the relative attributes of voxels at tissue interfaces (e.g. brain-CSF). The attributes of each voxel were further modulated by a "shading" factor which was a function of the relative directions of the density gradient and a simulated light source. This means that the shading of each voxel was controlled by a gradient which was related to the local gray level gradient (22) by a continuous non-linear transformation. An additional 3-D cue (depth-encoding) was added by suppressing the intensities of distant voxels by a quadratic function of the distance to the observer. A 3-D view was created by summing the intensity contributions from all the voxels along rays which were projected through the volume onto a viewing plane. The contribution of each voxel was determined by its own intensity and by the opacity of foreground tissues. It took approximately one minute to render each 3-D view.

PET images were obtained with a machine which was built in our institution after the design of Ter-Pogossian et al (23). Data were acquired 40 minutes after the patient was given an intravenous injection of 5-10 mCi (185-370 MBq) of 2-fluoro-deoxy-D-glucose Images were made of 5 or 10 contiguous axial sections with 14 mm thickness (Fig 4). The actual resolution of these images was 8-12 mm although the data were displayed as 100 x 100 matrices of 2.5 mm pixels. These images represented maps of the metabolic rate in the brain. We have also experimented with the use of SPECT images instead of PET scans; this will be reported elsewhere.

The PET images and edited MR scans were then correlated by a retrospective image registration program, described in detail elsewhere (16,17). Each set of data was used to create a coarse model of the brain surface which described the position and orientation of the head. The interscan coordinate transformation was then derived by finding the operations (translation, rotation, independent rescaling along each axis) which were required to fit one model to the other. The PET data were then resampled along the parasagittal MR planes to produce registered pairs of PET and MR images. Both sets of data were interpolated to create volumes of isotropic 1.2 mm voxels, which were reformatted into thin axial sections (Fig 5a,b).

These registered axial images were used to construct a 3-D computer model of brain surface metabolism. Specifically, the highly resolved MR images were used to define rim-like masks which delineated superficial brain tissue; 3-D views of brain metabolism were derived by projection of the average PET activity within the masked volume. In order to create these masks, each MR image was converted into a binary array by assigning the value zero (or one) to pixels having intensities above

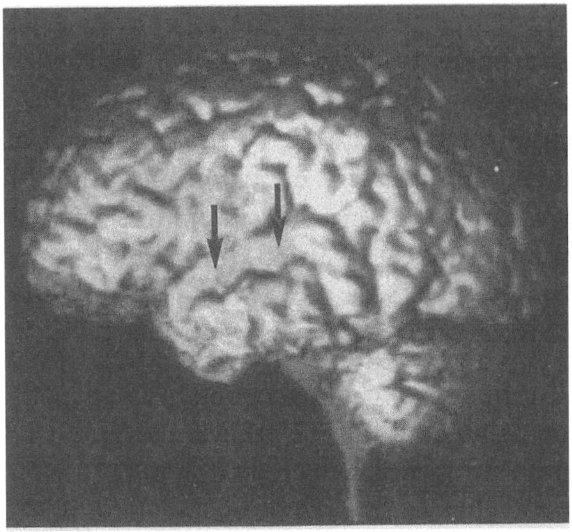

Figure 3. Three-dimensional view of the brain surface created by volume-rendering a stack of edited MR images, including the one in Fig 2c. No anatomic abnormalities are evident; there is good visualization of the superior temporal gyrus (arrows), which is associated with hearing.

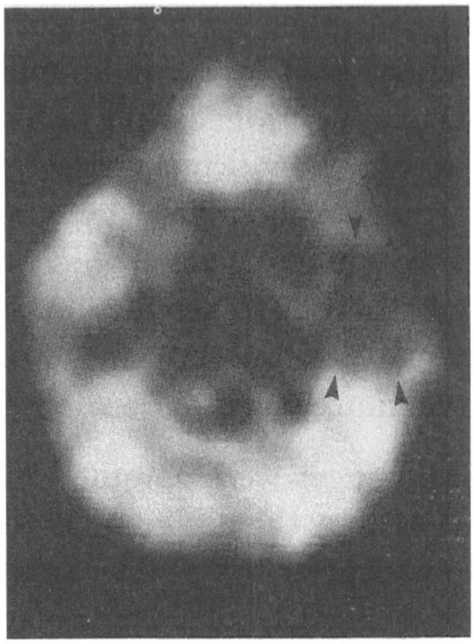

Figure 4. Axial PET image of brain metabolism of the patient in Figs 2-3. Note the ill-defined hypometabolic area (arrowheads) in the left temporal lobe. The poorly-resolved PET data does not show the relationship of this suspected seizure focus to the superior temporal gyrus, which is not visualized.

(or below) a user-selected "brain" threshold. This image, which depicted brain as a signal void against a background of unit intensity, was then blurred several times by averaging it over 3 x 3 neighborhoods. The original image was subtracted from the blurred version to create a rim-like image of the superficial brain tissue. This was converted into a binary mask by assigning unit (or zero) values to pixels having intensities above (or below) a user-selected threshold value (Fig 5c). The choice of this threshold, together with the chosen amount of blurring, determined the breadth of the rim, usually taken to be 5-8 mm. An image of metabolism in a rim of superficial brain was formed by applying the MR-derived mask to the registered PET image (Fig 5d). The entire stack of PET "rims" was then subjected to unshaded volume-rendering in order to generate a projection view showing the <u>integral</u> of brain metabolism along rays through the rims. The projection operation was done under control of an opacity distribution, which was constructed by assigning uniformly low (10-15%) and high (100%) opacities to the mask "material" and to the "empty" space inside the rim, respectively. The 100% opacity of

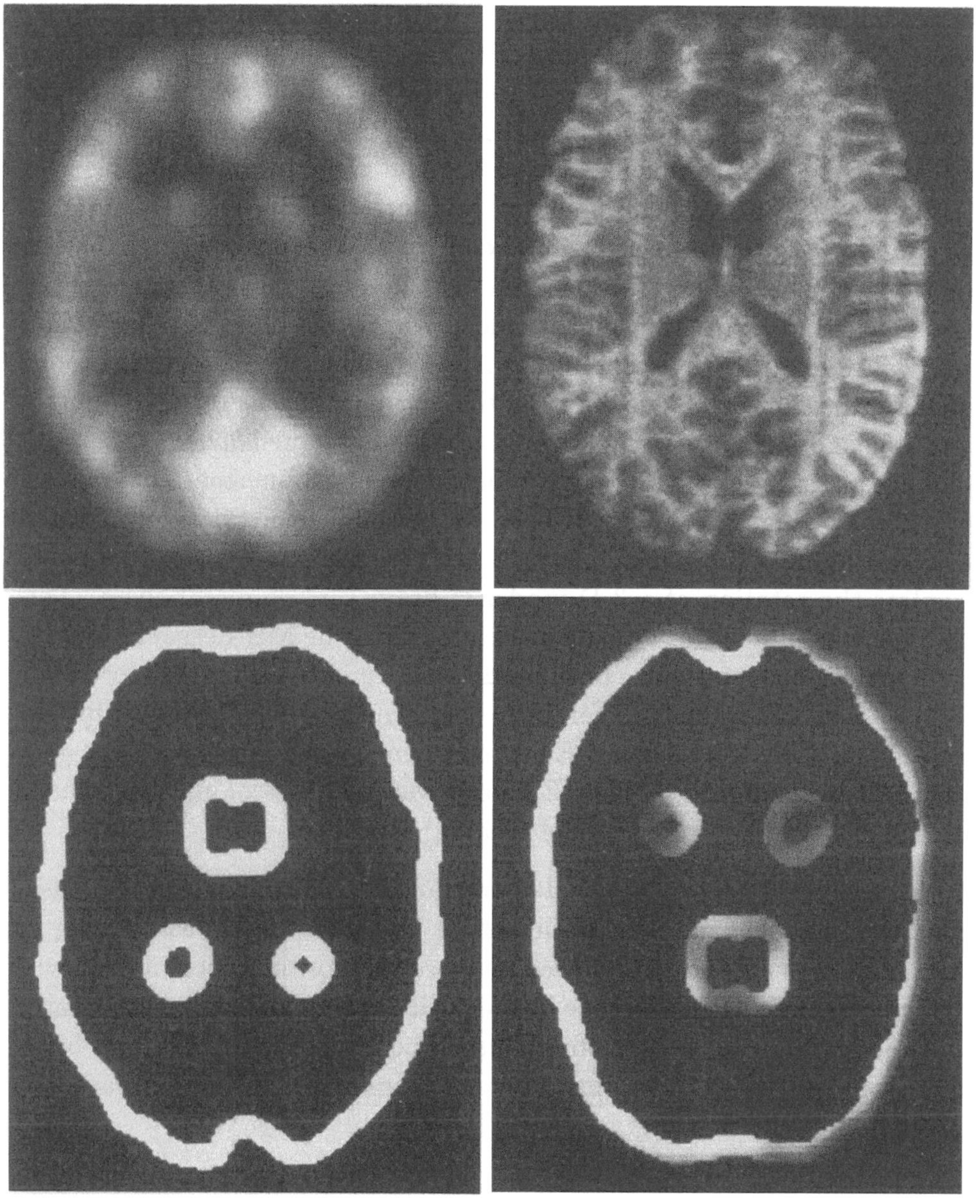

Figure 5. Steps in the MR-guided processing of the PET data, required to create a 3-D model of brain surface metabolism for the patient in Figs 2-4.
a) (top left) PET image of a thin axial section through the ventricles, obtained by interpolation and reformation of the original PET data.
b) (top right) MR image of the same section as a, formed by retrospective registration, interpolation, and reformation of MR data.
c) (bottom left) Rim-like mask of superficial and periventricular brain tissues, obtained by processing the image in b.
d) (bottom right) Metabolic activity in superficial and periventricular areas of brain.

the central zone prevented the far side of the brain from making "shine-through" contributions to the projection image. The small opacity of the rim itself served to reduce contributions of far side pixels which might "shine through" the curved ends of the rim; this also produced a slight increase in the relative contributions of the most superficial layers of the rim. A second projection image was created by volume-rendering the uniformly intense masks under control of the opacity distribution described above. The resulting view was a measure of the number of pixels contributing to the integral along each ray. The first projection view was divided by the second in order to display the average metabolic rate at each point on the near side of the brain. This was converted into a color image by means of a look-up table which assigned "cool" and "hot" colors (e.g. green and red) to hypometabolic and hypermetabolic areas, respectively (Fig 6). Thirty-two or 64 of these projections, separated by equal angles, could be "played back" in a digital "movie loop" to simulate the rotation of a 3-D model of brain surface metabolism. This PET-derived model was already registered with the MR-derived model of brain surface anatomy since both models were created from registered sets of cross-sectional images. Therefore, they could be fused by multiplying each channel of every color PET projection by the corresponding gray scale MR projection; i.e. the PET data was used to "paint" the shaded surface of the MR-derived model (Fig 7a).

We have also used morphological operations (24) in order to create masks from MR images. First, the brain surface was smoothed by subjecting an axial section to a dilation, followed by an equal erosion. Then, a mask was formed by taking the difference between this image and an image created by applying additional erosions. This procedure had the virtue of being more well-controlled than the previously-described technique based on thresholding and blurring.

Results

We have used MR data to construct 3-D computer models of the brains of over 30 patients and volunteers. In almost all cases (14,15) it was possible to identify the pre-central, post-central, left inferior frontal, and left superior temporal gyri which are usually associated with movement, sensation, speech, and hearing. Superficial lesions were visible as distor-

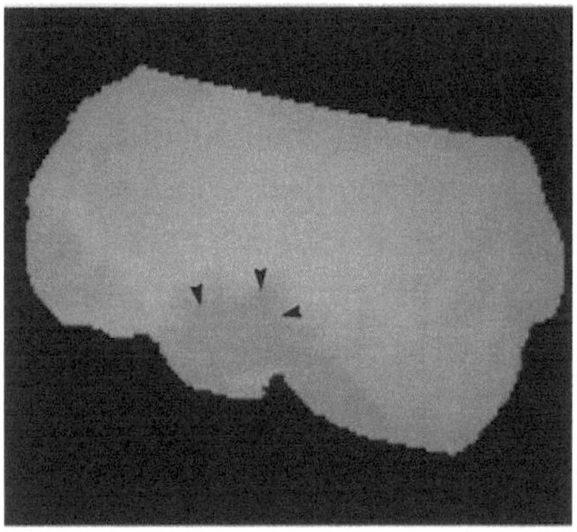

Figure 6. Image of the average metabolic rate at the brain surface, obtained by normalized projection of rim-like PET images such as Fig 5d. Note the hypometabolic area (green patch marked by arrowheads) in the left temporal lobe.

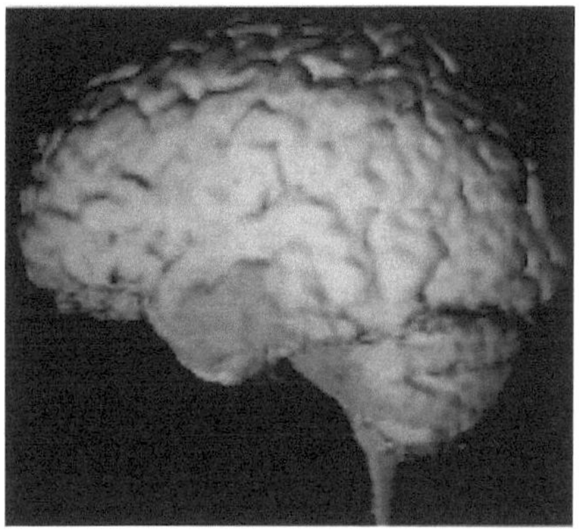

Figure 7. (a) Integrated 3-D view of brain anatomy and metabolism, obtained by fusing the MR and PET models in Figs 3 and 6. It is apparent that the superior temporal gyrus is spared by the hypometabolic lesion.

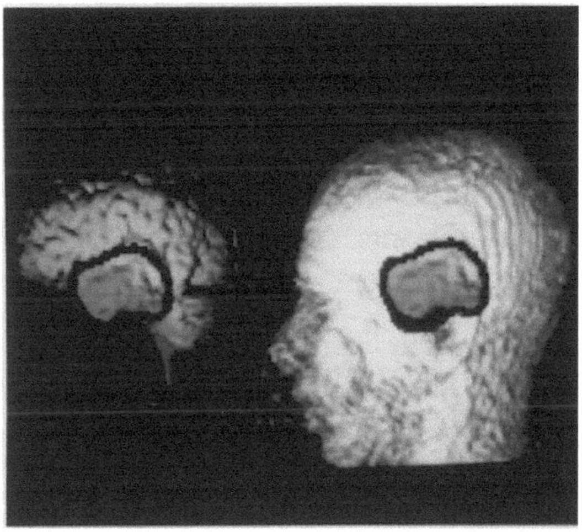

Figure 8. (a) Preoperative simulation of craniotomy to be performed on the patient in Figs 2-7. A mouse-controlled cursor was used to outline the patch of brain surface to be exposed. Then, the computer performed an instantaneous "craniotomy" on the model of the patient's skin surface.

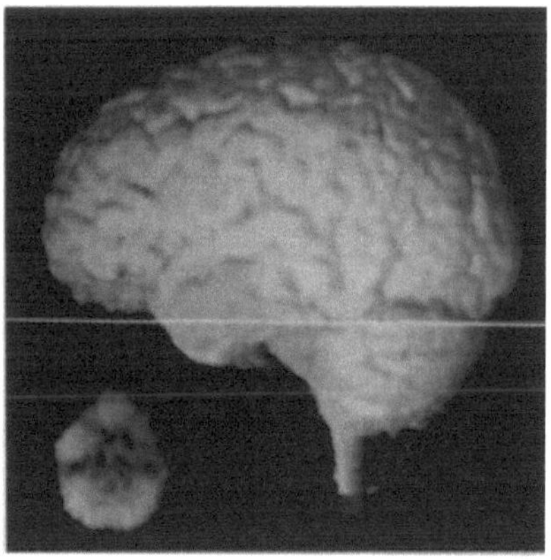

Figure 8. (b) "Plane-picker" program for simultaneous inspection of 3-D and cross-sectional images. As a mouse-driven line cursor (left panel) was moved through the integrated 3-D model, cross-sectional PET (and/or MR) images of the selected section were updated at the bottom of the screen.

tions of the brain surface. As described in detail elsewhere (15), these 3-D models were used to predict surgical anatomy in 16 cases, and in every case there was qualitative agreement between these predictions and intraoperative observations. Some 3-D models suggested that cortical lesions were flatter or more concave than was actually the case at surgery. This artifact was due to the fact that these lesions had abnormally low signal, which was between the signal levels of brain and CSF. These voxels were assigned low brain content and high CSF content by the look-up table for tissue characterization. Since such tissues were rendered as low intensity and low opacity materials, they were nearly invisible in the final 3-D views.

Studies of phantoms and human subjects (17) have shown that MR and PET images can be retrospectively registered to within a few millimeters. The accuracy of this procedure is limited by the relatively poor resolution of the PET images. We also encountered other sources of registration error but were able to circumvent them. For example, some cortical lesions produced different distortions of the coarse MR and PET surfaces which were matched during the derivation of the interscan coordinate transformation. Typically, this occurred when a region of severely hypometabolic brain was absent on the PET scans but had normal anatomy in the MR images. In some cases we excluded these areas from the MR and PET surfaces to be matched so that the computer did not skew the interscan coordinate transformation in a futile attempt to fit the discordant shapes to one another. In other cases, this problem was avoided by basing the registration on a fit between skin surfaces, derived from unedited MR images and PET transmission images.

All cortical metabolic abnormalities, which were identified on cross-sectional PET images, were clearly depicted by the integrated 3-D models of MR and PET data. However, some spurious metabolic "lesions" were also evident on the surface of the combined models. These were most frequently seen on the cerebellum, brain stem and inferior surface of the brain (Fig 7a). Many of these artifacts were due to image registration errors which caused the MR-derived masks to cover varying amounts of brain and bone in the PET images. Other artifacts were related to interpolation and reformatting of the original thick-section PET data which were affected by partial volume averaging. Fortunately, these false positive "lesions" were easily identified and rejected by noting their absence

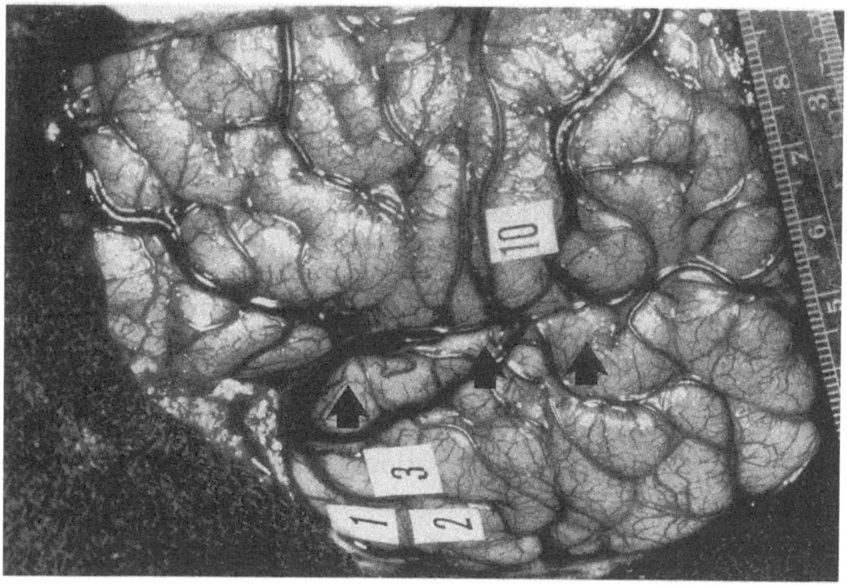

Figure 7. (b) Intraoperative photograph of the brain of the patient in Figs 2-6. Electrocorticography showed a seizure focus (labelled 1,2,3), which spared the superior temporal gyrus (arrows) as predicted by 3-D reconstruction.

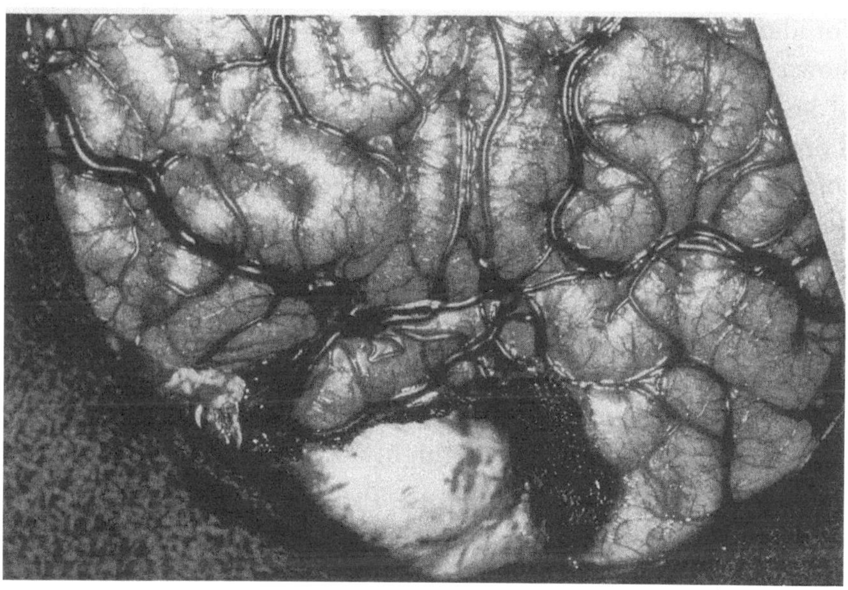

Figure 7. (c) Photograph of the brain, immediately after resection of the seizure focus in the inferior left temporal lobe. The patient has had an uncomplicated seizure-free post-operative course.

on the original PET cross-sections. A potential source of additional artifacts was the spurious distortion of the MR-derived masks by cortical lesions. Since the masks were derived by thresholding MR images, they could be artifactually flat or concave in cortical regions with abnormally low signal. In principle, such a mask could lead to a 3-D PET model which depicted the average metabolic rate in the subcortical tissues at that location; in practice we did not observe this phenomenon.

Other reports (21) describe our clinical experience with integrated 3-D models of the brain. In ten patients with brain tumors or epilepsy, 3-D models of MR and PET data were used for preoperative localization of metabolic abnormalities with respect to the gyral anatomy of the brain. In each case, intraoperative observations (visual inspection, electrocorticography, and pathological examination) were in qualitative agreement with predictions based on 3-D reconstruction (Figs 7b,c).

Discussion

MR-derived 3-D models of brain anatomy provide the only non-invasive means of identifying specific gyral convolutions on the brain surface. We have shown (15) that such models can help neurosurgeons to plan resections of lesions which are near the motor and sensory strips and other critical areas. In some cases, it is also helpful to display the depth and shape of a lesion by rendering it as an opaque colored object within a translucent brain (15,19). The technology described in this paper makes it possible to use such anatomical models to localize abnormalities seen on poorly-resolved PET scans. The interpretation of the PET data is enhanced by mapping it onto the MR model which serves as a "roadmap" or individualized atlas of the patient's brain. This is particularly helpful for planning surgical treatment of medically-intractable epileptics. In these cases, the MR and PET images are highly complementary, showing normal anatomy and abnormal function, respectively. Furthermore, surgery simulation software can be used to design an incision which will result in optimal exposure of anatomic and metabolic features of the underlying brain surface (Fig 8a).

3-D models should be used to enhance the interpretation of cross-sectional gray scale images, not to replace them (25). We have found that it is optimal to interact with a single-screen display of 2-D and 3-D images from both modalities (Fig 8b). MR and PET cross-sections of internal anatomy and metabolism are easier to interpret when their location is displayed with respect to a multimodality 3-D model of the brain surface.

The multimodality 3-D models suffer from certain limitations which should be recognized. For example, they may be degraded by false positive "lesions" induced by registration and interpolation errors stemming from the poor resolution of PET data. The use of color to encode PET data is also problematical since the appearance of the color display can be sensitive to the choice of a color look-up table. For these reasons, the 3-D views should only be used to localize metabolic abnormalities which are clearly identified by careful analysis of PET cross-sections. We are designing a 3-D/2-D "pointer" to facilitate such correlations between the two types of images. As one mouse-driven cursor is moved across the brain surface on a 3-D (or cross-sectional) image, a second cursor will automatically traverse the corresponding path on a cross-sectional (or 3-D) image on the same screen.

As described in the previous section, the MR-derived models of anatomy may also be artifactually concave or flattened at sites of cortical lesions. This was due to the fact that our "single-channel" tissue characterization algorithm confused abnormally dark cortex with a mixture of CSF and normal brain. We can remedy this by using a multispectral tissue characterization scheme (19, 26-28), which can distinguish between three or more tissues (e.g. normal brain, CSF, abnormal brain).

We are now extending this work in several directions. A 3-D workstation is being designed so that all of the above-mentioned display capabilities will be available to physicians in the operating room and other clinical settings. Physicians will be able to interact simultaneously with the data and with the patient instead of merely inspecting hard copies of 3-D views as is now the case. For example, it is not difficult to imagine how such a workstation could be used in conjunction with an interactive stereotaxic system (29-30). We have also developed preliminary programs for using MR "angiography" data to build 3-D models of the isolated intracranial vasculature (20) which can then be embedded in a multimodality 3-D model. Neurosurgical planning should benefit from a comprehensive 3-D display which is capable of showing the brain surface, brain metabolism, lesion anatomy, the vascular tree, and the skin surface.

Acknowledgments

This work was supported in part by Siemens Medical Systems, Inc., the Cancer Research Foundation, the Brain Research Foundation, and the Technology Commercialization Corporation.

References

1. Herman GT, Liu HK. Three dimensional display of human organs from computed tomograms. Proc Comp Graph Im 9:21 (1979).
2. Drebin RA. Volume rendering. Comp Graph 22:65-74 (1988).
3. Levoy M. Display of surfaces from volume data. IEEE Comput Graph Applic 8:29-37 (1988).
4. Demonstration of Voxel Flinger 3-D imaging system, commercial exhibit at the annual meeting of the Radiol. Soc. of N. America, November, 1989 (Reality Imaging Corporation, Solon, Ohio).
5. Pizer SM, Fuchs H, Levoy M, et al. 3D display with minimal predefinition. In Lemke HU, Rhodes ML, Jaffe CC, and Felix R (eds): Computer Assisted Radiology. Berlin: Springer-Verlag, 1989, pp 723-736.
6. Cline HE, Dumoulin CL, Hart HR, Lorenson WE, Ludke S. 3-D reconstruction of the brain from magnetic resonance images using a connectivity algorithm. Mag Res Imag 5:345-352 (1987).
7. Chuang KS, Udupa JK, Raya SP. High-quality rendition of discrete three-dimensional surfaces. Technical report MIPG 130. Department of Radiology, University of Pennsylvania, July 1988.
8. Drebin RA, Magid D, Robertson DD, and Fishman EK. Fidelity of three-dimensional CT imaging for detecting fracture gaps. J Comput Assist Tomogr 13:487-489 (1989).
9. Vannier MW, Marsh JL, Warren JO. Three-dimensional CT reconstruction images for craniofacial surgical planning and evaluation. Radiology 150:179-184 (1984).
10. Fishman EK, Drebin B, Magid D, et al. Volumetric rendering techniques: applications for three dimensional imaging of the hip. Radiol 163:737-738 (1987).
11. Kennedy DN, Nelson AC. Three dimensional display from cross-sectional tomographic images: an application to magnetic resonance imaging. IEEE Trans Med Imag MI-6:134-140 (1987).
12. Hoehne KH, Riemer M, Tiede U, et al. Volume rendering of 3D-tomographic imagery. In de Graaf CN, Viergever MA (eds): Information Processing in Medical Imaging. New York: Plenum Press, 1988, pp. 403-412.
13. Stimac GK, Sundsten JW, Prothero JS, et al. Three dimensional contour surfacing of the skull, face, and brain from CT and MR images and from anatomic sections. AJR 151:807-810 (1988).
14. Levin DN, Hu X, Tan KK, et al. Surface of the brain: three-dimensional MR images created with volume-rendering. Radiol 171:277-280 (1989).

15. Hu X, Tan KK, Levin DN, et al. Three-dimensional magnetic resonance images of the brain: application to neurosurgical planning. J Neurosurg 72:433-440 (1990).
16. Pelizzari CA, Chen GTY. Registration of multiple diagnostic imaging scans using surface fitting. In Bruinvis I, van der Giessen P, van Kleffens H, et al (eds): The Use of Computers in Radiation Therapy. Amsterdam: Elsevier, 1987, pp. 437-440.
17. Pelizzari CA, Chen GTY, Spelbring DR, et al. Accurate three-dimensional registration of CT, PET, and/or MR images of the brain. J Comput Assist Tomogr 13:20-26 (1989).
18. Levin DN, Pelizzari CA, Chen GTY, Chen C-T, Cooper MD. Retrospective geometric correlation of MR, CT and PET images. Radiol 169:817-823 (1988).
19. Levin DN, Hu X, Tan KK, et al. Integrated 3-D display of MR, CT, and PET images of the brain. Proc Nat Comput Graph Assoc 1:179-186 (1989).
20. Hu X, Tan KK, Levin DN, et al. Volumetric rendering of multimodality, multivariable medical imaging data. In Upson C (ed): Proc of Chapel Hill Workshop on Volume Visualization. Chapel Hill, NC: U of N Carolina, 1989, pp. 45-49.
21. Levin DN, Hu X, Tan KK, et al. Integrated three-dimensional display of MR and PET images of the brain. Radiol 172:783-789 (1989).
22. Hoehne KH and Bernstein R. Shading 3D images from CT using gray-level gradients. IEEE Trans Med Imag MI-5:45-47 (1986).
23. Ter-Pogossian MM, Ficke DC, Hood JT, Yamamoto M, Mullani NA. PETT VI: a positron emission tomograph utilizing cesium fluoride scintillation detectors. J Comput Assist Tomogr 6:125-133 (1982).
24. Serra J. Image analysis and mathematical morphology. New York: Academic Press, 1982, pp. 43, 50.
25. Hoehne KH, Delapaz RL, Bernstein R, Taylor RC. Combined surface display and reformatting for the three-dimensional analysis of tomographic data. Invest Radiol 22:658-663 (1987).
26. Vannier MW, Butterfield RL, Jordan D, et al. Multispectral analysis of magnetic resonance images. Radiol 154:221-224 (1985).
27. Koenig HA, Laub G. Tissue discrimination in magnetic resonance 3D data sets. In Schneider RH, Dwyer SJ (eds): Medical Imaging II: Image Formation, Detection, Processing, and Interpretation. Bellingham, Washington: International Society for Optical Engineering, Vol. 914, 1989, pp 669-672.
28. Cline HE, Lorensen WE, Jolesz F, and Kikinis R. 3D segmentation of brain tissue using connectivity. Mag Res Imag 8 (suppl 1):155 (1990).
29. Mosges R, Schlondorff G. A new imaging method for intraoperative therapy control in skull-base surgery. Neurosurg Rev 11:245-247 (1988).
30. Vannier MW, Gayou DE, Drebin RA. Applications in diagnosis and reconstruction methods for generic serial sections. Proc Nat Comput Graph Assoc 1:215-226 (1989).

Computer Assisted Radiation Therapy Planning

W. Schlegel

German Cancer Research Center, Institute for Radiology and Pathophysiology, Im Neuenheimer Feld 280, D-6900 Heidelberg, FRG

Abstract

The recent advances in 3D medical imaging are the basis to develop a new generation of computer assisted tools for radiotherapy planning. The precise extraction of therapy relevant information from 3D images, 3D display of anatomical structures, more accurate placement of beams and the calculation and display of dose disributions in three dimensions are expected to contribute to enhanced local control of tumours and to help to reduce side effects. The paper reports the current state of the art in radiotherapy planning and gives recommendations for future research.

Keywords

3-D medical imaging, treatment design, treatment simulation, radiation therapy, computer assisted treatment planning

1.Introduction

The goal of radiotherapy is to eradicate a tumour without causing severe damage to healthy tissue. Computer assisted radiation treatment planning (RTP) provides a tool to find the optimum configuration of radiation beams for the treatment of an individual patient.

In clinical practice, radiation treatment planning traditionally has been based on two-dimensional sectional representations of the treatment of the three-dimensional patient. However, the shape of the tumor and adjacent uninvolved anatomic structures can vary substantially from one section to another. The sectional 2D-planning approach may lead to inaccuracies in localizing internal organs as well as tumor and target volume which still can be considered as one of the largest source of errors in the whole radiation treatment procedure.

NATO ASI Series, Vol. F 60
3D Imaging in Medicine
Edited by K. H. Höhne et al.

Today, the situation is changing due to several technical developments, among which the advent of whole body X-ray computed tomography (CT) and of magnetic resonance imaging (MRI) are the most important ones. With the availability of CT and MR imaging devices, the ability to visualize anatomical structures in detail has been tremendously enhanced.

In order to take full advantage of these new imaging developments within the field of radiation therapy, new tools have to be developed and evaluated which e.g. permit the easy and precise extraction of the therapy relevant information from 3D images, 3D display of anatomical structures, accurate placement of beams, the display and evaluation of rival treatment plans and the display of dose distributions in three dimensions [Suit,1988].

The radiotherapy process can be separated into different independent steps, which are shown in Fig. 1. In order to reach more safe and effective treatment techniques, computer support of all steps of radiation therapy is indispensible. In the following chapters, the way how radiation therapy planning is carried out today will be described, current problems will be analyzed recommendations for future research and development work within the field of computer assisted radiation therapy will be given.

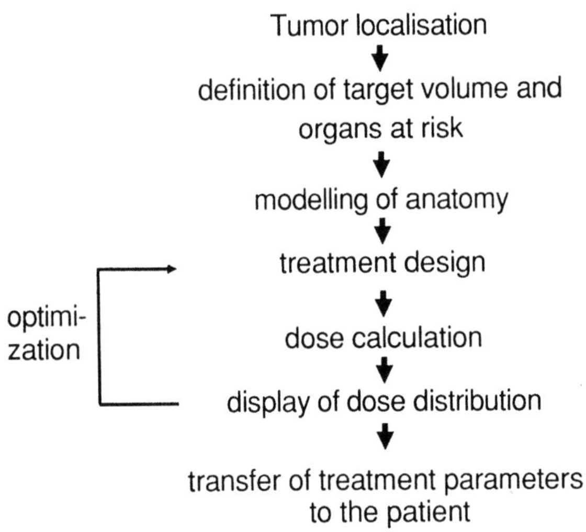

Fig. 1: The steps involved in radiotherapy planning

2. The Chain of Radiotherapy

2.1 Tumour localization, definition of target volume and organs at risk

The radiotherapeutic planning process starts with the detection of the location, form and extension of the tumour. After the identification of tumour tissue, the radiotherapist has to define the target volume, that means the volume which has to be completely covered by the therapeutic dose. The problematic nature of target definition is due to the fact, that tumour margins and safety margins have to be detected depending on histology and surrounding anatomy (esp. organs at risk). There are targets with well defined margins, blurred margins or even invisible margins (e.g. after subtotal or total tumour resection).

Tumour tissue, which is not included into the target volume, will be underdosed with high probability; organs at risk which are not included during this process will probably not be considered in the evaluation of treatment plans and eventually be overexposed. Errors, unaccuracies and neglections which are made during this first steps dramatically influence the effectiveness of treatment planning and the outcome of the whole radiotherapeutic process. These errors are irreversible and can not be corrected for after the patient is irradiated.

Current Problems. Therapeutic imaging requirements may be different from diagnostic requirements. While diagnostics concentrates to the perception and classification of tumor tissue, it is extremely important for radiotherapy to know the precise geometrical properties of the shape, extension and location of the tumor volume under therapeutic conditions.

Most of the current systems do not support the therapy relevant steps of tumor localization. Methods of digital image manipulation, image processing, image segmentation, locating, viewing and comparing of many images asocciated with a patient are not being supplied (or insufficient).

Recommendations for future research. Methods of digital tissue characterization are needed and have to be included in order to assist during the phase of determination of the tumour staging and grading [Schad, 1989].

The physiological and/or biochemical basis of MRI suggests that it will be more useful than CT for displaying tumor volumes for certain radiotherapy patients, most notably in the central nervous system [Bradley, 1984]. However, there may be a severe lack of geometrical accuracy of MRI [Schad, 1987b]. This lack of accuracy is counter to the most important need

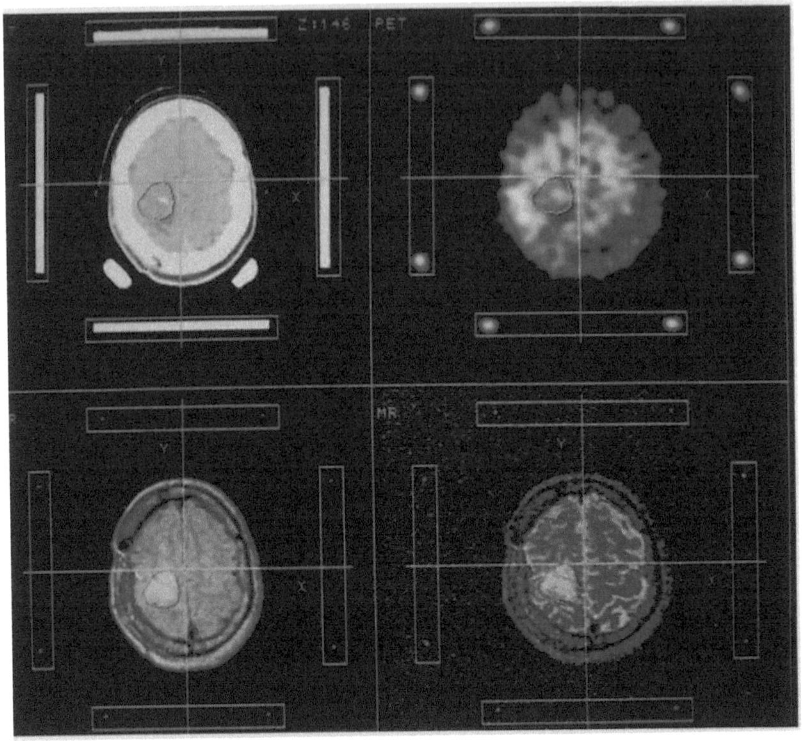

Fig. 2: 3D stereotactic image correlation study showing corresponding sections (upper left: CT, upper right: PET, lower left: T1-weighted MRI, lower right: T2-weighted MRI) of a patient with a brain tumour [Schad, 1987a]

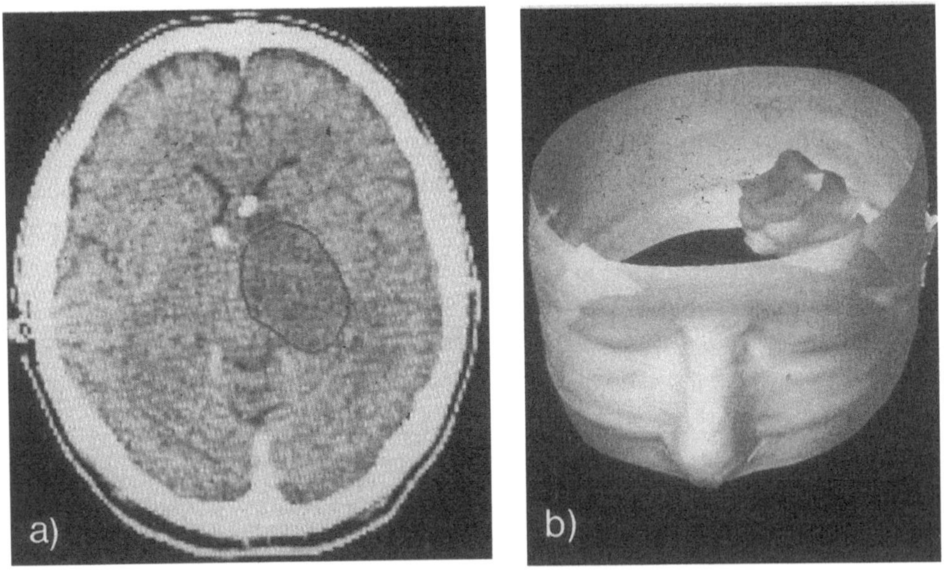

Fig.3: Manual segmentation of a brain tumor in a CT-image (Fig. 3a) and anatomic modelling of the target volume and the head surface (Fig. 3b).

of precision radiotherapy planning. Because electron density information is not available from MRI is a second, very major impediment to the usefulness of MRI for RTP (when it is used alone). Since the use of CT planning avoids the latter two limitations, then the only quantitative way to make use of the exceptional imaging capabilities of MRI is to integrate MRI with CT-assisted RTP [Fraas, 1987, Schad, 1987a]. Therefore, 3D-image correlation methods have to be developed, which allow to combine MRI-, CT-, PET- and DSA-imaging within the process of tumour localization (Fig. 2).

2.2 Modelling of anatomy

Tissues and organs which will be irradiated have to be included into the treatment planning process and have to be modelled in 3D. Such volume surfaces are the body surface, inhomogeneous anatomical structures (bone, lung, air cavities), the target volume (including the tumor volume, the tumor, adjacent or neighbouring structures of possible tumor infiltration or metastasation, and the tumor dependent safety region), organs at risk.

The commonly used method for extracting 3-dimensional information is manual segmentation of CT-images (Fig.3).

problems. The number of images comprising a 3D-image may vary from twenty to sixty. As many as ten or perhaps even more structures may need to be outlined on each of these images. This is an enourmous effort to be carried out for each patient. There are no tools available to make this process easier and faster.

Recommendation for Future Research. In order to shorten the time consuming process of anatomical modelling, as automated as possible 3D image segmentation and modelling tools have to be developed .

2.3 Treatment Design

Once the anatomical structures of interest have been delineated and modelled in 3D, the best possible directions from which to aim radiation beams at the target volume must be selected. To concentrate radiation dose at the target volume while sparing other regions, multiple beams of radiation directed from various angles at the target and permitting the best possible separation between the target volume and critical structures have to be used. The so-called "Beam's Eye View" technique [Goitein, 1983] provides an effective computer graphic tool for selecting optimum directions, seizes and shapes of beams by projecting 3D anatomical information into a coordinate system which is defined according to the geometry of the beam. The result is the representation of contours outlining the patients anatomic structures perspectively in three dimensions as if being viewed from the source of radiation along the central ray of the beam.

Problems. Because of their 2D nature, most of the available RTP systems do not supply this possibility of connecting patient geometry and beam geometry. The few systems which have already 3D capability are far away from having sufficient interactiveness in order to allow real time optimization of beam geometry.

Recommendations for Future Research. When 3D image data of the patient exists, this part of therapy simulation could be performed to a large extent within the computer. With the help of 3D computer graphics, it is possible to match the shapes of the beams to the target volume from every incidence angle, to ensure that critical organs will not be touched by primary beams over the whole length of the irradiation fields.

Especially for this phase of therapy planning, the importance of an appropriate user interface can not be overemphasized. Some special requirements should be considered:

In order to realize the potential of the new possibilities that 3D computer graphics offers to the radiotherapy simulation process, the best way is for the software to implement a superset of the functions used in conventional practice while retaining the ambiance of the traditional methods. For the process of treatment design, a system has to be created, which faithfully reproduces the function and the feel of a physical simulator. In this way a system can be constructed which an experienced radiotherapist can use with essentially no retraining [Sherouse, 1987].

2.4 Dose calculation

After the definition of the treatment techniques, the irradiation dose within the patient has to be calculated due to the physical laws of interaction of radiation with matter. Numerous algorithms have been developed in the past, which allow to calculate the irradiation dose within the patients body, but traditionally only in single sections. Prerequisite for these calculations is the knowledge of the physical beam characteristics of the irradiation units, the accurate patients geometry and the distribution of electron densities within the human body. The individual patient related parameters (geometry and electron densities) can be extracted with high accuracy from CT-data only.

Current problems. Most dose calculation methods are either inadequate (causing dose calculation errors up to 10-15% in dose) or much to slow for practical use [Schlegel, 1985, Schlegel, 1989].
Recommendations for future research. More adequate models for dose calculation in radiotherapy have to be developed. Especially divergence of the primary beam components and variations of the scatter components due to inhomogeneities of the human body, due to irregular shaped fields and due to compensators have to be considered with much higher accuracy. The speed of dose calculations has to be increased dramatically e.g. by using parallel processing techniques [Werner, 1987].

2.5 Evaluation of treatment plans

In order to confirm that a certain treatment technique will lead to the desired result (complete dose coverage of the target volume, sparing of organs at risk), dose calculation results have to be displayed and evaluated.
Current problems. Also for an experienced radiation oncologist it can be sometimes difficult to select the best of several competing treatment plans, even if consideration is given only to the dose distribution in one or a few planes. The problem becomes far more difficult if the entire spatial distribution of the radiation dosage is to be considered in judging the merits of a given plan. In this case the radiation oncologist can be overwhelmed by the amount of data to assimilate if the analysis must be done on a slice by slice basis.
Current system rarely have the ability to reconstruct CT data and display isodoses in sagittal, coronal or arbitrarily obliqued planes (as e.g. shown in Fig. 4). Such techniques can be very valuable in understanding 3 dimensional dose distributions as can techniques such as dose volume histogram analysis. Furthermore, the lack of suitable methods to simultaneously display 3D dose distributions superimposed on the relevant anatomy has probably been a major factor in delaying the routine use of 3D therapy treatment planning.
Recommendations for future research. The problem of 3D display of target volumes, organs at risk, neighbouring anatomical structures together with the 3D dose distribution has to be solved according to the needs of radiotherapy.

The uncertainties of the treatment planning (e.g. caused by dose calculation uncertainty) and of the treatment itself (caused by the inability of

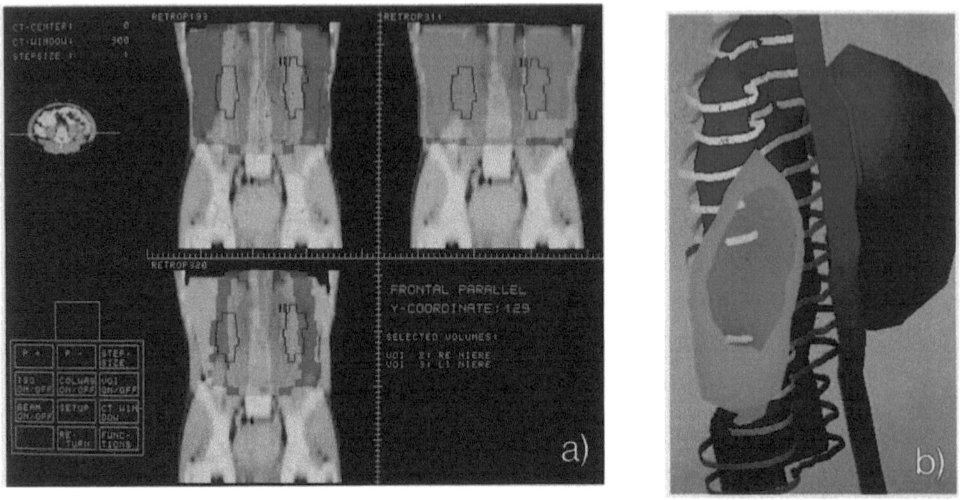

Fig.4a: Display of a medical workstation for evaluation of 3D-treatment plans. Fig. 4a shows Coronal reconstructions of 3 rival plans for a patient with a target volume including lymphatic metastases. The kindeys have been modelled as organs at risk. The overlayed coloured areas represent the radiation dose (blue= 20%-40%, light blue=40%-60%, green=60%-80%, pink=80%-100%). Upper left corner: treatment plan for a biaxial moving beam technique, upper right corner: treatment plan for a fixed beam technique, lower left corner: treatment plan for a conformation technique using a multileaf collimator.

Fig.4b shows the 3D shaded surface representation of the treatment plan with the conformation irradiation technique. Blue= spinal cord, red=target volume, green= kidneys, white ribbons= 80% isodose region. A clipping plane has been selected in order to show the interior surface of the left kidney. A penetration of the left kidney by the 80% isodose can be recognized.

Fig.5: 3D shaded surface representation of a treatment plan for a brain tumour. Red= target volume, blue=80% isodose level, green = 50% isodose level. A partial underdosage of the target volume can be recognized.

reproduction of patient position or by patient movement during the treatment) have to be estimated and visualized.

Quantitative evaluation of dose distributions by calculating mimimum and maximum dose levels, integral doses or dose volume histograms in the target volume or organs at risk are the minimum demand [Becker, 1989]. Objective functions have to be defined in order to describe the quality of a treatment plan.

Knowledge based systems are needed to assist in the evaluation, e.g. concerning doses tolerated by certain healthy tissue structures, e.g. mapping of morphology and radiotolerance to individual 3D patient images.

2.6 Optimization

During a treatment planning session, the definition of the treatment technique, dose calculation and evaluation of treatment plans have to be repeated, until an acceptable plan has been found.

problems. The optimization of treatment plans is currently performed by an interactive "try and error" method. Cumbersone interaction and extremely long dose calculation times often prevent the user to perform many iteration steps and may lead to an inadequate plan.

Recommendations for future research. In order to reach the optimum treatment for an individual patient, it is essential, that the time for such an iteration loop will be as short as possible. The need for short calculation times (with the aim of real-time response in dose calculation) urgently requests the use of parallel processing in radiotherapy planning. Geometrical pre-optimization definition of objective functions and fast dose calculation have to be combined to a self-optimizing treatment planning system, which only needs minor interactive corrections by the radiotherapist.

2.7 Transfer of treatment parameters to the patient

After treatment planning, it is of great importance that the parameters which have been defined during the treatment planning phase can be accurately reproduced while the patient is actually irradiated.

Current problems. Current systems scarcely support the radiotherapist as far as this transfer of treatment planning parameters to the patient is concerned. Radiotherapists commonly use anatomical landmarks within the images as a geometrical reference, which often cannot be accurately repro

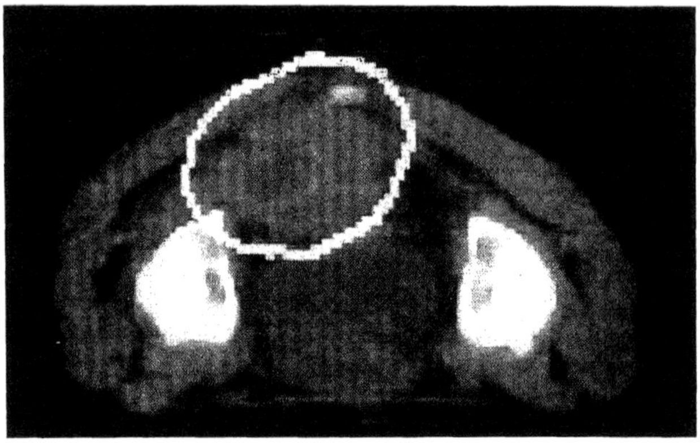

Fig.6: CT-section of a patient with a target volume in the pelvic region (manual segmentation of the target volume).

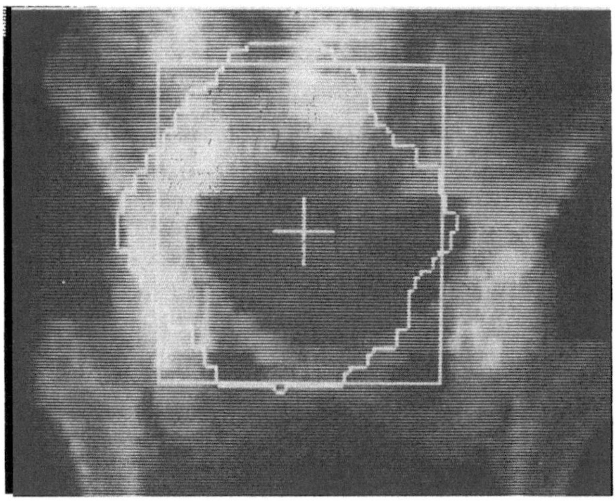

Fig.7: A fully divergent radiograph of the patient with the pelvic tumor reconstructed from CT-data (irregular white margins = target volume, white rectangle = geometric border of the radiation beam). In order to check the positioning of the precalculated radiation beam, this image can be directly compared with the simulator X-ray film.

duced and therefore in daily routine cause errors up to several cm [Reinstein, 1987, Blanco, 1987].

Recommendations For Future Research. Following the technique of stereotactic localization and positioning used in neurosurgery, image correlated 3D coordinate systems have to be defined, which can be accurately reproduced, either on the base of artificial or favourably anatomical landmarks.

Correlation of digital reconstructed radiographs (either from CT or MRI) with X-ray simulation radiographs and treatment verification films could be used for the solution of this problem (as e.g. shown in Fig. 5).
A completely unsolved problem is the detection and correction of movement of the patient during treatment. Computer vision and computer controlled treatment techniques should be investigated and applied.

3. Conclusions

Within the scientific community, it has been recognized that the most important developments in clinical medicine in the next decade will probably be in diagnostic radiology. Future techniques will make it possible to see better deep in our body. This will improve the feasibility of exactly delimiting the tumour and also improve the feasibility of achieving local control. For many types of tumours local control will remain important and the improved imaging may increase the possibility of using modern precision therapy to achieve local control of the tumour [Einhorn, 1987].

The fact, that the integration of the already existing superior imaging methods into clinical practice of radiotherapy is very slow may be due to the problems, that the involved computer assisted methods are very laborious and time consuming up to now and in many points don't meet the real practical requirements of radiation therapy. For future research and development within the field it is essential, that clinicians, particularly radiotherapists, perceive the methods of computer assisted radiation therapy as helpful tools and not as obstacles. I hope that this paper may be a small contribution to direct future research towards this goal.

4. References

Becker, G., Lohrum, R., Werner, T., Bürkelbach, J., Németh, G, Boesecke, R., Schlegel, W., Lorenz, W.J.: Presenation and evaluation of 3D dose distributions in radiotherapy planning. In: Computer assisted Radiology, CAR'89.(H.U.Lemke, ed.),pp. 254-261, Berlin-Heidelberg-New-York: Springer 1989

Blanco, S., López-Bote, M.A. and Desco, M.: Quality assurance in radiation therapy: Systematic evaluation of errors during the radiation execution, Radioth. Oncol. 8 256-261, (1987)

Bradley, W.G., Waluch, V., Yadley, R.A., Wycoff, R.R.: Comaprison of CT and MR in 400 patients with suspected disease of the brain and cervical spinal cord. Radiology 152, 695-702,(1984)

Einhorn, J: Recent Developments in Oncology and Prospects for the Future. In: Cancer Research and Treatment Today: Results, Trends and Frontiers (Lapis, Eckhardt, eds.), pp. 125-138, Basel, Karger: 1987

Fraas, B.A. and McShan, D.L.: 3-D Treatment Planning: Overview of a clinical planning system. In: The use of computers in radiation therapy (Bruinvis, I.A.D., ed.) pp. 273-276, Amsterdam: North Holland 1987

Goitein, M., Abrams, M., Multi Dimensional Treatment Planning : I. Delineation of Anatomy. International J. of Radiat. Oncol. Biol. Phys. 9, 777-787, (1983)

Reinstein, L.E. and Meek, A.G.: workshop on Geometric Accuracy an Reproducibility in Radiation Therapy. Int. J. Radiation Oncol. Biol. Phys. 13, 809-810, (1987) .

Schad, L., Boesecke, R., Schlegel, W., Hartmann, G., Sturm, V., Strauß, L. and Lorenz, W.: Three dimensional image correlation of CT-, MR- and PET studies in radiotherapy treatment planning of brain tumours.J of Comp. Ass. Tomography 11, 948-954, (1987a)

Schad, L., Lott, S., Schmitt, F.,m Sturm, V. and Lorenz, W.J.: Correction of spatial distortion in MR imaging: A prerequisite for accurate stereotaxy. J of Comp. Ass. Tomography 11, 499-505, (1987b) .

Schad, L., Brix, G., Zuna, I., Härle, W., Lorenz, W.J. and Semmler, W.: Multiexponential proton spin-spin relaxation in MR imaging of human brain tumours. J of Comp. Ass. Tomography 13 , 577-587, (1989)

Schlegel, W., Sturm, V.: Computer assisted Radiotherapy, Computer Assisted Stereotaxy. In: Tutorial Notes of CAR'89 (H.U.Lemke, ed.),pp. 99-110, Berlin: AMK 1989

Schlegel, W.: The Use of Computers in Radiotherapy Treatment Planning. In: Tutorial Notes of CAR 85. (H.U.Lemke, ed.),pp. 47-68, Berlin: AMK 1985.

Sherouse, G.W., Mosher, C.E.: User interface issues in radiotherapy CAD software. In: The use of computers in radiation therapy (Bruinvis, I.A.D., ed.), pp. 429-436, Amsterdam: North Holland 1987

Suit, HD, Becht, J, Leong, J, Stracher, M, Wood, W, Verhey, L, and Goitein, M: Potential for improvement in radiation therapy. Int. J. Radiation Oncol. Biol. Phys. 14 777-785, (1988)

Werner, T., Herrmann, G., Schlegel, W., Lorenz, W.J.: Parallel processing in radiotherapy treatment planning. In: The use of computers in radiation therapy (Bruinvis, I.A.D., ed.), pp. 21-24, Amsterdam: North Holland 1987

CAS – a Navigation Support for Surgery

Ludwig Adams[1], *Joachim M. Gilsbach*[2], *Werner Krybus*[1], *Dietrich Meyer-Ebrecht*[1], *Ralph Mösges*[3] and *Georg Schlöndorff*[3]

[1]Lehrstuhl für Meßtechnik, RWTH Aachen
Templergraben 55, D-5100 Aachen, FRG
[2]Neurochirurgische Klinik, RWTH Aachen
Pauwelstraße, D-5100 Aachen, FRG
[3]Klinik für Hals-, Nasen- und Ohrenheilkunde und Plastische Kopf- und Halschirurgie
RWTH Aachen, Pauwelstraße, D-5100 Aachen, FRG

Abstract

Computer Assisted Surgery (CAS) is a new navigation support for skull base surgeons. The combination of 3D coordinate measurement techniques, voxel processing methods and pseudo-3D image presentation supports preoperative planning of therapy, pathfinding during the operation itself and postoperative therapy control. For this purpose, the surgeon employs a hand-guided electro-mechanical 3D coordinate digitizer to locate points of interest within the operative field. The coordinates measured this way are correlated with a voxel model of the object gained by a preceding CT examination. With a prototype system the accuracy of this method has proven to be better than ±1 mm. The system has been successfully applied in more than 100 ENT operations and ten neurosurgical procedures. A similiar system was tested two times in the field of radiotherapy for the computer assisted placement of afterloading probes.

1. Scenario

In the time before the introduction of instrument navigation, an aircraft pilot had to give up his mission whenever weather conditions worsened to a degree where visual orientation did not allow for secure landing at his destination airfield. This is the situation of today's surgery. Even modern imaging techniques such as X-ray or magnetic resonance computer

NATO ASI Series, Vol. F 60
3D Imaging in Medicine
Edited by K. H. Höhne et al.

tomography do not offer anything else than the possibility of preoperative "map briefing". Now, CAS – Computer Assisted Surgery – shall introduce instrument navigation into the surgeon's daily practice by providing three-dimensional position measurement techniques for the operation theatre. The surgeon's map will be a pseudo-three-dimensional presentation of the patient taken down from CT images. A measuring system capable of determining an instrument's position with respect to the patient will be the navigation guide. This system will also serve as a "flight-simulator" for realistic preoperative planning.

Two major advances can be achieved by CAS: Before starting the operation, the surgeon can "walk" through the 3D image of the proposed operation area thus gaining a realistic impression of the region and its neighbourhood. During the operation the surgeon can yield position information by simply pointing at an unidentified structure with the tip of an electro-mechanical 3D coordinate digitizer. The respective CT sectional views are then displayed on the CRT in three perpendicular planes with an indication of the position of the coordinate digitizer. By means of the application of this orientation aid, safety and quality of the operation will be increased, time consuming procedures can be avoided and novel procedures become possible in surgery.

2. State-of-the-Art

In the course of operations at the skull base or in the region of the ears and the paranasal sinuses, intra-operative orientation can be extremely difficult due to individual variations of the patients' anatomy. The surgeon visually controls the position of the surgical instrument. Before and during the operation, several diagnostic imaging techniques are at his disposal for his orientation: X-ray films or CT image sequences, which have been prepared by preceding investigations, are displayed on lightboxes. However, because of lacking correlation between those images and his visual impression, even the experienced surgeon is sometimes not sure of the exact position of his instrument. In exeptional cases, intra-operative X-ray screening is applied. Radiation load and two-dimensionality, however, restrict application and benefit of this method. A continuous and reliable information concerning the position of the surgical instrument with respect to the anatomical structures is not possible this way. The same applies for neurosurgical operations of deep seated introcerebral lesions, for which no landmarks are available.

Up to now computers have been used for intra-operative navigation support only in neurosurgery. There, stereotactic operations are done with the help of a localisation frame and an instrument holder [Schad 1987]. In the course of the operation, the position of a pathological brain process related to the fixed frame is presented on CT images which were prepared by means of preceding CT examinations. Then, an instrument holder is firmly connected to the frame. By moving the instrument along the straight path of the fixed support, without further X-ray examinations, a sample of tissue is taken at the desired position, a local lesion is set or a radiating substance is implanted. It is the neurosurgeon's task to perform the linear displacement of the instrument along the path calculated by a computer. Disadvantages of the stereotactic method are the necessity to put the patient with a frame into the CT-scanner and the restricted handling of the instruments in the fixed frame during the operation.

In order to improve the stereotactic procedure, to position and to advance an instrument, devices have been developed which automatize the linear displacement of the instrument by the application of robots. In the first place, the experiment of Yik San Kwoh [Kwoh 1987] has to be mentioned. More recent developments in the field of stereotactic surgery also come from German neurosurgeons. In this connection, the group from Munich around Struppler and Lipinski [Lipinski 1987] is to mention. Their system permits the well-directed elimination of a defined area of grey matter. However, continuous operating, as it is necessary at the skull base, is not possible in this way since manual inputs and intra-operative electro-physiological measurements require interruptions of the operational process. In the course of these examinations, the patient's skull must remain fixed.

Stereotactic destroyment of brain lesions is also the subject of the group around Sturm at the DKFZ in Heidelberg, F.R.G. A traditional robot, as proposed by Doll [Doll 1987], has not been applied so far for practical intra-operative purposes. Neither this approach permits extensive manipulation as it is necessary when operating the paranasal sinuses or removing sealed-off tumours from the skull base. A new method presented by Roberts [Roberts 1986] does without the rigid linkage between patient and instrument. Its practical use, however, is limited on account of its low accuracy (deviations of up to 5 mm).

All systems applied so far suffer from the following shortcomings:

- they only work punctually,
- the point of interest is approached along a linear path,
- there is a rigid linkage between the localisation frame fixed to the skull and the displacement of the surgical instrument,
- the displacement of the instrument is neither feeded back nor visually presented in the context of the operative field.

3. A New Approach

The development of the CAS system at the Aachen University of Technology [Mösges 1988] aimed at predominantly supporting ENT surgery. For this purpose, the CAS methods had to satisfy the following requirements:

- the object to be operated and its environment shall be continuously presented on a display system,
- the position of the surgical instrument shall simultaneously be faded into the display presentation of the operative field,
- a frame attached to the skull shall be avoided and the surgeon shall be permitted to change the position of the skull during the operation.

The above requirements were fulfilled by combining 3D presentation methods with 3D coordinate measurement techniques. The surgeon uses a stylus which is fixed to an electro-mechanical 3D coordinate digitizer to tip onto points within the operative field of which he needs to know the position. Before the operation, a 3D data model of the object has to be prepared based on a complete set of computer tomograms. The measured coordinates are then correlated with the computer model to display three perpendicular sectional views of the object with faded-in position markers.

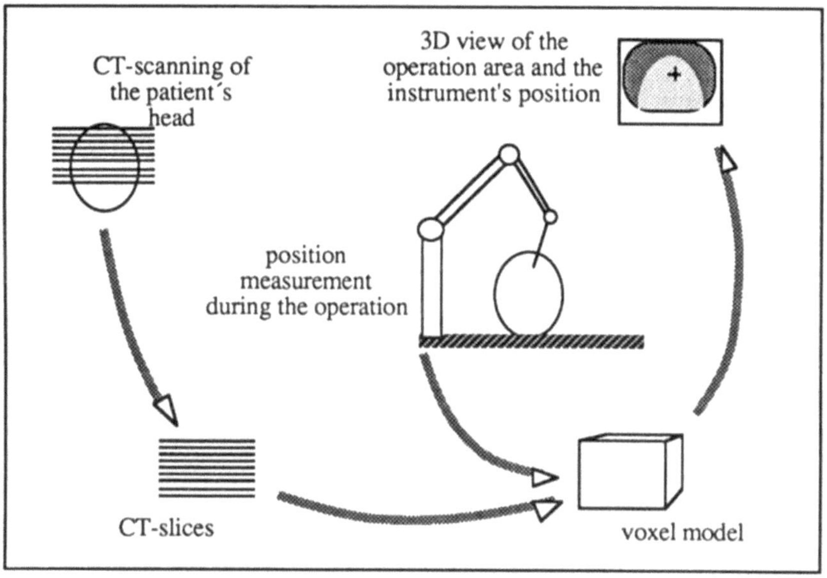

Fig. 1: Principle of Computer Assisted Surgery

Fig. 1 schematically shows the individual steps which are necessary for the application of the developed system. At first a CT investigation is performed (usual parameters in the ENT surgery: 50-70 sections, 2 mm slice thickness, 0.8-1.2 mm pixel size). In order to correlate the coordinate system of the image with that of the object, a sufficient number of reference points has to be defined. The reference points must be visible on the CT images and likewise be accessible during the operation. For this purpose radiopaque markings are attached to the patient's skull. They are afterwards replaced by colour markings. At the beginning of the operation as well as after every displacement of the skull during the operation, a calibration of the system has to be carried out. This is achieved by tipping the markings with the stylus. Their coordinates are identified with the coordinates of the markings in the CT images in order to establish the correlation between the position of the patient's skull and its voxel model. After the calibration procedure, the position of the stylus' tip in the operative field and its presentation on the screen are visually tied together, i.e. the display image dynamically moves to the corresponding three perpendicular sectional views of the voxel model.

4. About the Implementation

In the following we describe the steps which are necessary to fade in continuously the position of the 3D coordinate digitizer into the display presentation of the object. The coordinates given by the coordinate digitizer refer to the coordinate system within the operation theatre ("world coordinates"). The relation to the voxel model of the patient's skull in the computer is established by a transformation equation on account of which the coordinates given by the coordinate digitizer are transformed into coordinates within the voxel model. The equation given below allows all possible rotations, translations and scalings between the two coordinate systems.

$$\begin{pmatrix} x_{ct} \\ y_{ct} \\ z_{ct} \end{pmatrix} = \begin{pmatrix} m_{11} & m_{12} & m_{13} \\ m_{21} & m_{22} & m_{23} \\ m_{31} & m_{32} & m_{33} \end{pmatrix} * \begin{pmatrix} x_w \\ y_w \\ z_w \end{pmatrix} + \begin{pmatrix} x_0 \\ y_0 \\ z_0 \end{pmatrix}$$

The 12 factors of the above equation can be determined by correlating four points to both the coordinate systems. This is performed by the calibration procedure. Before the operation, the coordinates of the four reference points have to be determined within the coordinate system of the voxel model. Then, the coordinates of the four reference points at the patient's head have to be measured. This happens by tipping the four coloured points with the coordinate digitizer's stylus. By means of these two coordinate data sets, the parameters of the coordinate transformation can be calculated. During the operation, all coordinates given by the coordinate digitizer are transformed into the coordinate system of the voxel model. After this transformation the coordinates of the digitizer's stylus have to be superimposed on the CT images of the operative field in such a way that the surgeon can draw rapid and reliable conclusions concerning the three-dimensional situation on account of the representation of the object on a two-dimensional display.

5. Image Presentation

For the presentation of three-dimensional objects, there are two basically different approaches [Udupa 1986]. In the case of the first approach, the boundaries of anatomical structures (organs, bones, blood vessels etc.) are determined on the given two-dimensional CT images. Then, the so gained structures are translated into a 3D model of the object's surface which can be presented as if viewed from any desired point of view. The three-dimensional impression is conveyed to the observer on account of an appropriate shading [Oswald 1985, Herman 1987, Pizer 1986]. The disadvantage of this method is the necessity of defining the boundaries of the objects. The process of segmentation leads to a reduction of the informative content of the CT images to the surfaces of organs, bones etc.

In the case of the second group of methods, the complete information content of the primary data is maintained. As a basis for the presentation, the entire set of 3D data is employed rather than only the surfaces of objects [Höhne 1987]. Each pixel of a CT image is understood as a volume element (voxel) of a cube of voxels which represents the object space. Visual representations of the interior of this space can be presented by appropriate sectional views across the volume. Within these sectional views, also the original CT sectional views may appear.

After several experiments and thorough discussion concerning the best understanding of the display presentation, we decided to present the operative field by three perpendicular sectional views through the operative volume (Fig. 2). This type of presenting a 3D object on a 2D medium is well known from mechanical design. Incidentally the surgeon will also get the accustomed presentation of CT images this way. In addition to the usual *axial* CT presentation he also gets now the presentation of the corresponding perpendicular layers (*sagittal* and *frontal* display). An interpolation between given CT-sections is not necessary because the size of the voxels (1 x 1 x 2 mm) is small enough for a smooth image representation. The intersection of the three sectional views corresponds with the position of the coordinate digitizer within the operative field and is marked by a cross-hair cursor on each sectional view.

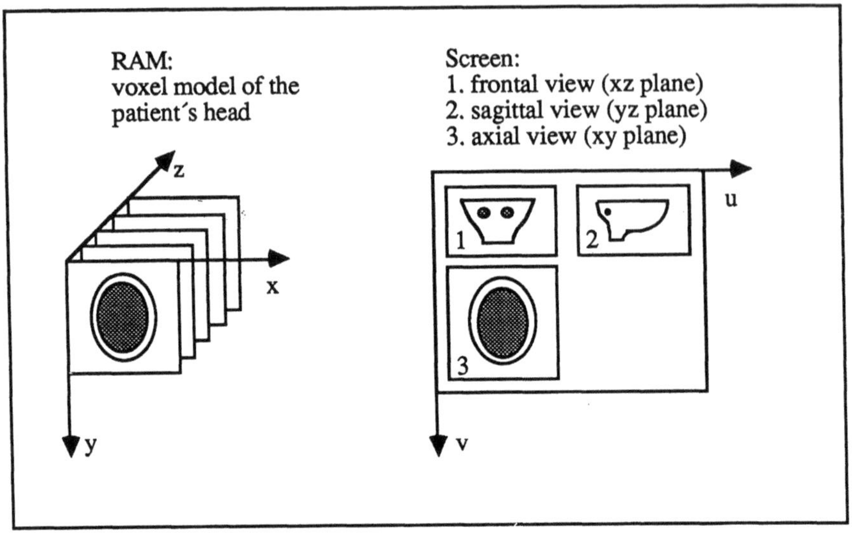

Fig. 2: Relation between volume model and display image

The following equation describes the relation between the image of the display and the sectional views within the voxel model:

$$\begin{pmatrix} x \\ y \\ z \end{pmatrix} = \begin{pmatrix} x_0 & \frac{\partial x}{\partial u} & \frac{\partial x}{\partial v} \\ y_0 & \frac{\partial y}{\partial u} & \frac{\partial y}{\partial v} \\ z_0 & \frac{\partial z}{\partial u} & \frac{\partial z}{\partial v} \end{pmatrix} * \begin{pmatrix} 1 \\ u \\ v \end{pmatrix}$$

By means of this transformation equation, sectional views through the voxel model can be calculated under any desired angle. This is necessary if structures of objects are not distinctly presented on account of the given three perpendicular sectional views.

Fig. 3 gives an impression about the image presentation of the present prototype system. The three-dimensional understanding is conveyed to the oberserver on account of the fact that the sectional views can be rapidly moved forwards and backwards along any desired axis. The possibility to present parts of the volume with an appropriate shading offers an additional aid for depth orientation. Fig. 3 shows the combination of a pre-calculated 3D image and a 2D cross section which is constructed on the fly. The generation of the required 3D images takes about 30 seconds because we restricted the representation to the two perpendicular viewing directions without perspective.

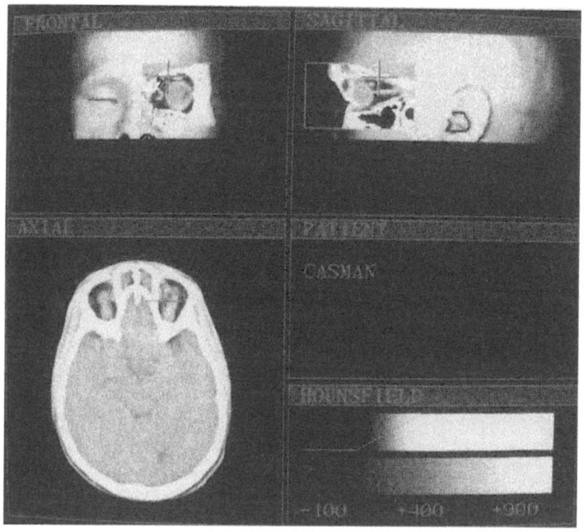

Fig. 3: CAS display representation

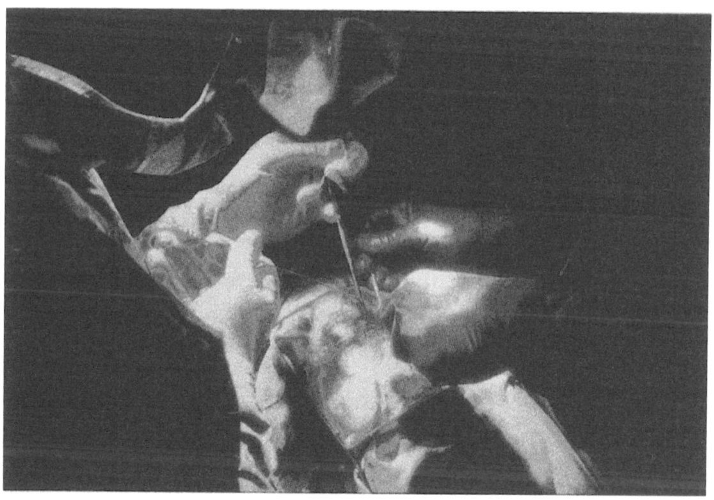

Fig. 4: 3D position measurement during an ENT operation

6. 3D Coordinate Measurement

Although it seems cumbersome to employ a mechanical device for the coordinate measurement before and in the course of an operation, at a first attempt a mechanical approach was followed rather than a more sophisticated and evidently less reliable non-mechanical approach. Electro-mechanical 3D coordinate digitizers have been developed for the application in mechanical engineering. Their principle of construction looks similar to an industrial multi-joint robot arm. The individual arm segments of the device are connected among each other by means of rotary joints. The angles of each of the rotary joints are measured by means of rotary encoders so that the coordinates of the tip of a stylus mounted to the arm can be calculated.

Based on this principle appropriate measuring arms were designed and manufactured by Ph. Moll, *Institut für Bekleidungsmaschinen* and *A. Mennicken*, Aachen. They were equiped with six optical increment encoders for shaft angle measurement. The pulse signals of the rotary encoders are evaluated by 16-bit counters. A dedicated 68008-microcomputer calculates the position of the stylus from the angle measurements and the arm lengths. The coordinates are transmitted to the host computer via an RS-232 link. The accuracy of this device proved to be better than ±.5 mm in a measuring volume of .3 m^3.

7. Computer Hardware

Fig. 5 schematically shows the configuration of the presently employed computer system which is based on the VMEbus. The CPU modul is an *Eltec* Eurocom-5 equiped with MC 68020 processor and MC 68681 coprocessor. The most important task of the CPU is the generation of the voxel model and the transformation of the coordinates of the 3D coordinate digitizer. By means of an 8" floppy drive (compatible with *DEC* RX01), CT-data can directly be read in (so far CT image data are provided by the *Siemens* CT systems “Somatom-DR” or “Somatom-2”). Two hard disks are used for mass storage. The voxel model is generated within a 16Mbyte RAM. Its capacity is sufficient for the storage of 128 CT images. A 1024 by 768 pixel display with 8 bit per pixel plus a 2 bit graphic overlay is refreshed non-interlaced at up to 70 Hz by a memory-mapped display memory. As usual the CT images are displayed in gray shades whereas grahic and alphanumeric information – such as the coordinate digitizer's position or pull-down menus – is superimposed in arbitrary colours via a colour look-up table.

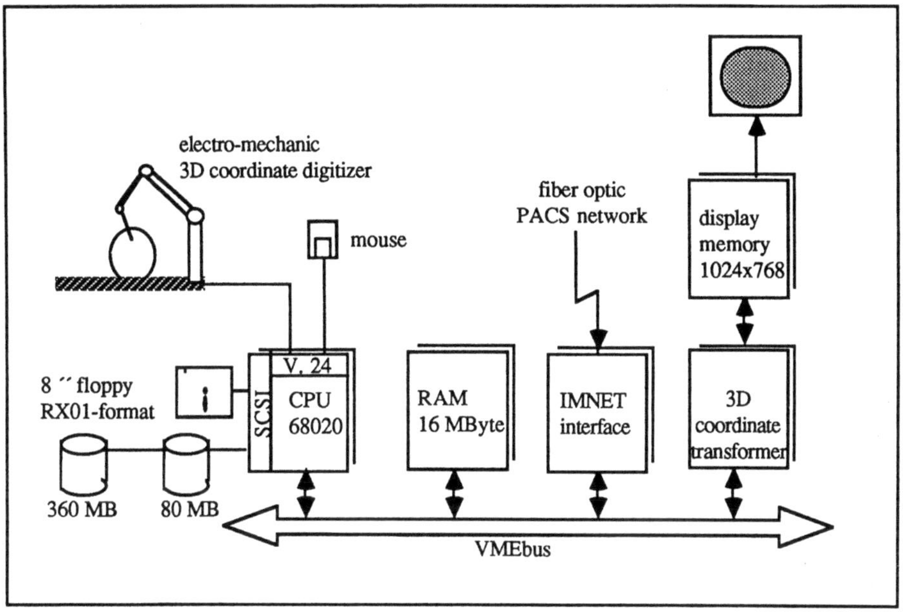

Fig. 5: Hardware of the CAS system

For the generation of the three perpendicular sectional views, a special hardware, the *3D coordinate transformer,* was developed which relieves the CPU from this task. It manages the complete transformation of the voxel model into the three sectional views. The CPU calculates a parameter set from the coordinates of the stylus' tip which is given to the coordinate transformer. This data set includes all parameters necessary for the presentation. For every pixel of the destination image the transformer then calculates the corresponding coordinates within the voxel model. Afterwards the voxel is read out of the RAM into the coordinate transformer. An internal look-up table, which compresses each pixel from

16 to 8 bit, offers a dynamic change of contrast and brightness of the presented sectional views. The destination image pixel is written into the display memory without any further bus access. The coordinate transformer is equipped with three "image resampling sequencers" TMC 2301 of *TRW*. It has two complete VMEbus interfaces so that it is compatible with the VMEbus to both sides. The CPU may also access the display memory directly. In this case the transformer is switched to a transparent mode. The transformer only needs 90ms for the generation of a 512^2 image independent of the specific transform coefficients (angle of the sectional views, zoom factors etc.). For the complete process (measuring, transformation, image construction), a repetition rate of 8 per second was achieved.

8. Experience and Outlook

Up to now, we have operated 100 ENT patients and ten neurosurgical patients with the help of the CAS prototype (Fig. 4). The operations have related to injuries, inflammations and tumours in the region of the paranasal sinuses, the skull base, the orbit, the fossa pterygopalatina and tumours of the centronervous system. The CAS application has proved succsessful particularly in the following domains of application:

- surgery of the paranasal sinuses,
- surgery of the orbit,
- osteosynthesis of facial fractures,
- identification and extraction of bone fragments and
- tumour surgery.

It turned out that already the ameliorated planning of the operation essentially increased its safety. Each CAS application confronted the surgeon with particular problems. So, its continuing application led to experiences which were helpful for the technical improvement of the system. Recently, the CAS equipment has also be successfully applied to "afterloading" which is a special process in radiotherapy. CAS helped to achieve an accurate positioning of probes of radiating matter. Another CAS application in radiotherapy could be to reproduce and supervise the positioning of the patient. There the accurate positioning is of particular importance when irradiating in the head-neck-region. It is also planned to spread the use of CAS to further areas of surgery. One will be endoscopical surgery, i.e. the determination of the position of an endoscope's tip or an endoscopical instrument. ■

9. Acknowledgement

The CAS project is financially supported by the *Landesminister für Wissenschaft und Forschung von NRW* under grant no. IV.B7-400.098.86, by the *Bundesminister für Forschung und Technologie* under grant no. 01KN88012 and by *Schlumberger Stichting Fund for Science, Technology and Research.* The authors are solely responsible for the content of this article. – The authors hope that the results of their work will not be used for military purpose. To the best of their knowledge they will not participate in any military application.

10. References

Schad, L. et al., "Magnetic Resonance Imaging in the Stereotactic Treatment Planning of Brain Tumours," *Proc. CAR '87*, H.U. Lemke et al (eds.), Springer 1987, p. 343

Kwoh, Y. S., Hou, J., Jonckheere, E. and Hayati, S., "An Improved Absolute Positioning Accuracy Robot for Stereotactic Brain Surgery," *Proc. CAR '87*, H.U. Lemke et al (eds.), Springer 1987, p. 899

Lipinski, H. G., Bitk, P., Struppler, A., "Computerized Stereotactic Neurosurgery," *Proc. CAR '87*, H.U. Lemke et al (eds.), Springer 1987, p. 348

Doll, J., et al., "The Use of an Industrial Robot as a Stereotactic Guidance System,"*Proc. CAR '87*, H.U. Lemke et al (eds.), Springer 1987, p. 374

Roberts, D. W., et al., "A frameless stereotactic integration of computerized tomographic imaging and the operation microscope," *J. Neurosurgery*, Vol. 65, October 1986

Mösges, R., Schlöndorff, G., "An new imaging method for intraoperative therapy control in skull-base surgery," *Neurosurgery Review*, Vol. 11, Berlin/New York 1988, pp. 245-247

Udupa, J. K., "Display of 3D information in discrete 3D scenes produced by computerized tomography," *Proc. IEEE*, Vol. 71, No. 3, pp. 420-431

Oswald, H., "A medical workstation for three-dimensional display of computed tomogram images," *Proc.CAR '85*, H. U. Lemke et al (eds.), Springer 1985

Herman, G., Udupa, J. K., "Display of three-dimensional discrete surfaces," Proc. *Soc. of Photo-Optical Instrumentation*, Vol. 283, pp. 90-97

Pizer, S. M. et al., "3D shaded graphics in radiotherapy and diagnostic imaging," *Proc. Computer Graphics 1986*, National Computer Graphics Association, III (1986), p. 107

Höhne, K. H., Riemer, M., Tiede, U., "Viewing Operations for 3D Tomographic Data," *Proc. CAR '87*, H.U. Lemke et al (eds.), Springer 1987

THREE-DIMENSIONAL IMAGING: CLINICAL APPLICATIONS IN ORTHOPEDICS

Elliot K. Fishman, Derek R. Ney, and Donna Magid

The Russell H. Morgan Department of Radiology and Radiological Science, The Johns Hopkins Medical Institutions
Baltimore, MD 21205-2180, USA

ABSTRACT

Advances in three-dimensional imaging now provide the radiologist and referring physician with the ability to create accurate three-dimensional models of any part of the human body. The major area of clinical applications of three-dimensional imaging has been in orthopedics. This paper presents a review of the basic applications of 3D imaging in such areas as trauma, oncology, degenerative arthritis, and congenital disease. The techniques used for the generation of three-dimensional images is presented as well as a detailed description of some of the basic clinical applications of three-dimensional imaging. A preview of some of the more advanced and newer applications of three-dimensional imaging in regard to orthopedics is also reviewed.

Keywords: three-dimensional imaging, volumetric rendering, skeletal trauma

INTRODUCTION

Regardless of the computer hardware used, the reconstruction algorithm applied, or the friendlessness of the user interface, the value of three-dimensional imaging can only be judged by its use in the clinical arena. The development of 3D imaging must mirror the needs of the radiologist and the non-radiologist alike in providing solutions for defined problems. The construction of high resolution 3D images would be little more than an exercise in futility if their use did not meet any specific need.

Orthopedic Radiology and Orthopedic Surgery have recently enjoyed a rapid curve of change due to the introduction of new imaging technologies such as CT and MRI. Prior to the introduction of these cross-sectional imaging

NATO ASI Series, Vol. F 60
3D Imaging in Medicine
Edited by K. H. Höhne et al.

techniques, plain films, tomography and arthrography were the mainstay of patient evaluation. Although adequate in may conditions, standard radiography proved limited in many areas due to the limitations of the display form. Overlapping structures, the inability to visualize soft tissues, limitations due to difficulty in patient positioning and patient compliance are all source of the well known limitations of plain radiographs.

Computed tomography provided the ability to directly evaluate both the skeletal, muscle and soft tissue components of the body. The cross-sectional imaging overcame many of the limitations of standard radiographs and quickly became the study of choice in the evaluation of most musculoskeletal pathology. Yet, standard transaxial CT has specific limits, especially to the non-radiologist. The understanding of a series of closely-spaced CT slices is a challenge to the person untrained in cross-sectional imaging. The orthopedic surgeon understands his anatomy in three-dimensions. The hip, shoulder or spine are not represented well by a series of planar scans. Rather, in the surgeon's mind the ideal image should represent a true three-dimensional presentation of data and it is this model which will aid in making critical decisions, both pre-operatively and during the surgical procedure. Three-dimensional imaging provides a template for the surgeon to understand fully the extent of anatomy and optimizes care of the individual patient. This chapter will review some of the basic clinical applications of three-dimensional imaging in regard to orthopedics as well as future directions and areas which will have to be addressed in the future.

MATERIALS AND METHODS

All CT scans are done on either a Siemens DR3 or DRH scanner. The patient is positioned supine in the gantry, made comfortable as necessary with small cushions or towel rolls behind the knees, and instructed in the importance of not moving throughout the study. Paper tape is applied across the shoulders or hips, knees, and ankles to encourage compliance. Careful positioning ensures the availability of bilateral symmetry to enhance appreciation of subtle abnormalities. Routine scanning parameters are 125 kVp, 230 mAs, 3 sec, 4 mm collimation at 3-mm intervals. Most areas of interest (for example, shoulder, sacrum, hip, elbow) require 35 slices or less for an adequate volume of reconstruction.

The first step, the 2D or multiplanar reformatting, can be done on the Siemens scanner using the multiplanar reconstruction and display software (MPR/D, Siemens Medical Systems, Iselin, NY) or can be done on the freestanding PIXAR system (PIXAR, San Rafael, CA) as part of the Orthotool program (Body CT Imaging Laboratory, The Johns Hopkins Medical Institutions, Baltimore, MD). For MPR/D on the CT scanner, a selected reconstruction box of up to 127 x 127 pixels is chosen with a resistor pen. This box, superimposed over the serial slices, defines the reconstruction volume of interest. The coronal 2D image is created by interpolation of matched transaxial rows, and the sagittal from matched transaxial columns. We use a 3 pixel (approximately 2 mm) image thickness for each reconstruction. The series of 3 pixel slices in each plane is generated and stored as splash files. They can then be displayed as a series of 4:1 images, or "flipped through" rapidly in a serial display mode. If desired, to simplify orientation to the reconstructions, a reference transaxial image with superimposed axes can be displayed that will demonstrate the exact level, medial-to-lateral (for sagittals) or anterior-to-posterior (for coronals), that each displayed reconstruction represents (Fig. 1). All images are filmed at both soft-tissue (WW-421, WC-36) and bone (WW-1000), WC-150 or WW-2300, WC-220) window width and levels. Total multiplanar reconstruction time averages 6 to 8 minutes.

For volumetric 3D image generation, including the Orthotool 2D/3D option, the transaxial CT data are transferred on nine track magnetic tape to the freestanding PIXAR Image Computer (PIC), part of an imaging system consisting of the PIC interfaced with a host Sun Microsystem Workstation (Sun Microsystems, Mountain View, CA). The PIC's high-speed parallel architecture, large image memory (24 megabytes), 4 parallel bit-slice processor (for a total of 40 million instructions per second), and high-resolution (1024 x 768) color display allow it to handle the massive volume of transaxial CT data within a clinically reasonable time frame and with only minimal user interaction.

Once the CT data are transferred, data are interpolated in the Z axis to allow the X, Y, and Z dimensions of each voxel to be equal (X and Y, determined by field of view, usually do not correspond with Z, determined by interscan interval). This takes 10 seconds. The CT gray scale information in each voxel is translated into three color and one alpha (translucency) channel. The gray scale intensity information in each voxel is replaced by gels of varying assigned color and transparency. For example, muscle attenuation values are replaced with scaled shades of red, fat attenuations with yellow, and bone with whites and grays to simulate real anatomy (4). If the muscle is the primary focus of interest, it is assigned minimal alpha channel translucency; if bone is the focus, muscle is assigned high transparency so that it nearly "disappears" and does not impede visualization of underlying structures. The ability to retain the entire gray scale of intensity formation and to retain all the CT data ensures greater fidelity of the final 3D image to the original transaxial data and to the original anatomy.

A series of simulated 3D images is generated from the data. As a rule, serial images simulating side-to-side or spinal axis rotation, and head-over-heels or somersaulting axis rotation, are generated. For each 360° rotation, 72 images, each representing approximately 5° rotation from the previous image, are rendered and stored. These are displayed and animated as a video loop, creating the illusion of real-time rotation of the structure of interest in the chosen direction. A third series is often created in which extraneous structures are edited off, isolating the structure of interest for each rotation. For instance, the medial clavicles, manubrium, and sternoclavicular joints may be isolated by editing off the rest of the thoracic cage and contents; or the sacrum and sacroiliac joints may be isolated by editing off the anterior pelvic ring, hips, and lateral iliac wings. If preferred, each series can be reviewed as a series of static images, usually 4:1 format.

The Orthotool 2D/3D software runs on the same Sun/PIXAR system and uses the same interpolated CT dataset. This software allows the display of 2D and 3D images simultaneously, allowing easy cross-referral and orientation. Three 2D orthogonal planar slices (transaxial, coronal, and sagittal) are displayed simultaneously and can be roamed in real-time by the user. Reference axes indicate the level of each slice for simpler orientation (that is, the transaxial image has north-south axis line showing the level of the currently displayed sagittal image, and one east-west axis line showing the level of the displayed coronal). These axes shift in real-time as the user travels medial-to-lateral through the sagittal images and anterior-to-posterior through the coronal images. Three 3D images are displayed on the same screen (15). The Orthotool software can integrate the full-field datasets of up to 192 transaxial CT slices in near real-time.

DISCUSSION

In determining which clinical applications 3D would be useful one can apply several simple criteria and concepts;

1. The more complex the anatomy to be evaluated the more likely 3D visualization is likely to be needed.
2. The more extensive the pathology (i.e., complex fracture, congenital malformation, tumor extension) present, the greater the need for 3D visualization.
3. The more varying therapeutic options available to the clinician, the more important additional information is in arriving at the correct procedural choice.
4. The more difficult the clinical decision the more likely the 3D information will be needed.

These rules are based on our belief that the major role for 3D imaging is in clinical management decisions rather than in diagnosis. This is not to say that a 3D display will not allow one to make the correct diagnosis, but rather a conceptual notion where 3D should fit in the diagnostic scheme. The radiologist and/or clinician should carefully review the transaxial CT images/supplemented by the two-dimensional reformations into coronal, sagittal and/or oblique planes. In cases of trauma the presence or absence of a fracture and/or dislocation should easily be made. The 3D display is then reviewed to get a better understanding of the fracture and its true orientation. This information is particularly useful in the event surgical repair is attempted. However, it would be utter foolishness to only review the 3D images and disregard the original transaxial and reformatted 2D data.

With this in mind the major clinical applications of 3D imaging have been for trauma, congenital deformities, oncology and degenerative arthritis. The major anatomic areas evaluated have included the acetabulum and pelvis, shoulder, spine, knee, and ankle. Although the applications of 3D imaging will expand with more sophisticated hardware and software solutions the major current applications at present include:

Pelvic and Acetabular Fractures (Fig. 1-3)

One of the most useful applications of three-dimensional imaging has been in the evaluation of the traumatized pelvis. It has previously been shown that even in the best of circumstances, up to 70% of pelvic plain films may be suboptimal due to a number of factors including superimposed air, feces and foreign matter or suboptimal patient positioning or exposure (18). The areas of the pelvis most often obscured include the posterior pelvic ring, sacrum and SI joints and the acetabulum itself. Initial use of CT in the late 70's and early 80's found that with standard transaxial images alone there was an increased detection of fractures as well as definition of the fracture extent and localization of intraarticular fragments. The introduction of two-dimensional and three-dimensional imaging is an attempt to expand the information gained from the transaxial CT's into a more useful model that the surgeon can apply when making patient management decisions and in the actual surgical procedure. Although the actual information as to the extent of fracture is present on the transaxial images it is difficult to create a mental visualization of the more complex injuries. This is true for the radiologist, but especially true for the non-radiologist who is less trained in cross-section image interpretation. The referring physician (which in this case is the orthopedic surgeon) must be presented a set of information that can be used by him for the care of the patient. The statement of "perception without conception is empty, and conception without perception is blind" can be applied in these cases.

We previously compared the value of routine pelvic plain films with three-dimensional imaging in adequately visualizing acetabular fractures. In a series of 19 cases, we found that even if the plain x-rays were of good quality the optimal projection for defining the fracture was in more than half the cases not one of the typical views obtained in a pelvic series. This is not surprising as the number and extent of fracture possibilities and combinations is endless. It is only the ability to rotate a three-dimensional dataset around any axis for 360° can one be certain of obtaining the optimal view of that injury.

The Shoulder Joint (Fig. 4-5)

Conventional radiography of the shoulder is limited due to overlapping structures and inability to obtain true lateral views. Although many of the special views have been developed, they are often impossible to obtain in the traumatized patient. In addition, fractures in this region are very subtle and can be easily missed. This is particularly true of fractures of the glenoid labrum and scapular body. The use of three-dimensional imaging allows the easy visualization of the entire shoulder joint including the acromioclavicular joint and sternoclavicular joint.The two-dimensional images are particularly helpful in these cases in measuring degree and angle of displacement, although the three-dimensional images truly do best define the full extent of injury. The ability to edit images allows for visualization of only the scapula which is particularly helpful in detecting the subtle body of scapula fractures. In addition, the humerus can be removed from the joint allowing direct visualization of the glenoid labrum. The glenoid labrum is often associated with injury which, if missed, can lead to future shoulder dislocations.

Spine and Sacrum (Fig. 6-7)

The presence of fractures of the spine are related to the mechanism of injury. The majority of patients with spinal trauma in our institution are automobile accidents or falls. Sacral fractures are commonly associated with pelvic acetabular fractures. Involvement of the sacroiliac joint is not uncommon, particularly in ring-type fractures. 2D and 3D imaging provide the ability to look at the entire pelvis in a single view and ascertain the extent of injury in complex multidimensional fractures. Using the two-dimensional reconstruction and choosing axes along the curvature of the sacrum, one can bring into view the sacral foramina. Looking at the sacral foramina directly, allows one to ascertain the full extent of the injury, particularly in vertical shear fractures which often do involve the foramen. The two-dimensional views are also very good at looking small fragments avulused of the iliac or sacral sign of the SI joint during injury. The three-dimensional image itself provides the best understanding of the full extent of pelvic ring view. Both the somersaulting and pelvic rotations help provide an understanding of the complicated injuries.

The goal of 3D imaging in lumbar or cervical fractures is for a better visualization of the extent of injury and definition of adjacent structural involvement. In the lower lumbar spine, detection of a fracture and displacement can all be clearly defined on the 3D reconstruction. The value of the 3D images in these cases is to mainly to look at areas such as the posterior elements to determine if there is any compression present; this is optimally viewed on these images. The extent of injury of the spinal injury cord can also be addressed at the same time.

CONCLUSION

Three-dimensional imaging has matured from being an interesting technique to becoming a clinically important part of the radiologist's diagnostic armamentarium. With recent advances in both hardware and software technology, images are presently created with higher resolution and speed than was possible before. Although there are many applications which use 3D technology at present, as 3D imaging expands to other datasets (i.e., CT, SPECT) it's use will become more widespread with more clinical applications. In addition, applications involving surgical simulation, prosthetic testing and design, and radiation therapy planning will become an integral part of radiology. (Fig. 8-10)

REFERENCES

Destout, J.M., Gilula L.A., Murphy, W.A., Sagel, S.S.: Computed tomography of the sternoclavicular joint and sternum. Radiology 138:123, 1981

Deutsch, A.L., Resnick, D., Mink J.H., et al.: Computed and conventional arthrothomography of the glenohumeral joint: Normal anatomy and clinical experience. Radiology 153:603, 1984

Fishman, E.K., Magid, D., Mandelbaum, B.R., et al.: Multiplanar (MPR) imaging of the hip. RadioGraphics 6(1):7, 1986

Fishman, E.K., Magid, D., Ney, D.R., et al.: Three-dimensional imaging and display of musculoskeletal anatomy. J Comput Assist Tomogr 12(3):465, 1988

Fishman, E.K., Magid, D., Robertson, D.D. Jr., et al.: Metallic hip implants: CT with multiplanar reconstruction. Radiology 160:675, 1986

Gill, K., Bucholz, R.W.: The role of computerized tomographic scanning in the evaluation of major pelvic fractures. J Bone Joint Surg 66A(1):34, 1984

Guyer, B.H., Levinsohn, E.M., Fredrickson, B.E., et al: Computed tomography of calcaneal fractures: Anatomy, pathology, dosimetry, and clinical relevance. AJR145:911, 1985

Heger, L., Wulff, K.: Computed tomography of the calcaneus: Normal anatomy. AJR 145:123, 1985

Heger, L., Wulff, K., Seddiqi, M.S.A.: Computed tomography of calcaneal fractures. AJR 145:131, 1985

Kuhlman, J.E., Fishman, E.K., Ney, D.R., Magid, D.: Complex shoulder trauma: Three-dimensional CT imaging. Orthopedics 11(2); 1561-1563, 1988

Magid, D., Fishman, E.K.: Computed tomography of acetabular fractures. Semin US CT MR 7(4):351, 1986

Magid, D., Fishman, E.K., Brooker, A.F. Jr., et al.: Multiplanar computed tomography of acetabular fractures. J Comput Assist Tomogr 10(5):778, 1986

Magid, D., Fishman, E.K., Robertson, D.D. Jr., et al: Assessing the post-operative hip: Computed tomography with multiplanar reconstruction. Contemp Orthop 15(3):17, 1987

Naam, N.H, Brown, W.H., Hurd, R., et al: Major pelvic fractures. Arch Surg 118:610, 1983

Ney, D.R., Fishman, E.K., Magid, D., Kuhlman, J.E.: Interactive real-time multiplanar CT imaging. Radiology 170:275-276, 1989

Pennal, G.F., Tile, M., Waddell, J.P., Garside, H.: Pelvic disruption: Assessment and classification. Clin Orthop 151:12, 1980

Phillips, W.A., Schwartz, H.S., Keller, C.S., et al.: A prospective, randomized study of the management of severe ankle fractures. J Bone Joint Surg 67A(1):67, 1985

Scott, W.W. Jr., Fishman, E.K., Magid, D.: Acetabular fractures: Optimal imaging. Radiology 165:537, 1987

Seltzer, S.E., Weissman, B.N.: CT findings in normal and dislocating shoulders. J Can Assoc Radiol 36:41, 1985

Smith, R.W., Staple, T.W.: Computerized tomography (CT) scanning technique for the hindfoot. Clin Orthop 177:34, 1983

Tile, M.: Fractures of the acetabulum. Orthop Clin North Am 11(3);481, 1980

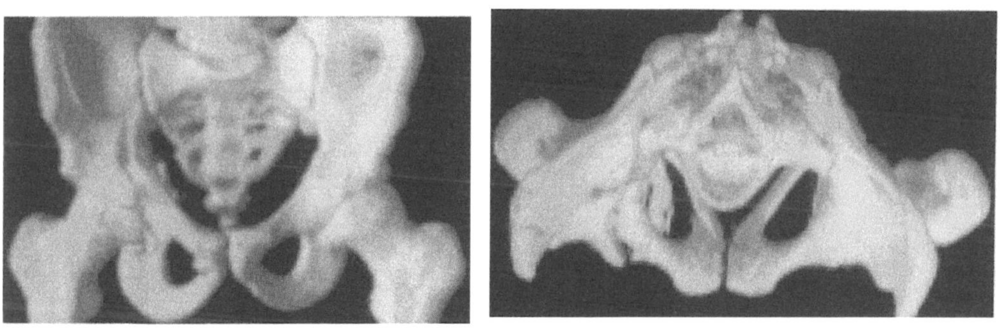

Fig. 1 (a-b)

Complex right acetabular fracture extending into iliac wing with displacement of medial acetabular wall clearly seen.

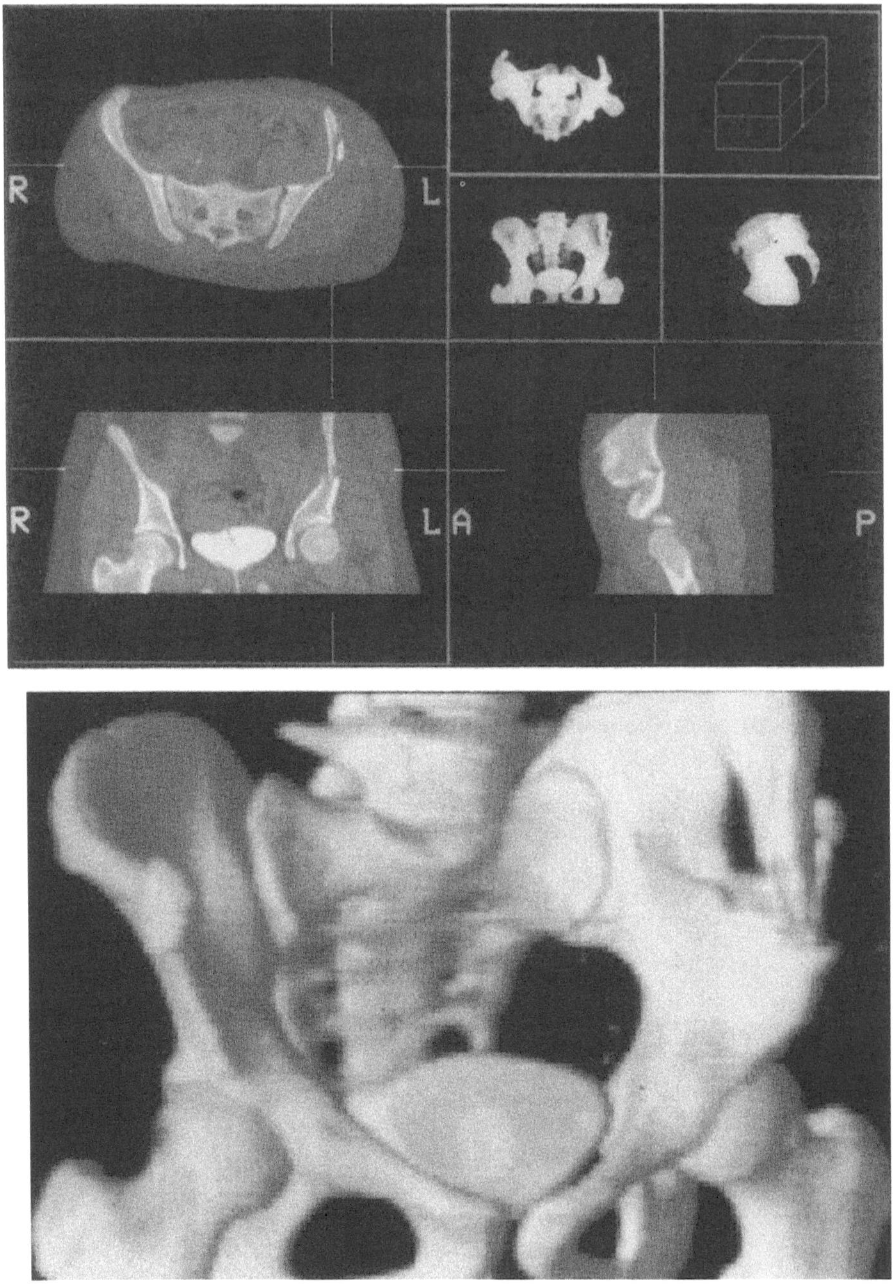

Fig. 2 (a-b)

Fracture of both iliac wings demonstrated. Contrast in bladder noted as 3D study was part of trauma examination of the abdomen.

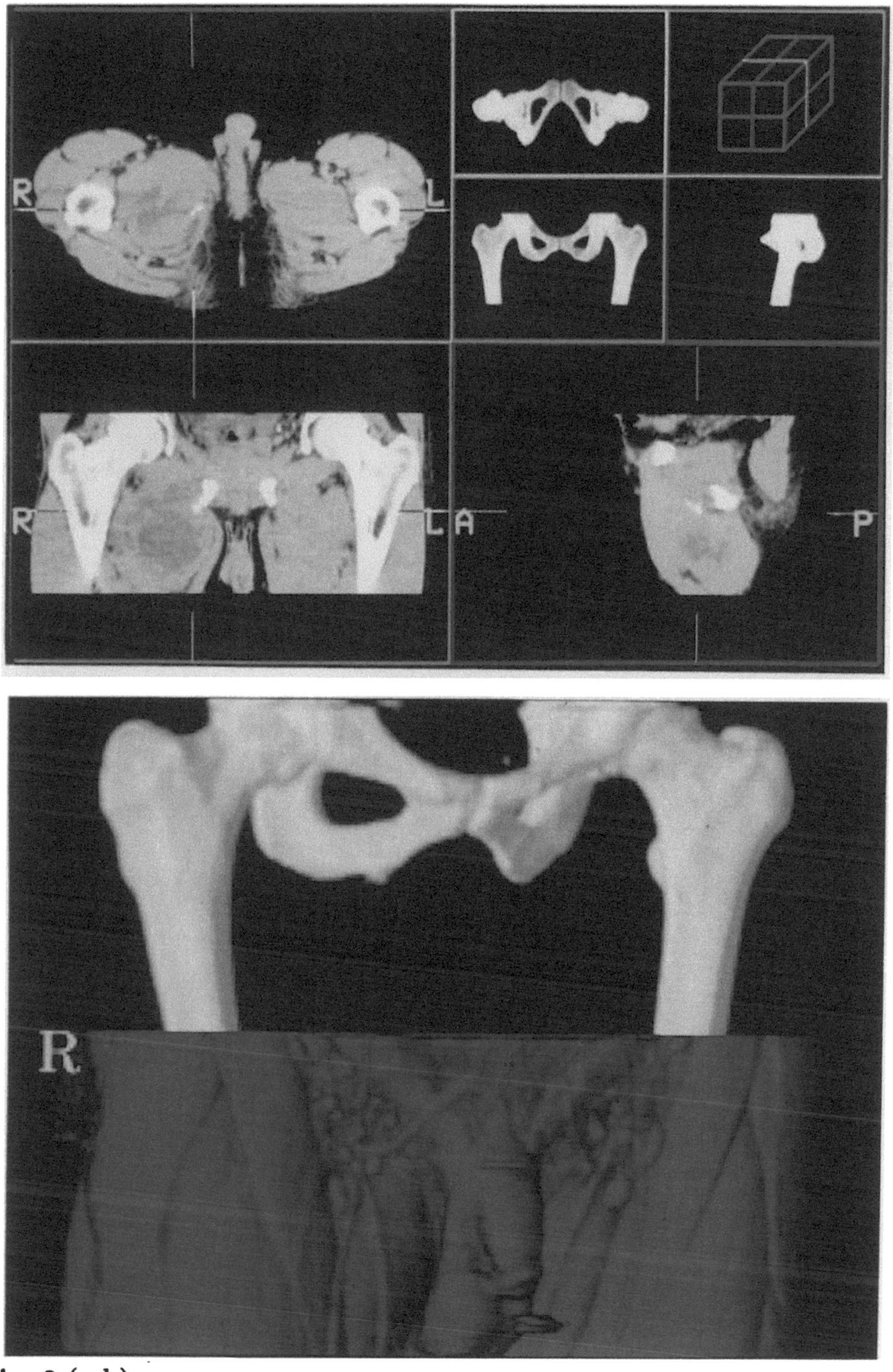

Fig. 3 (a-b)

25 year-old-male with right thigh mass felt to represent a sarcoma.
(a) avulsion fracture due to adductor muscle. Note bone fragment on coronal view.

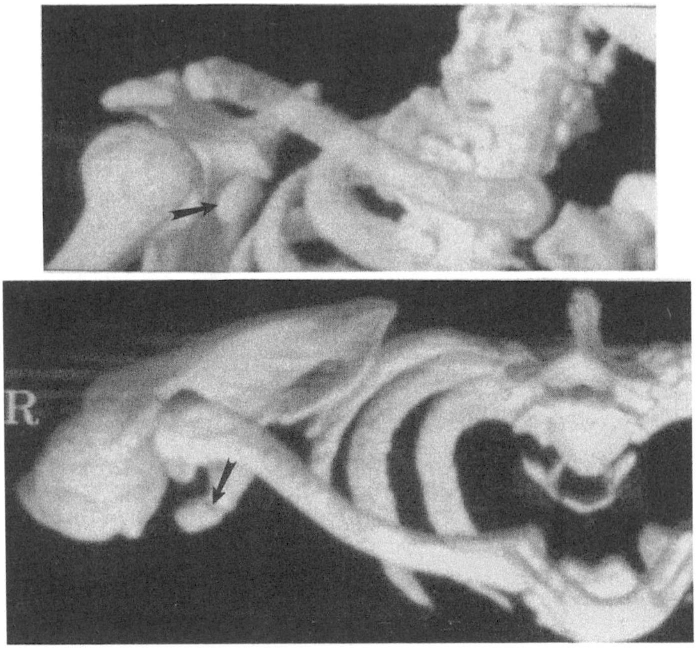

Fig. 4 (a-b)

55 year-old-male following blunt trauma to shoulder. 3D opaque reconstructions demonstrate fracture of corocoid process (arrow). Relationship of clavicle to acromion and sternum clearly defined.

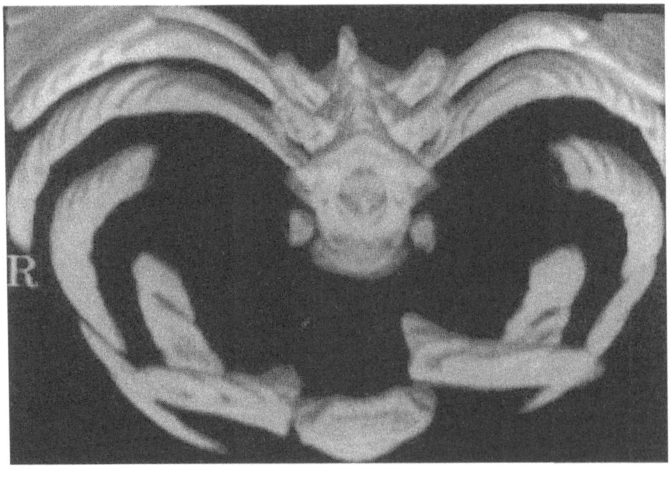

Fig. 5

Thoracic inlet view of posterior left sternoclavicular dislocation.

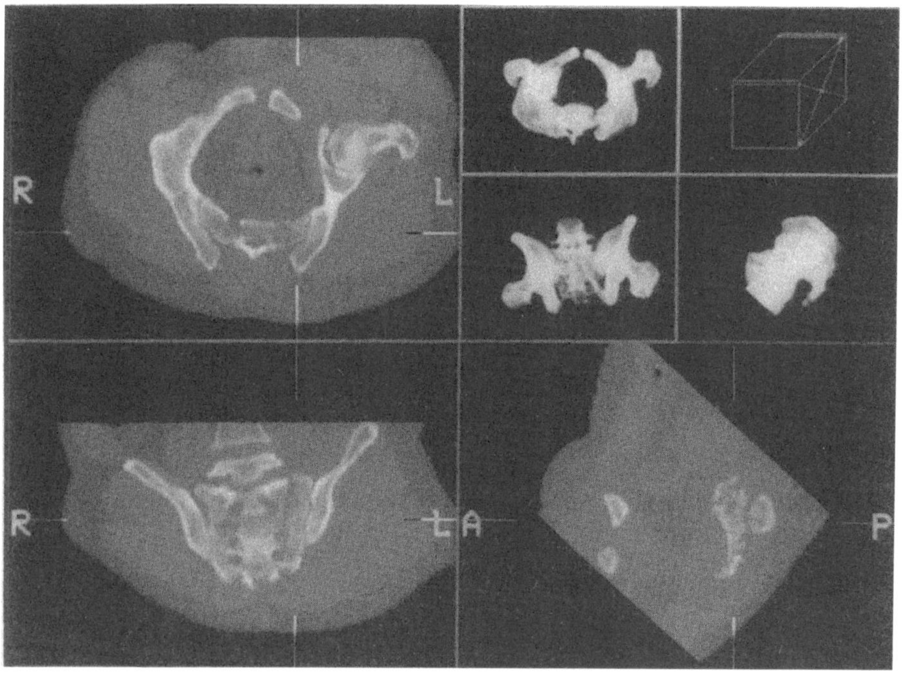

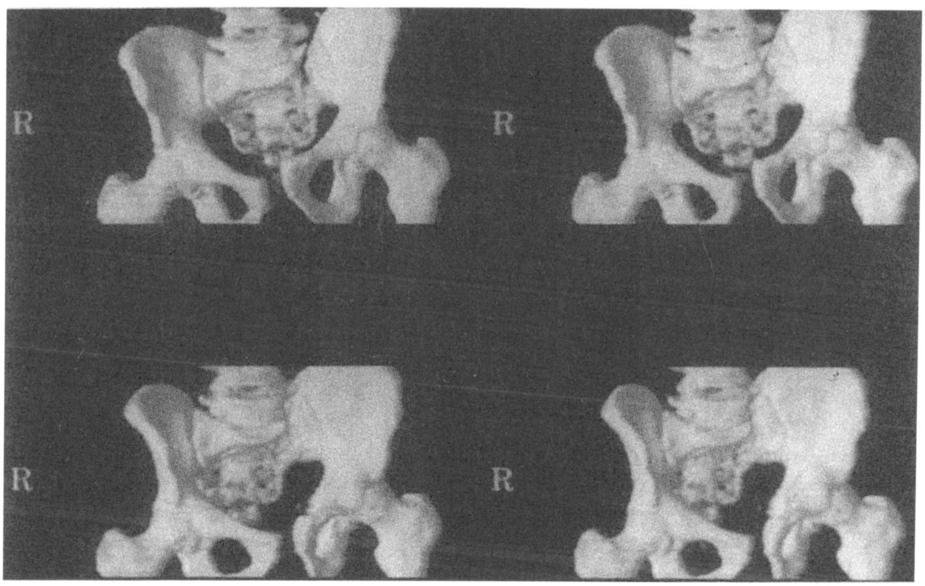

Fig. 6 (a-b)

Fracture of sacrum extending from S1 through S4 on left side. Fracture line extends through foramen best seen on oblique view. 3D views best demonstrate diastasis of symphysis and displacement upward of left hemipelvis.

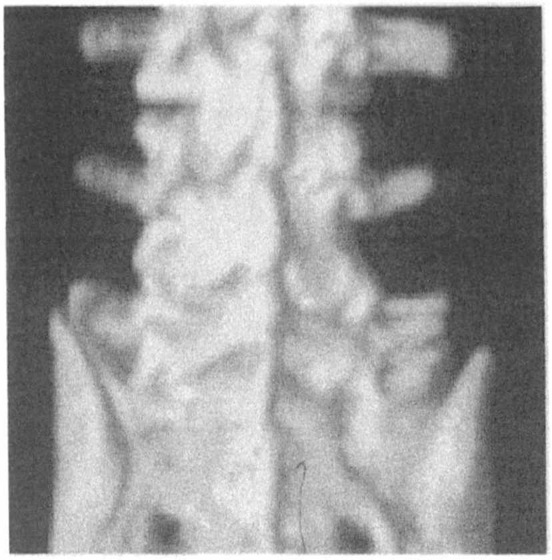

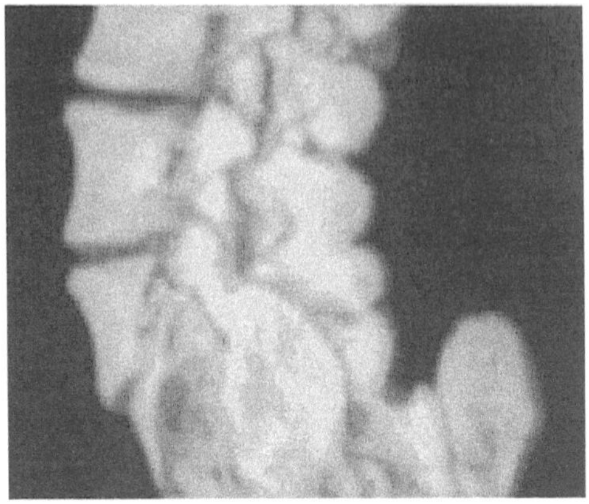

Fig. 7 (a-b)

3D reconstruction demonstrates spondylosis of L5 due to prior fracture.

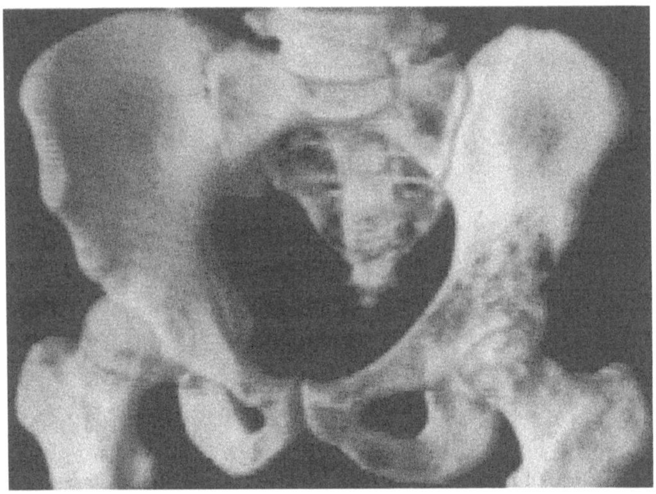

Fig. 8

Metastatic testicular tumor to left acetabulum and ileum. Infiltration of tumor clearly defined on the study. The 3D images were used for radiation therapy planning.

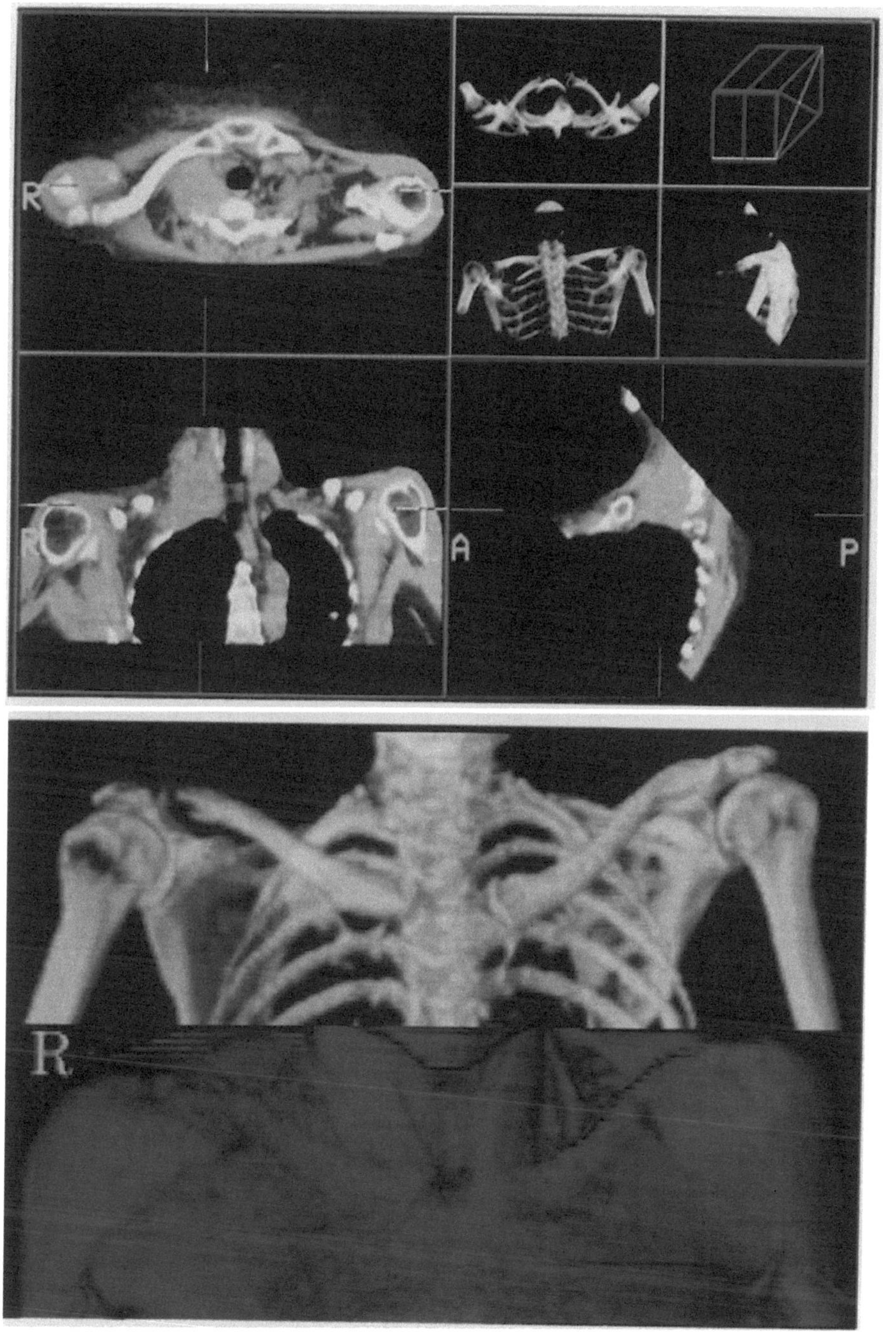

Fig. 9 (a-b)

2D and 3D reconstructions demonstrate recurrent lymphoma in right supraclavicular and cervical region. Tumor involved nerve roots at C6-C7 and C7-T1 resulting in brachial plexxopathy.

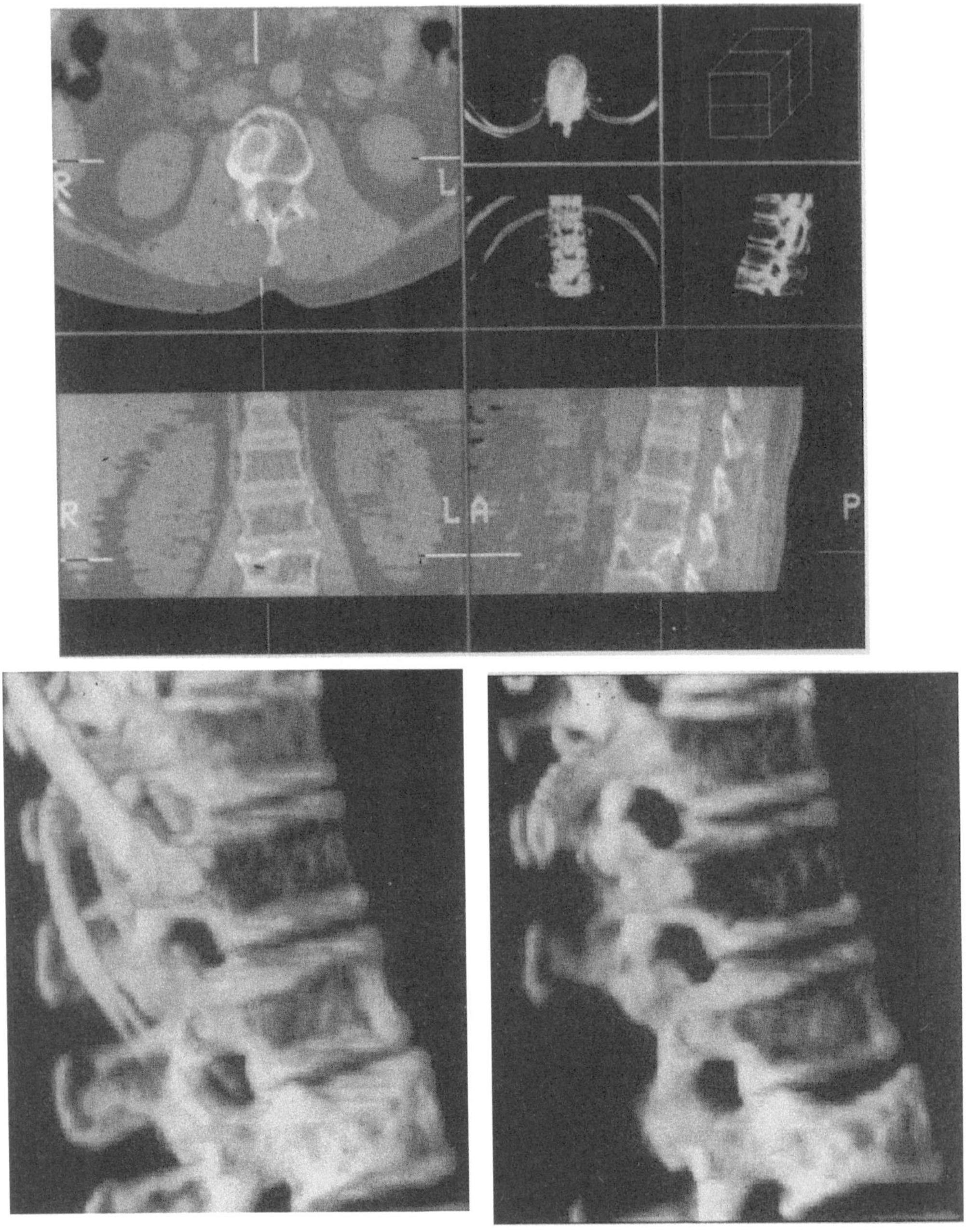

Fig. 10 (a-c)

2D and 3D reconstructions demonstrate collapse of L3 due to infiltration by multiple myeloma.

3D Morphometric and Morphologic Information Derived From Clinical Brain MR Images

Ron Kikinis[1], Ferenc A. Jolesz[1], Guido Gerig[2], Tamas Sandor[1], Harvey E. Cline[3] William E. Lorensen[3], Michael Halle[4], Stephen A. Benton[4]

From the Department of Radiology, Harvard Medical School and
Brigham and Women's Hospital, Boston, MA (1)
Institute for Communication Technology, Image Science Division,
ETH-Zürich, CH-8092 Zürich, Switzerland (2)
General Electric Corporation, Schenectady, NY (3)
Spatial Imaging Group, Media Laboratory,
Massachusetts Institute of Technology, Cambridge, MA (4)

Supported in part by NIH grants P01 CA41167, 5 K04 NS011083, and 2 P01 AG04953 and a grant from NYNEX. G. G. is supported by the Swiss National Science Foundation grant number 4018-11082. R. K. is recipient of a grant of the Swiss National Foundation.

Abstract

Data from conventional clinical MR brain images were processed using multi-step computerized segmentation as well as 3D analysis and rendering techniques. The usefulness of so obtained morphometric information and morphologic display for the development of new concepts for diagnosis and follow up of diseases was demonstrated with data sets from patients with Alzheimer's disease, normal pressure hydrocephalus, multiple sclerosis and brain tumors.

NATO ASI Series, Vol. F 60
3D Imaging in Medicine
Edited by K. H. Höhne et al.

Introduction

The clinical usefulnes of MRI is well established. The information provided by this technique is however, not yet fully exploited. Full use of all the available information for characterization of normal anatomy as well as of pathological changes and their distribution is still not granted. This is due to the fact that anatomical respectively pathological information is distributed over multiple slices of cross sectional images and has to be segmetented. Once identified, such information can be displayed as 3D reconstruction using algorithms and hardware which have been developed during the last decade and are available. It would be clinically of interest to extend the usage of such data beyond qualitative display by providing quantization derived from 3D data sets.

Measuring volumes from 3D data sets (morphometry) was introduced quite recently. The accuracy of morphometric measurements as a function of data acquisition modes, spatial resolution and segmentation has to be addressed before clinical validation (Filipek 89, Jack 88). For this purpose, different phantoms have been used (Filipek 89, Jack 88). The availability of reproducible quantitative measurements is important for clinical puroses because it helps to reduce subjectivity in the diagnosis and allows to improve the clinical follow-up of many diseases.

Segmentation of 3D data sets is a prerequisite for morphometry and 3D display of morphology. In order to be useful in a clinical environment, the segmentation should use the regular clinical data sets, allow retrospective application and require only minimal user-interaction. We have developed several specific applications based on this approach:

- Determination of the characteristic volumes and 3D configurations of the cerebrospinal fluid (CSF) spaces in normals as opposed to patients suffering from various forms of cerebral atrophy or hydrocephalus.
- Determination of volumes which may be used as quantitativemeasure for follow-up of treatments in patients with tumors or after shunt operations of hydrocephalic patients.
- Determination of the extent of white matter lesions in multiple sclerosis (MS) and other forms of white matter diseases.
- Generation of 3D renderings of skin, brain, and vessels for surgical planning

We applied image processing techniques for the extraction of morphometric and morphologic information from 3D MR data sets. The results were used to quantitate the extent of pathological involvement and to reveal the characteristic spatial patterns of various diseases. The information obtained with image processing techniques is not otherwise accessible and can result in a new conceptual approach for the correct description and interpretation of diseases. It was our goal to explore the clinical usefulness of a combination of morphometric and morphologic information derived from 3D MRI data sets and to develop clinically relevant criteria for a better diagnosis and a better follow-up of patients.

Material and Methods

Patients and Data Acquisition

Forty-one subjects were included in the comparative study of volumetries: 19 patients suffering from either Alzheimer's disease or brain atrophy were compared with 10 normal age matched volunteers. In addition, 8 patients suffering from clinically defined normal pressure hydrocephalus as well as 4 patients after shunt treatment were evaluated. Another 22 patients suffering from multiple sclerosis were included for analysis of the distribution of their white matter lesions.

Data were acquired on a Signa 1.5 T MR Scanner (General Electric Co., Milwaukee, WI). An axial, multislice, long TR (2000-4000 msec) double echo (30 and 80 msec) spin echo (SE) data acquisition sequence with flow compensation gradients, 192 phase encoding steps and conjugate synthesis (half Fourier, Feinberg 86) was used. In order to obtain enough slices to cover the whole brain two interleaved acquisitions were performed. Interleaving allows large interslice gaps with less crosstalk while providing contiguous data sets. Larger gaps improve the signal to noise of the images. Our scheme allows to acquire 30 to 60 slices of 3 mm thickness with two echoes for every slice in 6-12 minutes (depending on TR) (Jolesz 90).

Image Processing

Image processing was performed on various computers and software packages. The concept of the segmentation algorithm is based upon a multi-step approach (Gerig 89). These steps included:

1) filtering for improvement of signal to noise
2) supervised classification of all voxels based on multivariate analysis
3) connected component labeling for grouping of classified voxels.

Not all the steps were necessary in each case (e.g. filtering is only necessary in data sets with insufficient signal to noise). Using this approach, the brain data sets were segmented into the following distinct categories: subarachnoidal and ventricular CSF, gray matter, white matter (WM), WM lesions (WML) and background. At this point, every voxel was assigned unambiguously to one of the categories. Files containing this information in form of labels for every voxel were generated.

Morphometric Measurements

The simplest form of quantitative analysis is to count all voxels with the same label (i.e. CSF) for the entire data set or within a region of interest. The data in table 1 were obtained using this method. In order to obtain estimates of the accuracy of this process a phantom was prepared, using agarose gel doped with $NiCl_2$. This allows to control T1 and T2 times independently from each other

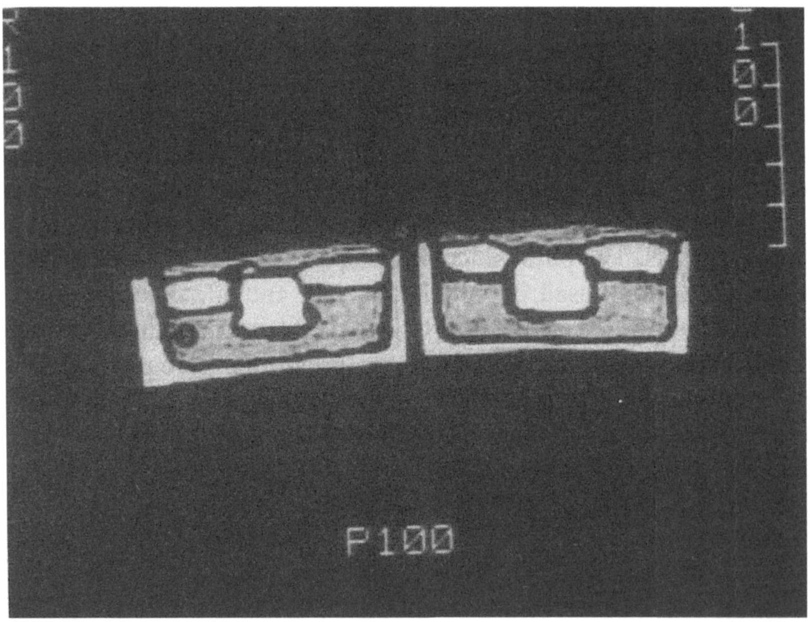

Fig. 1: Agarose Phantom. a) First Echo of a long TR series

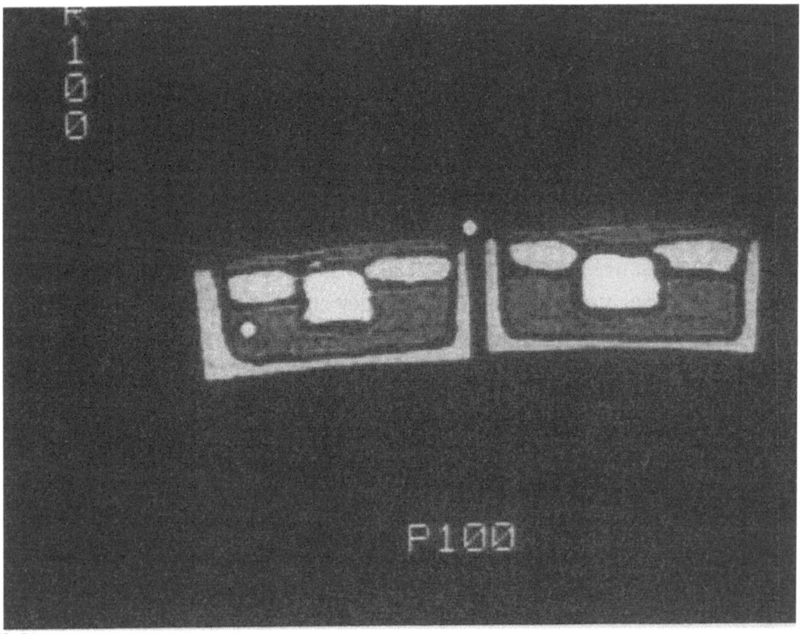

Fig. 1: b) Second Echo of a long TR series. The different compartments visible are made from agarose gels doped with $NiCl^2$ of varying concentrations. Thin test tubes containing D^2O doped water are visible as dots between the two phantoms and on either lateral side inside. As with the original, WM becomes darker on the second echo, grey matter is hyperintense to WM on both echoes and WML are brighter than fluid (CSF) on the first and isointense to fluid on the second echo.

(Mitchell 86). Such a phantom allows therefore to simulate the signal intensity changes of different tissues from first to second echo (see figure 1). Agarose gels are well suited for the creation of arbitrarily shaped phantoms for performance testing of complex image processing procedures.

3D Renderings

The label data sets can be used to generate three-dimensional (3D) renderings (Lorensen 87). These renderings allow visualization of different structures such as skin and brain surface, vessels, tumors, ventricles as well as the spatial distribution of WML. Structures can be displayed alone or in combination. Powerful computer hardware allows to perform such operations interactively.

Results

Table 1 gives an overview of the volumetric findings in a patient suffering from normal pressure hydrocephalus. Absolute volumes of different tissues have been determined after segmentation as described in material and methods. For the relative sizes, tissue volumes were normalized using the measured part of the intracranial cavity (ICC). In addition to the directly determined volumes, different ratios between the tissues can be formed. The ratio between subarachnoidal CSF space (SAS) and the ventricular CSF space can be used for differentiating different forms of hydrocephalus (e.g. in obstructive hydrocephalus, the ventricular size is increased, compared to the SAS).

Table 1: Example of Volumetry in a Case of Normal Pressure Hydrocephalus (NPH)

	Volume (ml)	Rel. Size
Total Intracranial Cavity (ICC)	1069.42	100
SAS	159.48	14.91
Ventricles	109.84	10.27
Grey Matter	506.98	47.41
White Matter	190.15	17.78
White Matter Lesion:	38.92	3.60
Vessels and Connective Tissue:	64.05	5.99
Brain Volume (G + W + WML)	736.06	68.83
ICC/Brain	1.45	
SAS/Ventricles	1.45	
Brain/CSF	2.73	
Brain/WML	18.91	

ICC = Intracranial Cavity, WML = White Matter Lesion, CSF = Cerebro Spinal Fluid, SAS = Subarachnoidal Space

Table 2: Summary of volumetric findings in different diagnostic groups

	Normal	NPH	Shunt	AD/BA
Number	10	8	4	19
Avg. Age	74	66	66	74
Brain Size	78	73	74	74
SAS	18	15	19	20
Ventricles	2	10	5	5

The numbers indicate the relative size of brain, subarachnoidal Space (SAS) and ventricles in normals, patients with normal pressure hydrocephalus (NPH), NPH patients after shunting (shunt) and patients with Alzheimer's disease or brain atrophy (AD/BA). The data was obtained retrospectively from routine data (spin echo, TR 2000-4000 msec, TE 30 and 80, slice thickness 3-5 mm).

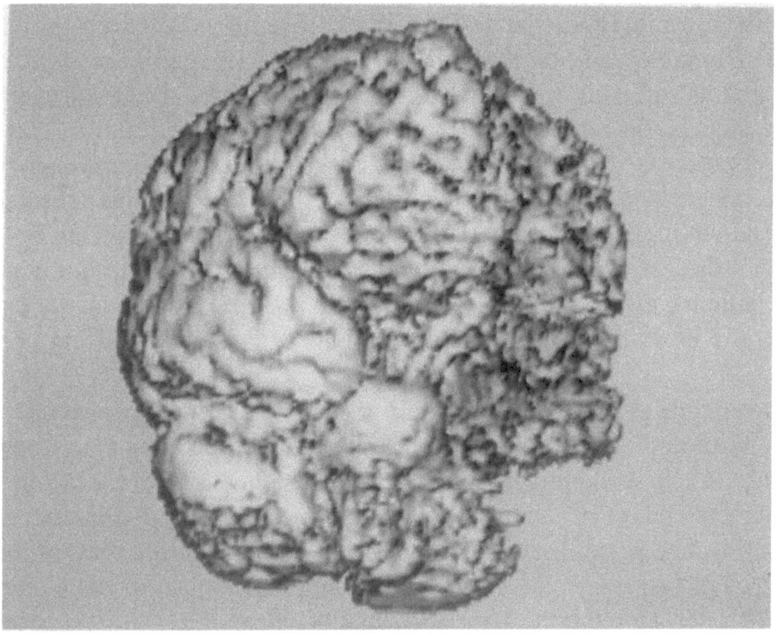

Figure 2

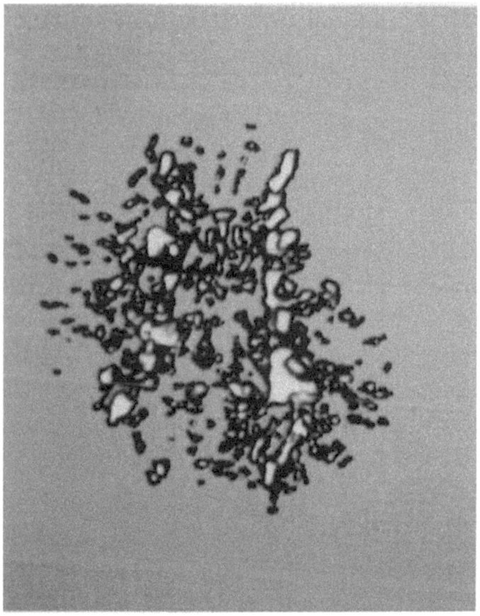

Figure 3a

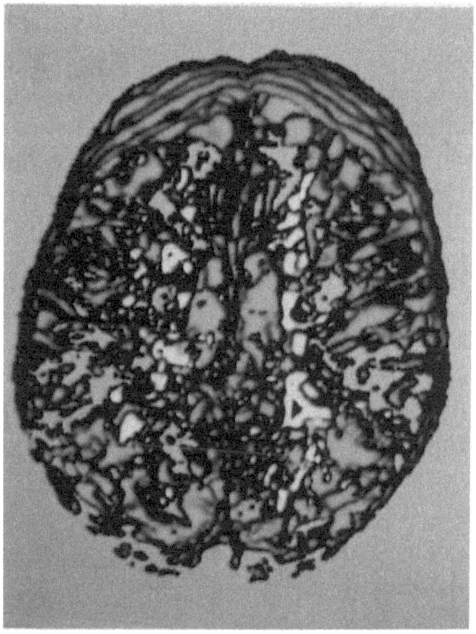

Figure 3b

Table 2 summarizes the relative sizes (determined in each individual as exemplified in table 1) of brain, SAS and ventricles in aged normals comparing them to patients suffering from various diseses. Compared to the normal subjects the brain size is decreased in all patients. In NPH the relative size of the ventricles is increased compared to all other groups. No relevant change in the brain size takes place after shunt operation in NPH patients. The ventricular size is decreased while the size of the SAS is increased. Patients suffering from either Alzheimer's disease (AD) or brain atrophy (BA) have volumetric findings similar to NPH patients after shunting. These preliminary results indicate that these patients could be distinguished from NPH patients and normals using volumetric findings.

3D renderings were obtained from the segmentated data (see example in figure 2). Although the number analysed cases is too small to draw final conclusions, the first examples show shapes which may be characteristic for normal subjects and NPH patients. Assessment of the extent and course of white matter disease such as multiple sclerosis (MS) requires identification of the white matter and of its lesions. Figure 3 depicts a brain, where parts of the white matter are exposed after subtraction of the gray matter.

In figure 4 white matter lesions in a patient suffering from MS were 3D reconstructed. Simultanous display of the CSF allows easier orientation. Combining different surfaces in a multicolored display allows a comprehensive evaluation of complex anatomical and pathological situations.

Figure 2: Grey matter and white matter surfaces. The original data for this frontal view of the brain of a patient was a 3 mm axial double echo spin echo data set. The segmentation ran fully automated after an interactive training for the multivariate classification. The grey matter is rendered in grey, the white matter is rendered in yellow. The surface of the white matter on the right side of the picture is visualized in this form for the first time. Its accessibility will provide new approaches to deal with white matter and its pathology. The pons and the pyramidal tracts are clearly visualized. The medial part of the temporal lobe shows the impression in the white matter surface caused by the hippocampus-amygdal complex. The grey matter surface allows recognition of landmark structures such as the sylvian fissure and the superior temporal gyrus. The ability to deal with grey and white matter separately is important for the diagnosis and follow-up in disease such as Alzheimer's disease and multiple sclerosis.

Fig. 3

White matter lesions and CSF in a patient suffering from multiple sclerosis. View from vertex. a) White matter lesions only b) white matter lesions in combination with the CSF spaces. The anatomical situation is easier to understand if the CSF spaces are rendered to provide an anatomical structure of reference. Clearly, the extent of disease is easier to appreciate from such renderings than from the cross sectional magnitude images. Currently, we are working on developing a quantitative description that goes beyond simple volumetry.

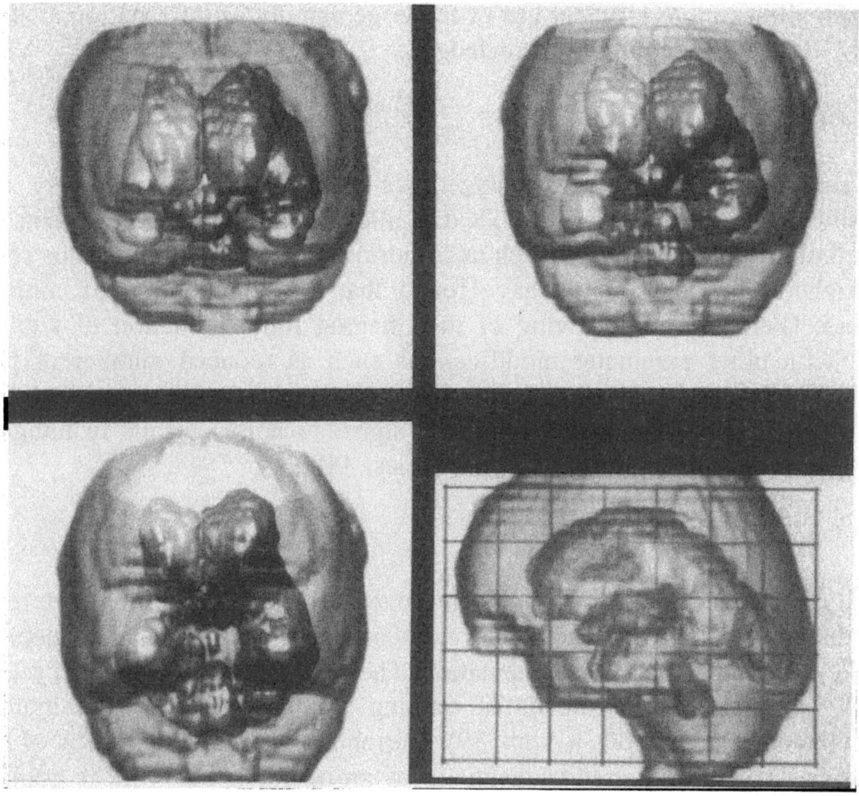

Figure 4: Ventricles in hydrocephalus. In this composite rendering the lateral ventricles of a patient suffering from NPH are displayed together with an outline of the intracranial cavity as a reference. The corpus callosal angle is visible in both frontal views. Depending to the exact viewing direction the angle changes, making it necessary to interactively adjust the view before the corpus callosal angle is determined.

Discussion

We have used various image processing techniques on conventional spin echo images to enhance some clinically significant features of 3D MRI data sets without having to obtain additional MRI pulse sequences. The application of the techniques reported here may improve and facilitate the correlation between clinical and MRI findings, the follow-up and the differential diagnosis between various types of WML, different types of tumors, and hydrocephalus. Both volumetric quantification of lesions (e.g. tumor, WML) and their spatial distribution within anatomic regions can be obtained from segmented images. Our approach allows more efficient use of the large amount of information available from MRI to advance clinical knowledge.

Data Acquisition

Data sets of 41 cases were analyzed for this study. The relatively short acquisition time for the axial long TR data allows to examine even patients with reduced ability of cooperation such as in normal pressure hydrocephalus (NPH) and Alzheimer's disease. We have found that the combination of conjugate synthesis (Feinberg 86) (halving of the imaging time, reduction of signal to noise) with other parameter modifications such as reduced number of phase encoding steps (also reduces imaging time and signal to noise) and interleaving of two series (more slices and improved signal to noise) results in images of good quality on our high field scanner (Jolesz 90).

Image Processing

The filter scheme used allows us to improve our spatial resolution (i.e. reduce slice thickness to 3 mm) and while maintaining sufficient signal to noise for visual evaluation as well as segmentation. The segmentation of the data sets can be fully automated after the initial training which requires 5-10 minutes of user-interaction (Gerig 89, Kikinis 89). Currently, the individual steps of these processes have to be started sequentially by an operator, but work is going on, to fully automate this part of the procedure.

Morphometric Measurements

Exact quantitative information can be obtained with full 3D anatomic definition demonstrating both the extent and spatial pattern of lesions. The determination of the volumes of anatomic objects is the most elementary form of 3D analysis. By counting all voxels which were segmented into a class and multiplying these numbers with the voxel-volume it is possible to obtain total volumes of anatomic objects such as brain, tumors, ventricles, the subarachnoidal space or WML (table 1). In order to estimate the error introduced by the whole procedure (measurement, segmentation, and connectivity), the same patient was examined at two different occasions with the same parameter settings. Another patient was examined at one occasion with two different parameter settings (slice thickness, TR, TE). The relative size of the different tissues differed less than 3%. This finding correlates well with values reported recently where absolute volumes were determined with an accuracy of few percent (-2 to +7% (Jack 88) and

4-10% (Filipek 89)). Using a cylindrical phantom, Filipek found a linear relation between slice thickness and error. Based on these findings we feel that the accuracy of morphometric measurements obtained in our environment are acceptable for the clinical usage. For most clinical applications the determination of relative sizes is sufficient (e.g. follow-up of tumor under therapy, or comparing two groups of patients). However, even the absolute measures obtained are adequately exact to be used clinically (e.g. stereotactic positioning).

3D Renderings

3D reconstructions can be used to visualize complicated anatomic situations such as the shape of the vascular tree, the ventricular system or the distribution of WML. In these examples the information is distributed over so many slices that it is virtually impossible to mentally reconstruct the objects. 3D representations of the anatomic and pathological structures reveal the full extent of disease processes and demonstrate the spatial pattern of the lesions, which may be pathognomonic in cases like MS or NPH.

Clinical results

Our approach was applied to various diseases of the brain in which quantitative morphometric and volumetric information, as well as spatial configuration or distribution of normal structures and/or lesions is essential. For comparison in larger groups of patients we normalized data sets using the volume of the intracranial cavity. This approach is valid for diseases in adults where the pathological changes occur only after the termination of skull ossification. Table 2 lists the results of such a comparison: In a patient presenting with the symptoms of dementia, gait disturbance and urinary incontinence the differential diagnosis includes NPH, Alzheimer's disease and brain atrophy. Only the NPH patients will profit from surgical implantation of a CSF shunt. Today, only about 60% of the patients receiving a shunt improve clinically. Therefore, it would be helpful to improve the diagnosis. The size of the brain relative to the intracranial cavity is reduced in NPH, Alzheimer's and brain atrophy patients. Characteristic patterns can be defiend for the volumes of the different CSF compartments for these diseases. Interestingly, the brain size doesn't recover after shunting operation in NPH patients. Instead, there is a shift of the relative volumes from the ventricles to the subarachnoidal space. The preliminary results in this example suggest that the size of the ventricles and the ratio between ventricular and subarachnoidal volumes is a differential diagnostic measure between NPH and

AD. No other noninvasive method is available today to make this differential diagnosis. The morphometric analysis can be used to study other focal and generalized atrophic processes such as in normal aging, MS, Alzheimer's disease, and schizophrenia. Such quantitative measures will allow a more precise follow-up in these patients. Following segmentation, 3D reconstructions of multiple anatomic surfaces such as skin, brain, and tumor surfaces can be generated. One benefit is the easy assessment of the distribution of white matter lesions (e.g. in MS). 3D renderings from over 20 patients were generated and used for differential diagnosis, for surgical and radiation therapy planning, for follow-up in patients with WML (such as in MS) and for teaching purposes (see figures 2-4). Many of these applicatons have already been evaluated in literature (Levin 88). However, most of them included many manual steps thereby increasing the time required for the processing of cases. We have started a prospective study for evaluation of the clinical significance of 3D reconstructions of WML for the diagnosis and follow-up in MS patients. In addition to simple display, 3D reconstructions can be used for the determination of additional morphometric measures (such as measuring angles and distances). Many linear measurements could be obtained from single reformatted planes. However, it is very difficult to identify the correct plane. Determination of these measures directly from the 3D renderings offers an elegant alternative. We have developed an application making use of this possibility: In pre-CT times pneumoencephalography (PEG) was the standard diagnostic procedure for suspected NPH patients. Radiologists used the general shape of the ventricular system as a measure for the differential diagnosis (LeMay 1970). In addition, the corpus callosal angle and the height of the lateral ventricles were introduced as morphometric measures. However, this invasive procedure requires to introduce air into the lateral ventricles and resulted in many complications in the NPH patients. PEG was abandoned after CT became available. MRI derived 3D renderings of the ventricles allow to determine these measurements noninvasively. A prospective study has been started to evaluate the different MR approaches (morphometric, morphologic, and physiologic information) for the prediction of shunt success in NPH patients. Crucial for the clinical usefulness of morphometric and morphologic information is its availability and therefore the time required for the processing of the data from cross sectional slices to 3D renderings. In our environment this time is between two to 8 hours depending on the complexity of the case and the number of interactive steps required. This means that we can process cases for research purposes. In order to be useful for the daily clinical program, the processing time will have to be reduced to less than 45 minutes. We expect to achieve this goal within the next six months. For bedside evaluation of 3D reconstructions by clinicans, and for portability in general it may be helpful to bypass the computers for display. We have started evaluating synthetic holographic stereograms for this purpose (Benton 85). The ability to generate 3D renderings fast can be exploited in yet another way. We plan to perform MR guided laser surgery using our segmentation and 3D techniques to guide the stereotactic positioning of a needle containing a laser glass fiber in patients whose heads are fixed to the MR scanner. This project requires exact absolute positioning, determination of angles and distances using both 3D renderings for determination of the entry point of the needle into the brain (preventing vital structures such as the central sulcus) and arbitrary positioned reformatted slices (for determining the position of the tip of the needle in the target structure). This approach will allow for the first time to evaluate the trajectory of a stereotactic device prospectively during the procedure. Because the image acquisition and the surgery take place in the same session without moving the patient, all the data sets will be registered and the coordinate system of the magnet can be used after proper calibration.

Conclusion

3D techniques have a great deal to offer in terms of improved accuracy of diagnosis, treatment planning, follow-up, and understanding of pathophysiology. The morphometric and morphologic 3D analysis and representation of medical images is useful not only for surgical planning, but also for diagnosis and follow-up in disease processes not requiring surgery. 3D morphometric information improves accuracy of clinical decisions by providing quantitative criteria to the clinicans. The optimized usage of the information available from cross-sectional images will improve the benefit of high-tech medicine to the patients.

References

Benton, S.A.: Display Holography: A SPIE Critical Review of Technology. SPIE. 532, 8-13 (1985)

Feinberg, D.A., Hale, J.D., Watts, J.C., et al.: Halving MR Imaging Time by Conjugation: Demonstration at 3.5 kG. Radiology. 161, 527-531 (1986)

Filipek, P.A., Kennedy, D.N., Caviness, V.S., et al.: Magnetic Resonance Imaging-based Brain Morphometry: Development and Application to Normal Subjects. Ann Neurol. 25, 61-67 (1989)

Gerig, G., Kuoni, W., Kikinis, R., et al.: Medical Imaging and Computer Vision: An Integrated Approach for Diagnosis and Planning. 11th DAGM Symposium Mustererkennung 1989, Hamburg, FRG, Oct 2-4, 1989. Fachberichte IFB 219, Springer Berlin:425-443

Jack, C.R,, Gehring, D.G., Sharbrough, F.W., et al.: Temporal Lobe Volume Measurement from MR Images: Accuracy and Left-Right Asymmetry in Normal Persons. J Comput Assist Tomogr. 12, 21-29 (1988)

Jolesz, F.A., Schwartz, R.B., LeClerq, G.T., et al.: Half Fourier Spin Echo Imaging in Routine Clinical Brain and Cervical Spine Protocols. Presented at the Society of Magnetic Resonance Imaging in Washington D.C. Feb. 24-28, 1990

Kikinis, R., Jolesz, F.A., Halle, M.W., et al.: Computer-assisted Analysis of Fluid Spaces from Routine MR Imaging Data. Radiology. 173(P), 367 (abstract) (1989)

LeMay, M., and New, P.J.: Radiological Diagnosis of Occult Normal-Pressure Hydrocephalus. Radiology. 96, 347-358 (1970)

Levin, D.N., Pelizzari, C.A., Chen, G.T.Y., et al.: Retrospective Geometric Correlation of MR, CT, and PET Images. Radiology 169, 817-823 (1988)

Lorensen, W.E., and Cline, H.E.: Marching cubes: a high resolution 3D surface reconstruction algorithm. ACM Computer Graphics 21(4), 163-169 (1987)

Mitchell, M.D., Kundel, H.L., Axel, L., et al.: Agarose as a tissue equivalent phantom material for NMR imaging. Magn Reson Imag. 4, 263-266 (1986)

List of Authors

Subject Index

NATO ASI Series F

Including Special Programme on Sensory Systems for Robotic Control (ROB)

Vol. 1: Issues in Acoustic Signal – Image Processing and Recognition. Edited by C. H. Chen. VIII, 333 pages. 1983.

Vol. 2: Image Sequence Processing and Dynamic Scene Analysis. Edited by T. S. Huang. IX, 749 pages. 1983.

Vol. 3: Electronic Systems Effectiveness and Life Cycle Costing. Edited by J. K. Skwirzynski. XVII, 732 pages. 1983.

Vol. 4: Pictorial Data Analysis. Edited by R. M. Haralick. VIII, 468 pages. 1983.

Vol. 5: International Calibration Study of Traffic Conflict Techniques. Edited by E. Asmussen. VII, 229 pages. 1984.

Vol. 6: Information Technology and the Computer Network. Edited by K. G. Beauchamp. VIII, 271 pages. 1984.

Vol. 7: High-Speed Computation. Edited by J. S. Kowalik. IX, 441 pages. 1984.

Vol. 8: Program Transformation and Programming Environments. Report on a Workshop directed by F. L. Bauer and H. Remus. Edited by P. Pepper. XIV, 378 pages. 1984.

Vol. 9: Computer Aided Analysis and Optimization of Mechanical System Dynamics. Edited by E. J. Haug. XXII, 700 pages. 1984.

Vol. 10: Simulation and Model-Based Methodologies: An Integrative View. Edited by T. I. Ören, B. P. Zeigler, M. S. Elzas. XIII, 651 pages. 1984.

Vol. 11: Robotics and Artificial Intelligence. Edited by M. Brady, L. A. Gerhardt, H. F. Davidson. XVII, 693 pages. 1984.

Vol. 12: Combinatorial Algorithms on Words. Edited by A. Apostolico, Z. Galil. VIII, 361 pages. 1985.

Vol. 13: Logics and Models of Concurrent Systems. Edited by K. R. Apt. VIII, 498 pages. 1985.

Vol. 14: Control Flow and Data Flow: Concepts of Distributed Programming. Edited by M. Broy. VIII, 525 pages. 1985.

Vol. 15: Computational Mathematical Programming. Edited by K. Schittkowski. VIII, 451 pages. 1985.

Vol. 16: New Systems and Architectures for Automatic Speech Recognition and Synthesis. Edited by R. De Mori, C.Y. Suen. XIII, 630 pages. 1985.

Vol. 17: Fundamental Algorithms for Computer Graphics. Edited by R.A. Earnshaw. XVI, 1042 pages. 1985.

Vol. 18: Computer Architectures for Spatially Distributed Data. Edited by H. Freeman and G. G. Pieroni. VIII, 391 pages. 1985.

Vol. 19: Pictorial Information Systems in Medicine. Edited by K. H. Höhne. XII, 525 pages. 1986.

Vol. 20: Disordered Systems and Biological Organization. Edited by E. Bienenstock, F. Fogelman Soulié, G. Weisbuch. XXI, 405 pages.1986.

Vol. 21: Intelligent Decision Support in Process Environments. Edited by E. Hollnagel, G. Mancini, D.D. Woods. XV, 524 pages. 1986.

NATO ASI Series F

Vol. 22: Software System Design Methods. The Challenge of Advanced Computing Technology. Edited by J. K. Skwirzynski. XIII, 747 pages. 1986.

Vol. 23: Designing Computer-Based Learning Materials. Edited by H. Weinstock and A. Bork. IX, 285 pages. 1986.

Vol. 24: Database Machines. Modern Trends and Applications. Edited by A. K. Sood and A. H. Qureshi. VIII, 570 pages. 1986.

Vol. 25: Pyramidal Systems for Computer Vision. Edited by V. Cantoni and S. Levialdi. VIII, 392 pages. 1986. *(ROB)*

Vol. 26: Modelling and Analysis in Arms Control. Edited by R. Avenhaus, R. K. Huber and J. D. Kettelle. VIII, 488 pages. 1986.

Vol. 27: Computer Aided Optimal Design: Structural and Mechanical Systems. Edited by C. A. Mota Soares. XIII, 1029 pages. 1987.

Vol. 28: Distributed Operating Systems. Theory und Practice. Edited by Y. Paker, J.-P. Banatre and M. Bozyiğit. X, 379 pages. 1987.

Vol. 29: Languages for Sensor-Based Control in Robotics. Edited by U. Rembold and K. Hörmann. IX, 625 pages. 1987. *(ROB)*

Vol. 30: Pattern Recognition Theory and Applications. Edited by P. A. Devijver and J. Kittler. XI, 543 pages. 1987.

Vol. 31: Decision Support Systems: Theory and Application. Edited by C. W. Holsapple and A. B. Whinston. X, 500 pages. 1987.

Vol. 32: Information Systems: Failure Analysis. Edited by J. A. Wise and A. Debons. XV, 338 pages. 1987.

Vol. 33: Machine Intelligence and Knowledge Engineering for Robotic Applications. Edited by A. K. C. Wong and A. Pugh. XIV, 486 pages. 1987. *(ROB)*

Vol. 34: Modelling, Robustness and Sensitivity Reduction in Control Systems. Edited by R. F. Curtain. IX, 492 pages. 1987.

Vol. 35: Expert Judgment and Expert Systems. Edited by J. L. Mumpower, L. D. Phillips, O. Renn and V. R. R. Uppuluri. VIII, 361 pages. 1987.

Vol. 36: Logic of Programming and Calculi of Discrete Design. Edited by M. Broy. VII, 415 pages. 1987.

Vol. 37: Dynamics of Infinite Dimensional Systems. Edited by S.-N. Chow and J. K. Hale. IX, 514 pages. 1987.

Vol. 38: Flow Control of Congested Networks. Edited by A. R. Odoni, L. Bianco and G. Szegö. XII, 355 pages. 1987.

Vol. 39: Mathematics and Computer Science in Medical Imaging. Edited by M. A. Viergever and A. Todd-Pokropek. VIII, 546 pages. 1988.

Vol. 40: Theoretical Foundations of Computer Graphics and CAD. Edited by R. A. Earnshaw. XX, 1246 pages. 1988.

Vol. 41: Neural Computers. Edited by R. Eckmiller and Ch. v. d. Malsburg. XIII, 566 pages. 1988.

NATO ASI Series F

Vol. 42: Real-Time Object Measurement and Classification. Edited by A. K. Jain. VIII, 407 pages. 1988. *(ROB)*

Vol. 43: Sensors and Sensory Systems for Advanced Robots. Edited by P. Dario. XI, 597 pages. 1988. *(ROB)*

Vol. 44: Signal Processing and Pattern Recognition in Nondestructive Evaluation of Materials. Edited by C. H. Chen. VIII, 344 pages. 1988. *(ROB)*

Vol. 45: Syntactic and Structural Pattern Recognition. Edited by G. Ferraté, T. Pavlidis, A. Sanleliu and H. Bunke. XVI, 467 pages. 1988. *(ROB)*

Vol. 46: Recent Advances in Speech Understanding and Dialog Systems. Edited by H. Niemann, M. Lang and G. Sagerer. X, 521 pages. 1988.

Vol.47: Advanced Computing Concepts and Techniques in Control Engineering. Edited by M. J. Denham and A. J. Laub. XI, 518 pages. 1988.

Vol. 48: Mathematical Models for Decision Support. Edited by G. Mitra. IX, 762 pages. 1988.

Vol. 49: Computer Integrated Manufacturing. Edited by I. B. Turksen. VIII, 568 pages. 1988.

Vol. 50: CAD Based Programming for Sensory Robots. Edited by B. Ravani. IX, 565 pages. 1988. *(ROB)*

Vol. 51: Algorithms and Model Formulations in Mathematical Programming. Edited by S. W. Wallace. IX, 190 pages. 1989.

Vol. 52: Sensor Devices and Systems for Robotics. Edited by A. Casals. IX, 362 pages. 1989. *(ROB)*

Vol. 53: Advanced Information Technologies for Industrial Material Flow Systems. Edited by S. Y. Nof and C. L. Moodie. IX, 710 pages. 1989.

Vol. 54: A Reappraisal of the Efficiency of Financial Markets. Edited by R. M. C. Guimarães, B. G. Kingsman and S. J. Taylor. X, 804 pages. 1989.

Vol. 55: Constructive Methods in Computing Science. Edited by M. Broy. VII, 478 pages. 1989.

Vol. 56: Multiple Criteria Decision Making and Risk Analysis Using Microcomputers. Edited by B. Karpak and S. Zionts. VII, 399 pages. 1989.

Vol. 57: Kinematics and Dynamic Issues in Sensor Based Control. Edited by G. E. Taylor. XI, 456 pages. 1990. *(ROB)*

Vol. 58: Highly Redundant Sensing in Robotic Systems. Edited by J. T. Tou and J. G. Balchen. X, 322 pages. 1990. *(ROB)*

Vol. 59: Superconducting Electronics. Edited by H. Weinstock and M. Nisenoff. X, 441 pages. 1989.

Vol. 60: 3D Imaging in Medicine. Algorithms, Systems, Applications. Edited by K. H. Höhne, H. Fuchs and S. M. Pizer. IX, 460 pages. 1990.

MIX
Papier aus verantwortungsvollen Quellen
Paper from responsible sources
FSC® C105338

If you have any concerns about our products,
you can contact us on
ProductSafety@springernature.com

In case Publisher is established outside the EU,
the EU authorized representative is:
Springer Nature Customer Service Center GmbH
Europaplatz 3, 69115 Heidelberg, Germany

Printed by Libri Plureos GmbH
in Hamburg, Germany